THE SCIENCE OF NATURE IN THE :

STUDIES IN HISTORY
AND PHILOSOPHY OF SCIENCE

VOLUME 19

The titles published in this series are listed at the end of this volume.

THE SCIENCE OF NATURE IN THE SEVENTEENTH CENTURY

Patterns of Change in Early Modern Natural Philosophy

Edited by

PETER R. ANSTEY

University of Sydney,
NSW, Australia

and

JOHN A. SCHUSTER

University of New South Wales,
NSW, Australia

 Springer

A C.I.P. Catalogue record for this book is available from the Library of Congress.

ISBN-13   978-90-481-6909-2 (PB)
ISBN-13   978-1-4020-3703-0 (e-book)

Published by Springer,
P.O. Box 17, 3300 AA Dordrecht, The Netherlands.

www.springeronline.com

Printed on acid-free paper

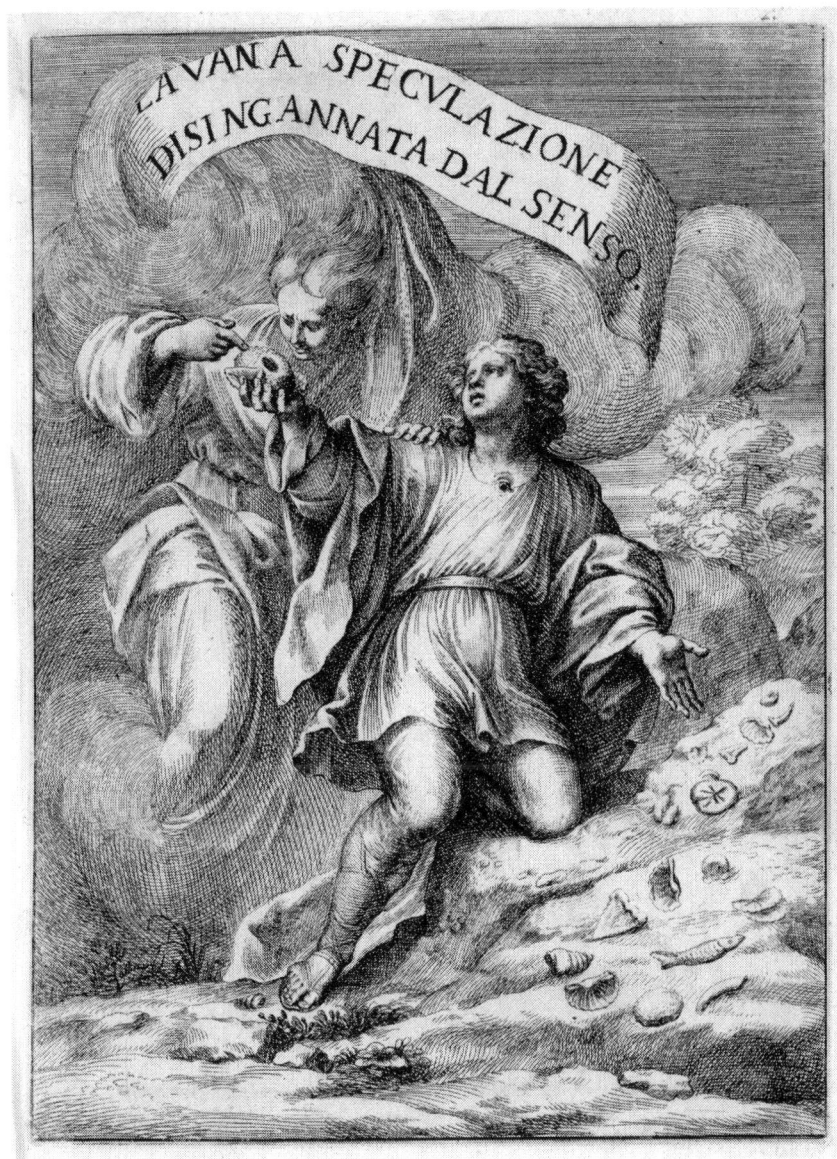

'Vain speculation undeceived by the senses'. Engraving from *La Vana Speculazione Disingannata dal Senso* by Agostino Scilla, Napoli, 1670, page opposite p. 168. Used with permission of the Bodleian Library, University of Oxford (Shelfmark Lister C 101).

CONTENTS

CONTRIBUTORS

PETER R. ANSTEY is Lecturer in the Department of Philosophy at the University of Sydney.

LUCIANO BOSCHIERO is a Fellow at the Department of History of Science and Technology, Johns Hopkins University.

H. FLORIS COHEN is Professor of History of Science at the University of Twente, Netherlands.

PETER DEAR is Professor of History and of Science & Technology Studies at Cornell University.

STEPHEN GAUKROGER is Professor of History of Philosophy and History of Science at the University of Sydney.

PETER HARRISON is Professor of History and Philosophy at Bond University, Gold Coast, Australia.

HELEN HATTAB is Assistant Professor in the Department of Philosophy and Honors College at the University of Houston.

JOHN A. SCHUSTER is Associate Professor [Reader] and Head of the School of History and Philosophy of Science, University of New South Wales, Sydney, Australia.

ACKNOWLEDGMENTS

Many debts have been accrued while compiling the papers in this collection. The conference on 'The Origins of Modernity 1543–1789' which gave rise to this volume was generously sponsored by the Faculties of Arts at the University of Sydney and the University of New South Wales and by the International Society for Intellectual History. Special thanks are due to Constance Blackwell for her support of the original conference in July 2002 and to Stephen Gaukroger for his guidance as the project has developed. Peter Harrison suggested the frontispiece which is used with permission of the Bodleian Library, Oxford.

We also acknowledge the permission of the Smithsonian Institution Libraries, Dibner Library of the History of Science & Technology, Washington DC, to reproduce the pictures in Helen Hattab's chapter and the permission of the Biblioteca Nazionale Centrale, Florence to use the pictures in Luciano Boschiero's chapter. Parts of H. Floris Cohen's chapter have appeared previously in French in 'Les raisons de la transformation: la specificité europeenné (trans. A. Barberousse) in *L'Europe des sciences. Constitution d'un espace scientifique*, eds M. Blay and E. Nicolaidis, Paris: Le Seuil, pp. 51–94.

PETER R. ANSTEY AND JOHN A. SCHUSTER

INTRODUCTION

One of the hallmarks of the modern world has been the stunning rise of the natural sciences. The exponential expansion of scientific knowledge and the accompanying technology that so impact on our daily lives are truly remarkable. But what is often taken for granted is the enviable epistemic-credit rating of scientific knowledge: science is authoritative, science inspires confidence, science is right. Yet it has not always been so. In the seventeenth century the situation was markedly different: competing sources of authority, shifting disciplinary boundaries, emerging modes of experimental practice and methodological reflection were some of the constituents in a quite different mélange in which knowledge of nature was by no means pre-eminent. It was the desire to probe the underlying causes of the shift from the early modern 'nature-knowledge' to modern science that was one of the stimuli for the 'Origins of Modernity: Early Modern Thought 1543–1789' conference held in Sydney in July 2002. How and why did modern science emerge from its early modern roots to the dominant position which it enjoys in today's post-modern world? Under the auspices of the International Society for Intellectual History, The University of New South Wales and The University of Sydney, a group of historians and philosophers of science gathered to discuss this issue. However, it soon became clear that a prior question needed to be settled first: the question as to the precise nature of the quest for knowledge of the natural realm in the seventeenth century. This collection is the product of the preliminary soundings made at the conference on that crucial prior question.[1]

The papers in this collection start from the premise that in the early modern period the central category for the study of nature was natural philosophy, or as Robert Hooke called it in his *Micrographia*, the Science of Nature. Any system of natural philosophy, whether a version of the hegemonic and institutionalised Scholastic Aristotelianism, or one of its challengers, concerned itself with a general theory of nature—that is, the nature of matter and cause, the cosmological structuring and functioning of matter and the proper method for acquiring or justifying knowledge of nature. To place the evolution of natural philosophy, and in particular the shifting patterns of its relations to other enterprises and disciplines, at the centre of one's conception of the Scientific Revolution is not novel, and more

---

[1] Six of the eight chapters in this volume ultimately derive from presentations at the Sydney Conference—those by John Schuster, Peter Dear, Helen Hattab, Peter Harrison, H. Floris Cohen and Stephen Gaukroger, the latter two having been plenary addresses. To these have been added related papers by Peter Anstey (who spoke in Sydney on another topic) and Luciano Boschiero.

P. R. Anstey and J. A. Schuster (eds.), The Science of Nature in the Seventeenth Century, 1-7.

scholars are realising the value of such a perspective, but neither is it obvious or agreed upon in the scholarly community.

Many older discussions, and some contemporary ones, are marred by a tendency to lump the culture of natural philosophising under an anachronistic label of 'science',[2] thus obscuring the possibility of speaking convincingly about the internal texture and dynamics of the culture of natural philosophy and its patterns of change over the period. If such anachronism truncates historical analysis by making the object of study 'science' from the first, a more recent, sophisticated and, to many, convincing approach is to read natural philosophy entirely out of the story of 'modern science' and its early modern origins. This has been done by identifying the large and encompassing culture of 'natural philosophising' solely with its dominant, institutionalised form, neo-Scholastic Aristotelianism. Thus, some recent scholars have defined the Scientific Revolution in terms of the end or demise of 'natural philosophy', supposedly followed by an equally abrupt triumphant origin of something called 'experimental science' or 'modern science'.[3] Over against both these pitfalls, this volume assumes that natural philosophy, understood as a large and contested field of systematic natural inquiry, encompassed Aristotelianism and its various challengers, their evolution and conflict over time. It is precisely this sort of understanding that has allowed some scholars to view the Scientific Revolution, so-called, as a process of conflict, co-optation and displacement amongst different natural philosophical claims.

This volume aims to cast more light on this approach. But, the focal concern of the papers in the collection resides in the deeper question of how claims were constructed and located in the field of natural philosophising. This is where our central theme takes shape: the issue of how natural philosophical claims were positioned in relation to other enterprises and concerns, taken variously to be superior to natural philosophy (such as theology); or cognate with it (other branches of philosophy, such as ethics or mathematics); or subordinate to it (as in the dominant Aristotelian evaluation of the mixed mathematical sciences, such as astronomy, optics and mechanics); or simply of some claimed relevance to it, as for example pedagogy or various of the practical arts. Taking on board the assumptions of sociologists of science and social historians of science when working on similar issues in later periods, we may straightforwardly assume that the positioning of natural philosophical claims in relation to other enterprises and concerns always involved two routine manoeuvres: the drawing or enforcing of boundaries and the making or defending of particular linkages (including efforts to undermine others' attempts at bounding and linking). The description of concrete examples of such

---

[2] H. Floris Cohen's massive survey of Scientific Revolution historiography (Cohen 1994) illustrates that the term 'natural philosophy' has been endemically present in the literature, but not systematically theorised, often serving as a synonym for 'science' or (some of) the sciences. Recent attempts to delineate the category of natural philosophy and deploy it in Scientific Revolution historiography include, Schuster 1990, 1995; Schuster and Watchirs 1990; Cunningham 1988, 1991; Cunningham and Williams 1993; Dear 1991, 2001; Harrison 2000, 2002 and his chapter in this volume; and Henry 2002.

[3] Shapin 1994 and Dear 1995.

machinations is one aim of the papers in the collection; the articulation of better general models and conceptions of such dynamics is another.

In the time period addressed by the collection, it was of course the case that the dominant Scholastic Aristotelianism tended to provide all players with the fundamental grammar for how such boundaries and linkages were to be made, since many natural philosophers, including some of the most dedicated advocates of alternative systems, had originally been scholastically trained. But, even amongst Aristotelians the topography of boundaries and linkages was not overly rigid and could be contested; for example, in shifting evaluations of the natural philosophical import of the definitely 'subordinate' mixed mathematical sciences. Moreover, advocates of natural philosophical alternatives to Aristotelianism could and did propagate different patterns of bounding and linkage. That is why, arguably, the process of the Scientific Revolution can be mapped in terms of the larger secular trends in these moves, and the dynamics that governed them, and that is also why we can focus on these developments in the cases of major non-Aristotelian natural philosophers studied in this volume, amongst them, Bacon, Galileo, Descartes, Beeckman, Kepler, Huygens, Boyle, and Newton.

In sum, then, the volume aims to offer a set of interrelated but distinct studies motivated by these concerns. Hence it takes a position on the historiography of the Scientific Revolution, stressing patterns of change in the continuing culture of natural philosophising, and it offers various, related suggestions for improving the concepts and tools used to study natural philosophy and its dynamics.

In the first chapter, H. Floris Cohen examines the simultaneous emergence in the early seventeenth century and the ongoing impact of three different, yet mutually complementary modes of acquiring knowledge of nature. With broad historiographical brushstrokes, Cohen shows how the mathematisation of natural phenomena, the fact gathering experimentalism of Bacon and his heirs and the re-emergence of ancient (though rival) explanatory models in natural philosophising blended and interlocked throughout the seventeenth century, culminating in the achievement of Newton's *Principia* and *Opticks*. For Cohen, the emergence of these three modes of 'nature-knowledge' is constitutive of the Scientific Revolution, and, in the tradition of Koyré and Westfall, he claims that the mathematisation of natural knowledge was the most decisive. However, even though the mathematisation of nature was to yield more long-term fruit, it was the re-emergence of Classical and Hellenistic explanatory models of the functioning and structure of nature that proved to be the rallying points for allegiances and the basis of polemics amongst natural philosophers of the early modern period. And the dominant explanatory model to emerge was a form of kinetic corpuscularianism as found in the writings of Gassendi, Boyle and others.

The most systematic and ambitious development of such an explanatory model of nature was René Descartes' and it is only fitting that his articulation of a cosmological system in terms of vortices should form the focal point of the next three chapters of the collection. Descartes' cosmology fulfilled the need for a credible mechanistic theory of the heavens on the demise of the Ptolemaic system and the modelling of his system on ancient hydrostatics and its relation to late

Renaissance work on mechanics provide central reference points for other contributions to this volume.

In chapter two John Schuster explores Descartes' often misunderstood concern with vortices; that is his vortex model for celestial motions and for light in the cosmological context. Schuster analyses the internal conceptual architecture of the vortical model, as well as its genesis out of Descartes' early attempts to construct a mechanical natural philosophy by both co-opting and in turn resynthesising the Scholastic 'mixed mathematical sciences' of mechanics, hydrostatics, optics and astronomy under the label of 'physico-mathematics'. He argues that whatever Descartes thought he was doing with vortices, it bears no relation to simplistic glosses routinely offered in the latter half of the seventeenth century and beyond, and that it constituted a serious and innovative, if ultimately flawed, cosmic hydraulics, or 'waterworld' for both light and celestial motion. In this way Schuster's chapter also illustrates some of the challenges, gambits and pitfalls that presented themselves to mathematically-oriented natural philosophical innovators of Descartes' generation. In particular, it exposes Descartes' complex debt to Beeckman, including Beeckman's own attempt, in the late 1620s, to produce a mechanist version of Kepler's radical program of a neo-Platonic synthesis of realist Copernicanism, with a new 'physics' of light and celestial motion.

Staying with Descartes' vortices, Peter Dear's chapter deals with the question of where, after all, did Descartes obtain his conception of vortical motion? Within what kind of textual, artisanal, or other context of practice (surely not that of 'mechanics' in its classical sense) did vortical motion appear as a topic for discussion? And how was the image or figure of the vortex supposed to clarify a new kind of physics for Descartes' readers? Dear discusses the most likely points of resonance that a philosophically-educated European of the period would have recognised in Descartes' use of vortical motion, and develops their implications for the disciplinary games that the 'mathematical' Descartes played in developing an alternative natural philosophy.

Yet despite the self-proclaimed novelty of his cosmological system, the causal explanations of Descartes' physics have affinities with the methods of the mixed mathematical science of mechanics. In her chapter, Helen Hattab explores the continuities and differences between the Aristotelian tradition in mechanics and Descartes' mechanistic view of causation and scientific explanation. She does this by focusing on the pseudo-Aristotelian *Quaestiones Mechanicae* (attributed to Aristotle, first printed in Latin in 1517) and a series of commentaries and other texts that took up its subject matter during the sixteenth century. These texts developed a form of explanation that, while not in contradiction with Aristotelian physics, nevertheless offered an alternative—one based on geometrical principles rather than the four causes. Her analysis provides then a basis for comparing the project of these Renaissance mathematical practitioners and humanists, who understood mechanics as the investigation into the causes of mechanical devices and wondrous effects, with Descartes' endeavour to apply the principles of mechanics to natural philosophy as a whole.

Thus it can be seen that Descartes' vortex theory provides an illuminating case study of the emergence of a new form of natural philosophising in the early to mid seventeenth century, and of some of the tactics he employed to exploit and transform the mixed mathematical sciences in the service of his natural philosophical gambits. Yet Descartes was not a lone player in these regards, nor were relations with mathematics and the mathematical sciences the only issues at stake. Stephen Gaukroger shows in his chapter on the autonomy of natural philosophy that some of the deep epistemological issues tied up with the emergence of natural philosophy as an autonomous discipline, and the concomitant realigning of disciplinary boundaries, can be best brought out by a comparison of some of the central players. To this end, Gaukroger provides parallel treatments of Galileo, Francis Bacon and Descartes to illustrate the protracted process by which natural philosophy extricated itself from a Christianised Aristotelianism and became established as a discipline in its own right. This process involved not only a rearticulation of the boundaries between natural philosophy and pre-established theological truths, but also careful manoeuvring by leading natural philosophers within their own intellectual and social milieux. It also involved deep epistemological issues concerning the relation between justification and truth as illustrated by the Renaissance debate over the immortality of the soul.

Gaukroger's concern with the relation between theology and natural philosophy becomes the focal point of the next chapter by Peter Harrison, who examines the origins and contours of the hybrid discipline of physico-theology in the seventeenth century. The early modern period witnessed the emergence of a number of hyphenated disciplines and modes of explanation. Physico-mathematics and physico-theology are perhaps the best known of these mixed disciplines, but in numerous works we also encounter physico-chemical, physico-medical, or physico-mechanical accounts of natural phenomena. All represent revisions of the traditional disciplinary boundaries inherited from the scholastics. Historians of the period have become increasingly aware of the significance of the introduction of mathematical principles into natural philosophy whether under the model of mixed mathematics or physico-mathematics. 'Physico-theology', however, is generally assumed to be simply a synonym for 'natural theology', and thus of marginal interest as a specific category in discussions of the identity of seventeenth-century natural philosophy. Harrison's chapter explores analogies between physico-mathematics and physico-theology, and suggests that the emergence of the latter discipline also sheds important light on the identity of early modern natural philosophy. In particular, he shows how some individuals dealt with the problematic issue of the extent to which theological concerns could have a legitimate place in natural philosophy. He therefore also addresses the broad question of the extent to which early modern natural philosophy was an inherently religious activity.

Of Floris Cohen's three modes of acquiring knowledge of nature: the mathematisation of natural phenomena; the emergence of kinetic corpuscularianism; and the rise of Baconian experimentalism, it is the latter which receives detailed treatment in the final two chapters of this collection, in ways indicative of our focus on patterns of change in a wide culture of natural philosophy in the seventeenth century. First Luciano Boschiero offers a case study of experimental work at the Accademia del Cimento in Florence, the first of the new scientific institutions, along with the Royal Society of London and Parisian Académie des Sciences, to embrace

Baconianism as its public legitimatory rhetoric. Exploring the academicians' attempt to resolve debates about the rings of Saturn, Boschiero reinforces the importance of focusing upon natural philosophy as a wide, complex and evolving field of natural inquiry. He shows the continued existence within the Accademia of personal and group agendas in natural philosophy that framed experiments and the accounting of their results, thereby promoting competition and tension amongst the members. This contrasts with the strict maintenance by the Accademia's Medici patrons of a uniform public rhetoric of inductivist experimental methodology, supposedly issuing in a consensually agreed harvest of atheoretical matters of fact. Boschiero concludes that recent concentration by some historians on this rhetoric, whilst correct and useful, has had the unfortunate, if often unintended consequence of occluding the continued natural philosophical theorising and conflict that marked the actual knowledge making practices inside the Accademia.

In the final chapter Peter Anstey traces, in the case of England, some often overlooked elements in the growth and triumph of Baconian discourse in the self understandings and public representations of natural philosophers. He shows that references to 'experimental philosophy', 'observation and experiment' and a rejection of 'speculative hypotheses' were commonplace in early modern English natural philosophy. Yet what is invariably overlooked is that these terms mark a fundamental distinction in discussions about natural philosophical methodology from the 1650s on. This is the distinction between experimental and speculative natural philosophy. Anstey argues that the experimental/speculative distinction provides the basic terms of reference by which early modern English natural philosophers understood their practice and theoretical reflections on natural philosophy. Robert Hooke's comment, from which the title of this book derives, captures the sentiment nicely.

> The truth is, the Science of Nature has been already too long made only a work of the *Brain* and the *Fancy*: It is now high time that it should return to the plainness and soundness of *Observations* on *material* and *obvious* things.[4]

Anstey claims that this distinction transcended disciplinary boundaries within natural philosophy and beyond to medicine and that it appears to have been set in sharper focus in the 1690s when English anti-hypotheticalism reached new heights and when a 'dumbed down' version of the Cartesian vortex theory was paraded as the paradigm speculative hypothesis. Furthermore, the distinction provides us with a hitherto neglected methodological context for the interpretation of Newton's notorious comments on the value and role of hypotheses in natural philosophy.

The eight studies in this collection were inspired by a shared but not doctrinaire commitment to exploring problems about the Scientific Revolution from the perspective of continuity and change in the culture of natural philosophy, rather than within the more usual narratives of the origin of 'modern science', either by *de novo* discoveries of method or fact, or by heroic defeat of older regimes of knowledge. The cumulative effect of the studies presented here certainly is not intended to be the provision of a definitive analysis of the early modern discipline of natural

---

[4] Hooke 1665, The Preface, b1.

philosophy and its linkages and boundaries with other intellectual and artisanal pursuits. Rather, it is hoped that the present studies will inspire further research into that complex set of relations and the process of disciplinary definition that natural philosophy underwent in the seventeenth century.

REFERENCES

Cohen, H. F. (1994) *The Scientific Revolution: A Historiographical Inquiry*, Chicago: University of Chicago Press.

Cunningham, A. (1988) 'Getting the game right: some plain words on the identity and invention of science', *Studies in History and Philosophy of Science*, 19, pp. 365–389.

— (1991) 'How the *Principia* got its name; or, taking natural philosophy seriously', *History of Science*, 24, pp. 377–392.

Cunningham, A. and Williams, P. (1993) 'De-centring the "big picture": *The Origins of Modern Science* and the modern origins of science', *British Journal for the History of Science*, 26, pp. 407–432.

Dear, P. (1991) 'The church and the new philosophy' in *Science, Culture and Popular Belief in Renaissance Europe*, eds S. Pumfrey, P. L. Rossi and M. Slawinski, Manchester: Manchester University Press, pp. 119–139.

— (1995) *Discipline and Experience: The Mathematical Way in the Scientific Revolution*, Chicago: University of Chicago Press.

— (2001) 'Religion, science and natural philosophy: thoughts on Cunningham's thesis', *Studies in History and Philosophy of Science*, 32, pp. 377–386.

Gaukroger, S. W., Schuster, J. A. and Sutton, J. eds (2000) *Descartes' Natural Philosophy*, London: Routledge.

Harrison, P. (2000) 'The influence of Cartesian cosmology in England' in *Descartes' Natural Philosophy*, eds S. W. Gaukroger, J. A. Schuster and J. Sutton, London: Routledge, pp. 168–192.

— (2002) 'Voluntarism and early modern science', *History of Science*, 40, pp. 63–89.

Henry, J. (2002) *The Scientific Revolution and the Origins of Modern Science*, 2nd edn, London: Macmillan.

Hooke, R. (1665) *Micrographia*, London.

LeGrand, H. E. ed. (1990) *Experimental Inquiries: Historical, Philosophical and Social Studies of Experiment*, Dordrecht: Reidel.

Olby, R. C., Cantor, G. N., Christie, J. R. R. and Hodge, M. J. S. eds (1990) *Companion to the History of Modern Science*, London: Routledge.

Pumfrey, S., Rossi, P. L. and Slawinski, M. eds (1991) *Science, Culture and Popular Belief in Renaissance Europe*, Manchester: Manchester University Press.

Schuster, J. (1990) 'The Scientific Revolution' in *Companion to the History of Modern Science*, eds R. C. Olby, G. N. Cantor, J. R. R. Christie and M. J. S. Hodge, London: Routledge, pp. 217–242.

— (1995) 'Descartes *Agonistes*: new tales of Cartesian mechanism', *Perspectives on Science*, 3, pp. 99–145.

Schuster, J. and Watchirs, G. (1990) 'Natural philosophy, experiment and discourse in the 18th Century: beyond the Kuhn/Bachelard problematic' in *Experimental Inquiries: Historical, Philosophical and Social Studies of Experiment*, ed. H. E. LeGrand, Dordrecht: Reidel, pp. 1–48.

Shapin, S. (1994) *A Social History of Truth: Civility and Science in Seventeenth-Century England*, Chicago: University of Chicago Press.

H. FLORIS COHEN

THE ONSET OF THE SCIENTIFIC REVOLUTION

Three Near-Simultaneous Transformations

This chapter deals with the radical transformation in modes of pursuing nature-knowledge that took place in Europe in course of a few decades around 1600.[1] My principal thesis is that this transformation involved three very different modes of acquiring knowledge about nature.

1. The mathematical, broadly 'Alexandrian' portion of the Greek legacy, after undergoing several centuries of reception and enrichment in Islamic civilization and then in Renaissance Europe, was turned, by Galileo and Kepler alone, into the beginnings of an ongoing process of mathematisation of nature, a process that was sustained and articulated through experimentation.

2. The broadly 'Athenian' portion of the Greek legacy, which consisted of four distinct, rival systems of natural philosophy with Aristotle's paramount, was replaced, at the instigation of Descartes and a range of other corpuscularian thinkers, by a natural philosophy of atomist provenance yet decisively enriched with a Galileo-like, mathematical conception of motion.

3. A quite specifically European-coloured mode of investigation bent upon accurate description and practical application that had started to emerge by the late fifteenth century began to consolidate around 1600, largely under the aegis of Francis Bacon's calls for a general reform of nature-knowledge, into an empiricist and practice-oriented form of experimental science.

What we are wont to call the Scientific Revolution consisted in these three by and large simultaneous transformations, plus an unprecedented amount of fruitful exchange amongst the resulting modes of investigation of nature over the remainder of the seventeenth century—a process leading up to and including Newton's *Principia* and *Opticks*. Thus, out of the revolution came three distinct modes of nature-knowledge of a kind the world had not seen before. Of these, the decisive mode was the program and practice of mathematisation of nature which was of universal import. Fact-finding experimentalism constituted a lesser mode, as yet very much coloured locally. Meanwhile the kinetic-corpuscularian mode of pursuit of nature was by far the most widely adopted at the time and certainly had more than local appeal, yet was of an essentially transient nature.

---

[1] Several portions of the present paper overlap with passages in other publications of mine, all written with distinct, non-overlapping audiences in mind, to wit, Cohen 2001, 2004a and 2004b. Also, some are lifted more or less verbatim from the book mentioned in the text to note 3.

9

P. R. Anstey and J. A. Schuster (eds.), The Science of Nature in the Seventeenth Century, 9-33.

Given this sort of analytic framework, if we wish to understand how modern science could arrive in the world, we must ask how, around 1600, these three almost simultaneous transformations could come about. From that, a further question emerges—how did such kernels of 'recognisably modern science'[2] manage to stay in the world once they arrived there? To address that question we must examine the built-in dynamics of the three modes of thought and concomitant practice thus produced, and the nature of their interactions in the course of the seventeenth century, as well as their differential societal appeal and anchoring. In the present chapter the focus is very much on the former question. In seeking to explain, at least in rough outline, the arrival in the world of basic elements of recognisably modern science, I present here a range of salient points that I treat at much greater length in the first half of a forthcoming book, provisionally entitled 'How Modern Science Came Into the World: A Comparative History'.[3] Inevitably, the three transformations that stand at the centre of my present argument are being painted in very broad brushstrokes, with many an issue which here I dispose of in a few sentences (if at all) taken up there at section- or even chapter-length. Consideration of many possible objections also falls to the wayside — the reader is invited to read what follows as the kind of 'ideal type' argument needed to draw so big a picture in so limited an amount of space.

1. CAUSES OF WHAT? A COMPARATIVE APPROACH

So much by way of introduction. Now, our first substantive task is to examine the principal components of European nature-knowledge on the eve of their radical transformation. In doing so, we need to take into account the diverse, cultural constellations in which these components found themselves over their respective life-times in a variety of distinct civilizations, notably those of Hellenism, of Islamic civilization, and of medieval and of Renaissance Europe. For if we fail to do this, we sacrifice the most significant source for subsequent causal analysis. The point, therefore, is to make our search for causes of the Scientific Revolution a *comparative* one. This, in my view, is indispensable if the causal investigation of the past is to avoid the indiscriminate piling up of an, in principle, unending array of antecedent events and circumstances. What was unique about what happened in Europe in the early seventeenth century can be brought out best by comparing it to what happened in other civilizations than the one that, through a typical blend of coincidence and causally linked chains of events, was indeed to create 'recognisably modern science'. However much Islamic civilization and medieval and Renaissance Europe surely differed, one thing they did have in common was that their respective pursuits of nature-knowledge only burst into life after they had taken up and sought to master what the Greeks had previously achieved. Their achievement then, provides our point of departure.

---

[2] Stillman Drake frequently used this felicitous expression in his books and articles on Galileo.

[3] Since in that book I list the scholarly resources drawn upon in every successive chapter, in the notes for the present paper I refrain from indicating more than just provenance of literal quotations.

2.TWO MODES OF NATURE-KNOWLEDGE IN ANCIENT GREECE

The key point regarding nature-knowledge in the Greek tradition is that it was pursued in two fundamentally different modes. One was mathematical science, which had its centre in Alexandria and which we shall therefore label here by that city's name for short; the other was natural philosophy, which (in the original period of school formation certainly) was centred in Athens. Both certainly went back to identifiable strands in pre-Socratic thought, yet developed from there in quite distinct, indeed, in almost fully separate ways. I propose to outline in seven points the nature of the contrast between the Athenian and the Alexandrian modes of pursuing knowledge of nature, and in order to give the reader a 'feel' for our otherwise somewhat abstract, successive points of contrast, I shall illustrate these points by reference to issues about the properties of sound.

Questions about sound appeared in all four Athenian schools of natural philosophy. Leaving Platonism aside (where it played a very subordinate role), we find Aristotelianism most concerned with a qualitative account of the perception of sound (details of which we also leave aside here), whilst the Atomists, as well as Stoics, concentrated on its mode of propagation. Atomists, in the wake of Democritus taking the world as made up of particles moving through void space, consequently took sound to be produced when atoms, pressed out of our throat or other vessel, on their flight through empty space strike our eardrum. Stoics, taking the world as made up of *pneuma*, a material/spiritual, air/fire-like substance in dynamic equilibrium, consequently took sound to be a disturbance of such an equilibrium reaching our sense of hearing the way wavelets produced by a stone thrown in a quiet pond propagate. In Alexandria, meanwhile, neither the production, the propagation, nor the perception of sound received special attention, but rather (in the wake of the Pythagoreans) the phenomenon of consonant sound, or, to be more precise, the empirical fact that the very musical intervals which strike us as sounding well are produced by strings of lengths in ratios of the first few integers (the octave (C–c) 1:2; the fifth (C–G) 2:3, the fourth (C–F) 3:4). Upon this observation they erected a mathematical discipline called 'harmonics', in which they examined properties of the 'harmonic', i.e., consonance-generating numbers. In this regard it stood opposed to a much more directly empirical analysis of music, emerging from Aristotelianism, centred not on harmonic relations but on the flow of the melody.

From this brief example, our set of seven pertinent contrasts may be gleaned.

1. In Athenian thought the central operation was explanation through the positing of first-principles; in Alexandrian thought, description in mathematical terms. First-principles of various kinds were put forward by a range of Athenian thinkers. What these first-principles held in common was, indeed, their being *posited*, with a blend of inner self-evidence and external, empirical illustration serving to underwrite their status as certain rather than probable knowledge. Such certainty was held to be both attainable and actually attained. Alexandrian thought had no use for any such first-principles. Its sole aim was to establish mathematical regularities without explanatory pretensions or underlying ontology; however, it did likewise lay claim to certain knowledge, with one of

the greatest representatives of Alexandrian thought, Ptolemy, deeming natural philosophical knowledge to be 'guesswork'.[4]

2. Athenian thought subsisted in four schools engaged in ongoing rivalry and (over time) alternating paramountcy. In addition another tradition, scepticism, opposed in principle the very possibility of the certain knowledge each claimed to have actually attained. Alexandrian thought appeared in one mode only, with the differences amongst its practitioners being solely over subjects examined and/or results arrived at.

3. In Athenian thought empirical phenomena appear as samples, chosen primarily in view of their capacity to illustrate the validity of the first-principles posited. In Alexandrian thought, empirical phenomena serve as individual points of departure for mathematical analysis. Each school of Athenian natural philosophy was ideally capable of explaining each and every natural phenomenon in terms of its own first-principles, which after all embrace the whole world, with no exceptions. In practice, however, empirical evidence served primarily to make the first-principles plausible (so, for example, in Stoicism empirical wavelets help us understand, by way of a profound analogy, what *pneuma* is, thus lending further credence to its existence and imputed properties). In Alexandrian thought, just as a vibrating string gave occasion to observe the numerical regularity of the consonances, other objects of sense, like beams in equilibrium or mirrors or lenses or planetary trajectories could give rise to mathematical analysis, provided they proved susceptible to such treatment.

4. In Athenian thought the aim, and the claim, was to gain a solid grasp of reality; in Alexandrian thought, real phenomena quickly vanished behind a process of ever increasing abstraction. The reality Athenian thought was seeking to grasp was our everyday reality, considered from a special point of view (this is true even of Platonism, so concerned to overcome everyday reality). By contrast, Alexandrian thought became ever more abstract the farther the process of mathematical idealisation went. Archimedes' proof of the law of the lever applies, not to real balance beams with real weights suspended, from which his analysis took its point of departure, but to straight lines to which numbers denoting weights have been assigned. Similarly, once the integer ratios for the consonances had been established, there was no trace of an inquiry into the nature of the vibrations produced by the string at its various lengths. In short, natural philosophy was about reality, grasped (with few exceptions) qualitatively; mathematical science about abstract entities treated with exactitude.

5. Athenian thought was comprehensive, Alexandrian piecemeal. The aim of Athenian thinkers was to grasp the whole; to explain the world or at the very least to understand that which gives the world the inner coherence they assumed identifiably to exist. The natural world was only a portion (in some cases, rather a subordinate portion) of all that had to be understood, in that the nature and mutual dealings of human beings, our place in the world, and how we can arrive at knowledge of all this in the first place, was likewise subject to the kind of

---

[4] Ptolemy 1984, Section I: 1.

understanding sought, called 'philosophy' for short. Alexandrian thought was none of these things. Investigators went about their researches one at a time, without positing or even seeking any necessary coherence between them, with the sole common thread being the mode of investigation applied, that is, the application of known mathematical theorems and properties.

6. Athenian thought spread out from four schools in Athens over the length and breadth of the Hellenistic, then the Roman world over a period of seven centuries, by means of coherent successions of teachers and disciples. Alexandrian thought, while fed from intellectual resources in a variety of Mediterranean cities (besides Alexandria also Rhodes, Perga, Syracuse), was and remained focused throughout on the Alexandrian centre and was cultivated on a more than individual scale for some two centuries; that is, by a few mostly isolated individuals, such as Ptolemy, for some two centuries more.

7. Any educated person could take part in philosophical debate, whereas to contribute to Alexandrian mathematical science required highly specialised skills. Philosophers in the Athenian mode filled an obvious social role in helping people make sense of the world at large; Alexandrian science could survive only for as long as the one powerful court that held a sustained interest in its doings persisted in its interest—which in antiquity was true only of the Hellenistic kings of Egypt, for reasons at which we can only guess.

3.MUTUAL ISOLATION

The next essential thing to grasp is how thoroughly the pursuit of these two distinct modes of nature-knowledge went ahead in mutual isolation. This applies both to practitioners (no philosopher was also a mathematical scientist or the other way round) and to contents. No Stoic or atomist thinker sought to link up his conception of sound with Euclid's account of consonant sound; no mathematical scientist sought to enrich that account with a notion of sound propagating by way of either wave-like processes or the emission of particles. The separation was not, to be sure, entirely rigid. Both Athenian cosmology and Alexandrian mathematical astronomy took their point of departure in the self-evident conception of a fixed, central earth, and there were a few more overlaps. More than that, on two specific occasions attempts were undertaken at reconciliation or even fusion. Toward the end of the Golden Age of Alexandrian mathematical science some results of mathematical astronomy and elements of Aristotelian cosmology were jointly put into an astrological synthesis held together by the basic tenet 'as above, thus below'. Further, Ptolemy's overwhelmingly mathematical work in planetary theory, in optics, and in harmonics testifies to an awareness of the gap between the two modes in that in each case he sought to bridge it. For example, he sought to reconcile Euclid's analysis of music in terms of the ratios of consonant intervals with Aristoxenos' Aristotle-inspired, perception-based account in terms of melodic flow. Such attempts by Ptolemy to infuse abstract mathematical analysis with some greater degree of 'reality content' look hardly less misconceived from our modern point of view than the attempt at astrological synthesis. After all, that modern point of view has irredeemably been shaped by the kind of 'mathematical realism' introduced by Kepler and Galileo, and at bottom reconfirmed ever since (on which more below, of course). But this is not the main point of these mistaken efforts at synthesis. Their main point is rather that the very effort to overcome the gap is witness to its presence throughout antiquity (and way beyond), with Ptolemy's very failure suggesting that there was no obvious or easy route toward doing a better job in this regard.

In sum, whereas in our modern era the big problem is to preserve quality in a world of quantity, in the intellectual legacy of the Greeks the issue—definable, obviously, only in retrospect—was quite the reverse. Not only in Greece, but everywhere humankind was living in what that most perceptive of historians of science, Alexandre Koyré, once called the 'world of the more-or-less'.[5] In that world, so hard for us to recapture nowadays, the problem for the mathematical sciences was rather how to find a place for quantity in a world of quality. Recall how extremely tenuous the connection with reality actually was even in these few mathematised bits and pieces of science. With the fictitious and/or purely numerical handling of planetary trajectories, musical intervals, and the like, the only remaining, somewhat solid points of connection between the empirical world and its mathematical treatment were the mirror, the five simple machines known to obey the law of the lever, and regularly shaped bodies floating in water; and even these were treated in a thoroughly idealised manner. With so little quantity introduced into so relentlessly qualitative a world, it should not come as a big surprise that no breakthrough toward mentally conceiving a world of quantity occurred at this point (which is not to say that such an event would have been wholly impossible). With a bow to Koyré's terminology once again, we might express the utterly marginal position of the mathematical branches of Greek nature-knowledge by stating that they formed little pockets of mathematical precision inside a world of the more-or-less, without there being any significant occasion to think that they might be turned into kernels of a new, entirely unheard-of 'universe of mathematical precision'. We can say, guided by hindsight, that the Greek heritage was inherently capable of such an outcome; we cannot say that such an outcome was bound to occur *either then or at any later time*.

4. AGENTS OF ACTUALITY

What, then, was required to turn what was potentially there into actuality? The primary answer is that, due to a range of wholly unrelated, world-historical events—military conquests mostly—the Greek legacy became subject to a range of *cultural transplantations* and thus gained the very sort of opportunities for creative innovation that have so often in history gone with the meeting, or the clash, of cultures. That is to say, potentials inherent in the Greek legacy now got chances to unfold, and every subsequent feat of cultural transplantation entailed such chances afresh. From the perspective of the creator-civilization, once fresh developments

---

[5] See his 'Du monde de l'"à-peu-près" à l'univers de la précision' in Taton 1966.

turn into tradition, they tend to stifle and become routine. But, considered from the viewpoint of the receiver-civilization, the very effort required to master and appropriate a tradition foreign to one's own ways, may set free energies to go ahead and enrich or even, *under particularly propitious circumstances*, radically transform it.

5.THREE RECEPTIONS DISTINGUISHED

No such transformation did occur in the course of the *first* reception of the Greek legacy, which took place in Islamic civilization. In mathematical science as in natural philosophy, the legacy was adopted, expounded and creatively extended. In mathematical science it was enriched with new theorems here, new geometric tools there, and with syncretist efforts and shifts of emphasis in natural philosophy amidst their continuing rivalry. This process left intact not only these two overall frames and modes of thought as the Greeks had produced them, but also, once again with very few exceptions, the intellectual as well as social chasm between them.

Nor did any large-scale transformation occur in the course of the *second* and far less complete reception of the Greek legacy, which took place in medieval Europe. This reception was really an exception in so far as one of the four schools of natural philosophy (Aristotle's, of course), right from the start became so dominant that it either drove its three immediate rivals as well as the Alexandrian mode underground, or scholasticised portions of them to the point that they were almost unrecognisable.

Nor, during its earlier stages in the sixteenth century did any large-scale transformation occur in the *third* reception of the Greek legacy, which was overall a much more balanced one like its Islamic counterpart had been. In the first place, full rivalry in natural philosophy returned as Platonic, Stoic, and atomist conceptions along with their sceptical nemesis, were restored through textual transmission and in scholarly debates. Furthermore, mathematical scientists in the Alexandrian mode, by means of a similar restoration of texts and theorems and proofs, sought to regain, both intellectually and socially, such terrain as had been occupied by their counterparts in the worlds of Hellenism and Islam, but had been lost during the reign of the schoolmen. Thus in Renaissance Europe mathematical humanists like Regiomontanus or Maurolyco soon found themselves in a situation such as had confronted earlier Islamic mathematical scientists like Thabit ibn Qurrah or Ibn al-Haytham. They moved beyond the sheer recovery of proofs and theorems, through the hesitant reconstruction of some material, which over the centuries appeared to have gotten irretrievably lost, and eventually became involved in even more hesitant attempts at improvement of portions of the inherited archive.

It is important to realise that almost all this humanist activity was aimed at recovery of lost knowledge which was now about to be restored to its original integrity. What innovation actually took place in this regard was the unintended by-product of an essentially backward-looking business—the sense that all that could be known had once been known already was, if anything, more outspoken in Renaissance Europe than it had been in Islamic civilization. This is true of the modest extension of Archimedean theorems on equilibrium states accomplished by the end of the sixteenth century in Stevin's work, or in the school of Urbino that operated under Guidobaldo del Monte's patronage; it is no less true of the restoration on the grand scale of Ptolemaic planetary astronomy undertaken by Copernicus half a century earlier.

Once more, then, just as had happened in Islam, just so in late Renaissance Europe some increasingly creative yet retrospectively modest enrichment took place, while leaving both the overall frames of the Athenian and Alexandrian legacies and the chasm between them fully intact. The number of practitioners was quite considerably larger and a narrower geographical scope, the printing press, religious controversies, and a proliferation of princely courts gave a certain increased speed and urgency to ongoing debates. Yet by the turn of the seventeenth century, there was little reason to anticipate any major break in a by now familiar, perhaps already somewhat worn-out pattern of such a 'renaissance' of Athenian natural philosophy and of Alexandrian mathematical science. Or, was there not, in fact, something more going on, something that portended a break in that pattern?

Note here that not all pursuit of nature-knowledge in Renaissance Europe was aimed at restoration. A great number of books appeared at the time with the word 'new' in their lengthy titles (written mostly in the vernacular). It can safely be said that they had nothing to do with this recovery-oriented movement of Greek or Latin writing humanists, but tended rather to be contemptuously dismissed by them. Instead, works advertising their novelty, like a Spanish book on American herbs typically translated as *Joyfull Newes of the Newe Found World* (1565/1577) belong to another mode of pursuit of nature-knowledge altogether. Their programmatic insistence on novelty, to be sure, stood for a confidently future-oriented, dynamic approach to things much more than for unalloyed, fully genuine originality. This vigorous current of thought of a novel kind had begun to manifest itself by the mid-fifteenth century, along with the humanist replay of the Greek performance, yet separated from that movement by a considerable intellectual and also social chasm. Here, one does not see the mathematical handling of a restricted set of geometric or numerical figures thoroughly abstracted away from selected pickings of natural reality, as in Alexandrian science. Nor is one presented with assorted pieces of real-life evidence adduced to shore up empirically a set of comprehensive principles established beforehand, as in Athenian natural philosophy. Rather, one finds here a dedicated striving for life-likeness, for factual accuracy and for exhaustive description. This is what came to mark domains as varied as anatomy (Vesalius), plant description (the three German herbalists, Garcia de Orta), the cataloguing of planets and stars (Tycho), or geography (Pedro Nunes, an assortment of scholarly and/or commercial mapmakers). This thirst for facts accurately rendered, as strongly exemplified in the work of Leonardo da Vinci, was accompanied by a strongly practical orientation. Paracelsian iatrochemistry and other currents of natural magic under the banner of Hermes Trismegistos offer the most spectacular examples of this action-directed aspect of Europe's third mode of nature-knowledge. But, the linking of the pursuit of nature-knowledge to matters of current concern was equally exemplified in the widely expressed aspiration to apply mathematics to practical problems in perspective, fortification, and navigation. Thus arose a new kind of knowledge intermediate between the artisans' design of ingenious devices for practical use and the lofty abstractions of Greek provenance taught (albeit in a simplified manner) in all of the universities of Europe.

6.EUROPE'S COERCIVE EMPIRICISM

It is now time to make a distinction between the culture-transcending nature of both the Alexandrian and the Athenian modes of pursuit of nature-knowledge, and the much more locally determined nature of the motley of activities just surveyed— locally determined, in that our third mode can be seen to reflect certain specifically European values. What, then, was so specifically European about hosts of accurate descriptions, the application of some mathematics to artists' problems, the emergence of several other possible interfaces between nature-knowledge and the crafts, the universal claims raised for chemistry, and a magical philosophy bent on the conquest of nature?

The answer is, in the first place, that several other long-term processes going on at the same time in Europe stand clearly reflected in those activities. Artists were similarly concerned with finding new modes of naturalist representation, and it is not by chance that we find Vesalius' atlas or Brunfels' and Bock's herbals illustrated by contemporary men of art, or Leonardo even blurring any distinction whatever between art and the pursuit of nature-knowledge. The voyages of discovery, too, shine through our third mode at many a spot, as in Orta's extensive descriptions of herbs and plants in India or in Nunes' pioneering work on navigation, with the whole enterprise as such turning into a powerful symbol of a forward-looking stance generally. An ongoing concern with machine tools and their labour-saving capacities can further be seen at work behind the scenes of the ongoing *rapprochement* between the pursuit of nature-knowledge and the crafts. For example, they are behind Leonardo's painstaking analyses of how machine tools work so as to optimise their effective power, and they are behind Agricola's creative survey of current mining practice.

Modes of naturalist depiction; explorations of foreign lands and peoples, and the invention, importation, and employment of machine tools, were surely not absent from other civilizations at the time. European uniqueness does not of course rest in that; it was of a more *restricted* kind in that what happened elsewhere in fits and fashions turned into far more sustained enterprises in Europe. What began as a comparably limited exercise in naturalist depiction by men like Giotto and Duccio turned in the end into a sustained, and by and large desacralised art no longer bound up with stereotyped modes of depiction. Vasco da Gama's voyage to India did not remain what it originally appeared to be—the regional counterpart to incidental voyages like Ibn Battuta's or even to far-flung expeditions like those under Chêng Ho—rather it turned into an early link in a chain essentially unbroken until the last blank spot on the map of the earth had been filled in. Similarly, in Europe the invention and/or importation of machine tools did not remain a matter of sporadic and incidental activity, but came to display a dynamic characterised by unusually eager reception, very quick spread, and a comparatively huge impact upon daily life, especially so in the cities. With justice these phenomena have been termed 'Europe's love affair with the machine'.

Two major driving forces may in their turn be identified at work behind all this. Both were specific to this fairly small and in hardly any other way particularly advanced subcontinent. One such driving force was a consistent *dynamism* arising out of a number of mostly contingent circumstances and historical developments, such as Europe's general lack of commodities fit for commercial exchange (so that what was desired had to be taken), the ongoing tensions between the few forces of unity and the many forces of (both geographically and linguistically reinforced) division, leading all conjointly to a certain restlessness, a profound lack of the kind of self-sufficiency that had come to mark those contemporary civilizations further east. The other major driving force was, not so much Europe's monotheist creed itself, which it of course shared with others, but rather the particular turn Latin Christendom gradually yet ever more consistently took, under the sustained influence of Europe's very dynamism, in a this-worldly direction of labour conceived as a mode of worship, and of nature as a gift of God to humankind to be exploited more or less at will. The Christian message *need* not be read that way; it *can* be read that way, and (as one consequence of the kind of overall dynamism just pointed at) it *was* increasingly read that way, thus serving both to interiorise the values that went with these secular forces of dynamism, and to sanction them from sources already lying ready for centuries in its own doctrine. That this is indeed how these things hang together finds some confirmation in the circumstance that Protestantism stands clearly over represented numerically in Europe's coercive empiricism (with men like Tycho, Ramus, Mercator, Blaeu, Paracelsus, or all 'big three' German herbalists). Protestantism, after all, was the variety of Latin Christendom in which the separation between nature and the divine (with humankind duty-bound toward both) appears in particularly marked fashion. In contrast Protestants and Catholics by and large took part in proportional measure in the ongoing recovery of the Greek legacy in mathematical science and in natural philosophy.

In sum, the mode of pursuit of nature-knowledge we have now been considering can be seen to reflect ongoing developments in Europe (a sustained naturalism in art, a sustained enterprise of exploration abroad, a sustained sense of enchantment and corresponding practice with machine tools) that in their turn reflect profound driving forces of a definably specific, European nature.

With their local-cultural origin thus traced, the various activities so far considered under the bland label of 'third mode of nature-knowledge' may be defined in more pointed fashion. At first sight we are dealing here with a quite incoherent range of endeavours, made up of a quite disparate set of probings, approaches, and results tried out and/or attained in a broad range of quite diverse domains of contemporary concern stretching from magnets to plant stems. What held together its various components was an approach to natural phenomena governed all over the range by a broad empiricism adorned (or, if one prefers, disfigured) by a marked desire to gain some measure of control over phenomena: hence our introduction of the label 'Europe's coercive empiricism'.

This broad, control-oriented empiricism had a potential, already manifesting itself in some quarters by the late fifteenth century, *to consolidate into more narrowly targeted, deliberate experiment*. Not that the act of experiment was so novel *per se*. But what *was* new was experiments being carried out, not by and large incidentally as before in alchemy or in some optical work, but in well thought-out, progressive series. That is what, hidden from the world at large, Leonardo was engaged in—systematically to set up a coherent range of experiments to find out, for instance, how materials behave under deliberately varied conditions of friction. That is also what in the 1530s the Portuguese admiral João de Castro undertook when systematically exploring the workings of the compass and the behaviour of waters, or Vincenzo Galilei in the 1590s when testing strings of deliberately varied qualities and parameters for their consonance-producing capacities.

7.LIKELY PROSPECTS AROUND 1600 AND THE BREAK THAT MADE ALL THE DIFFERENCE

So far we have surveyed the state of nature-knowledge on the eve of its grand transformation in terms of structures more than of detailed contents, and we have also inquired about how that state had come about. All in all, we may conclude that in Europe during the period of the Renaissance a familiar pattern had re-emerged. The Golden Age of Islamic nature-knowledge had witnessed the reception of the Greek legacy, its partial enrichment in accordance with its own accepted principles, and also a few activities (notably, mathematical determination of the direction and times of prayer) stamped by definitely locally-situated circumstances. Up to the end of the sixteenth century, the equally Golden Age of European nature-knowledge, for all the numerous local variations it of course displayed, was not, from a structural point of view, composed differently at all. Just as the former had come to an end at some point, as episodes of flourishing nature-knowledge customarily did in pre-modern societies, which possessed no built-in motor drive for the continual advance of such activity, so the structurally similar flourishing of European nature-knowledge seemed bound to proceed in the same mould for a while, but eventually to come to its own, locally determined yet natural end. The historical fact that, against all precedent, it did not then or at any later time come to an end, but instead ushered in the so far unbroken era of unceasing scientific expansion in the midst of which we ourselves live, is the direct consequence of the radical break in the pattern that occurred around 1600. What, then, did that break consist in?

Consider that those excitingly dynamic, future-directed inventions and discoveries that were taking place all over western Europe from mid-fifteenth century onward—its *Joyfull Newes*, so to speak—were all part of that locally rooted brand of control-oriented empiricism we have just surveyed. From a contemporary point of view, it may well be that this European mode of coercive empiricism would have represented, amongst a host of remote possibilities, the least unlikely place for radical novelty to occur. Nor, as we know, would such an expectation have been entirely wrong—it is indeed true that by this time the third mode was in for transformation of a kind. Below we shall see in slightly greater detail that its already apparent potential for condensing into a consistently experimental approach to natural phenomena began, by 1600, to be realised with hastening speed and also (in the guise of the Baconian gospel) to be given a consistent rationale of its own. Still, the truly radical transformation, the real break in the pattern, occurred quite elsewhere. It occurred inside a movement directed not forward but backward, not toward an unknown future confidently faced and actively prepared, but toward gradual extension along lines already drawn by the Greek pioneers—toward enrichment, that is, of the Alexandrian legacy. It occurred at the hands of two men whose intellectual ancestry, while surely going back in part to a variety of ancient philosophies of nature, was profoundly rooted in the mathematical approach to phenomena undertaken in prominent fashion by their respective scientific forebears, Ptolemy and Archimedes. These two men were Johannes Kepler and Galileo Galilei. Mathematical humanists themselves, they rose to transcend their ancestry and to turn it into something almost unrecognisably different; that is, into the 'universe of mathematical precision'. And, the decisively novel element they brought to the Alexandrian legacy rested in their consistent *realism*.

Earlier we discussed the extremely tenuous link obtaining between the rare subjects (planetary trajectories, mirrors, the lever, etc.) treated the geometric way in Alexandrian science and what little empirical reality ultimately underlay those subjects. Additionally, we noted that the very rarity of such incidental, quantitative treatment well-nigh precluded a quantitative world being thought up. All this now changed, and very quickly so.

Kepler operated throughout from the conviction that God had created the world in geometric fashion, so as to conform to harmonic models founded upon the musical consonances. Out of this came as one by-product a conception of the planetary system simultaneously mathematical and (this was the new thing) physical in the sense that the elliptical orbits he eventually determined for the planets and the two other laws he discovered about them, were held by Kepler to depict the true state of the solar system, not just fictional gadgets fit to 'save the phenomena'. The link with reality was established by means of a blend of considerations about God's harmony expressed in the world; of physical (mostly Aristotelian) argument about the celestial forces he held to be operative; and of sustained, empirical checking of the calculated outcomes (most notably so in the celebrated rejection of his earlier theory due to a quite small yet, to him, sufficiently significant quantitative divergence).

Likewise Galileo operated throughout from a religion-laden conviction that the only language fit to decipher our world is that of mathematics. Out of this came a novel, really mathematical conception of motion expressed in (among other things) his view that bodies tend to retain their motion once acquired; that the motion of a body can only be judged relative to the state of motion of other bodies, and that a body may be subject to various motions at the same time. The prime examples Galileo elaborated were a mathematical analysis of free fall, with distance covered, time passed, and velocity acquired being correlated by means of ratios geometrically expressed, and of the trajectory of projectiles through his derivation of their parabolic shape. The link with reality was established by means of experiment. That is to say, Galileo acknowledged the chasm between the empirical, everyday phenomenon of free fall (which does not nearly display the feature of uniform acceleration Galileo ascribed to it in the abstract) and the mathematical, 'ideal' reality of bodies descending through an imagined space with all impediments removed. His achievement was to bridge the chasm, by artificially imitating—as well as could be done with available tools—that imagined, ideal reality in the 'real' reality of near-spherical balls sliding down well-polished, wooden grooves along an incline, and by checking the quantitative outcomes obtained from a water clock against the quantitative outcomes yielded by prediction in the abstract.

In short, Kepler mathematised nature in the sense that a subject previously treated mathematically in a fictional way was now turned real ('physical') for the first time, whereas Galileo mathematised nature in the sense that a subject previously treated in realist ('physical', i.e. natural-philosophical) fashion was now being treated mathematically for the first time. In so doing, the two men jointly ushered in the 'universe of mathematical precision', by which expression we mean the positing of an *ideal world* (whether as truly existent or by way of a handy tool for the scientist) taken to be devoid of all pertinent real-life impediments, approachable through preferably mathematical laws or models or theories, and linked to our everyday empirical reality by means of deliberate experiment set up in thought or in an artificially created material reality by way of a more or less close imitation of that ideal world.

To the extent that Kepler and Galileo embodied this idea of science (the latter overall more so than the former) they may with justice be called the first modern scientists in a sense in which even Archimedes or Ptolemy were not. Which statement at once raises two questions. One is how, given the necessarily fragmentary nature of their achievement (with only a few, though hardly the least significant portions of our world being subjected as yet to such idealising, mathematical treatment), the universe of mathematical precision has managed to expand ever further, rather than being nipped in the bud as quite conceivably it might have? Our answer to this question involves the identification of a threefold dynamic underlying the broadly forward movement of science over the remainder of the seventeenth century, as expounded at length in part III of my forthcoming book. The other question, which we do take up here, is whether this altogether quite sudden transformation of the Alexandrian legacy into the onset of the universe of mathematical precision took place through sheer chance, or whether some definite cause or causes may with some confidence be assigned to the event? Whence, in short, these two giant strides from Ptolemy to Kepler and from Archimedes to Galileo? Why then? Why there? Why at all?

8.CAUSES OF THE FIRST TRANSFORMATION

The answer to the question 'why at all?' has really been given already. The onset of the universe of mathematical precision out of the Alexandrian legacy in seventeenth-century Europe was neither a miracle nor a fully determined (let alone a foreordained) event. We take it to be established through our preceding survey that (1) the transformation rested as a yet to be realised possibility in the Alexandrian legacy in mathematical science, and that (2) at the start of the Islamic Golden Age that legacy had proven capable of transcending the confines of the civilization from which it sprang. It follows that, in principle, every single case of transplantation of that legacy was to offer chances afresh for the transformation actually to occur. There is no *inherent* reason for why no Galileo-like figure appeared in, say, the 5th/11th century to cap the ongoing enrichment of the legacy by outstanding men like Ibn al-Haytham or al-Biruni. Still, those fresh chances differed in every case, and that is why causal analysis of the specific transformation we are discussing here is to be done by way of pointing at factors less or more propitious for such an outcome to be actually realised.

Why medieval Europe hardly provided a favourable environment for the outcome is obvious at a glance. With the Alexandrian legacy partly falling to the wayside and partly being disfigured in scholastic categories, not even modest enrichment was feasible, with the very best in this regard being the incidental quantification of a few Aristotelian categories displayed in Oresme's work. In the remaining cases of Islamic civilization and of Renaissance Europe the ongoing enrichment of the Alexandrian legacy *went on at a high level of performance in both*, but eventually petered out in the former yet gave rise to wholesale transformation in the latter. Here then is the nub of the problem we seek to resolve, at least in a schematic fashion: What propitious factors emerge from a sustained comparison between the two?

A very elementary consideration has to do with sheer numbers. The more intensively work in a creative spirit was being done on the legacy, the greater the chance that someone would hit upon the unrealised potential hidden near its core. Here Europe had an advantage over Islamic civilization simply by virtue of being next, just as (if Europe had likewise neglected to grab its chances) some successor civilization to which the legacy might once again have been transplanted would have had a further chance. In addition, the European university system, with its comparatively huge turnover of people equipped with at least a nodding acquaintance with Greek thought, provided a both deeper and wider soil on which true mathematical talent could grow and on occasion flourish than in Islamic civilization. Kepler's career is not at all atypical in the European context. Half-way through his theological studies he received an appointment as a mathematics teacher. But such a trajectory seems most uncommon in Islam. Note carefully the quite limited scope of this consideration. It is not at all meant to rule out the possibility of someone achieving broadly what Kepler achieved over his lifetime in a non-European context; it is meant solely to illustrate that paths toward a career in mathematical science were in Europe more numerous and, above all, more varied than they had ever been anywhere else. And, the more numerous and varied those career paths were, the greater the chance that a genius might come forward to perceive what no one else had so far perceived.

What, then, did those two geniuses perceive? We have already called attention to the decisive innovation of their shared *realism* in linking mathematics to the empirical world in a novel way. Hence, our search for propitious factors ought to be directed, as indeed it has recently begun to do among historians of the Scientific Revolution, toward the question of what may have helped turn these two men into mathematical realists in the sense defined. Two *a priori* plausible resources of realism lay more or less readily available in the two other modes of pursuit of nature-knowledge current at the time, that is to say, in the practice-oriented empiricism of our third mode and in the comprehensive understanding of the real world sought for in natural philosophy. Inside those broad resources we shall now specifically probe half a dozen, partly overlapping stimuli toward realism in the sense defined, none of which had much of a counterpart in Islamic civilization. These possible stimuli are: Tycho's data; the tradition in practical mathematics Galileo encountered in Padua; the uniquely wide, both intellectual and social chasm obtaining in Europe between the mathematical scientist and the philosopher; Copernicus' peculiar brand of cosmological realism; Kepler's personal vision of 'the astronomer as a priest of God to the book of nature',[6] and, finally, Jesuit promotion of 'mixed mathematics' within an Aristotelian frame.

1. *Possible stimuli toward realism arising out of Europe's coercive empiricism.* For all our insistence that the mathematical humanists and the practitioners involved in the varied activities of our third, empiricist mode of nature-knowledge were operating in splendid isolation from one another, it remains true that some osmosis did take place here and there. The clearest case of such osmosis is provided by Kepler's creative usage of the data patiently accumulated by Tycho which, among other significant feats, enabled him to perceive, as no one else did, the full consequences of the dissolution of the heavenly spheres those data appeared to imply. There is further good reason to assume that Galileo, who was still at the end of his Pisan days a pure Archimedean of the Urbino school, was turned into the man to transform that legacy at least in some part by his exposure to the practical application of basic mathematical insights during his Padua years. This is not to say that no other ways to transcend the legacy might have been taken under other circumstances—a possible counterpart in Islamic civilization would surely have done it differently. It is only to say that the presence of an environment enriched by this mode of broad, control-oriented and accuracy-seeking empiricism may reasonably be taken to have contributed its share to how the Alexandrian legacy was in historical fact transformed. Whether, and to what extent, the near-simultaneous rise of validation-oriented experiment in its universe-of-mathematical-precision context and heuristic experiment in its Baconian context were fully independent events we shall seek to find out a few pages further down.

2. *Possible stimuli toward realism having to do with natural philosophy.* One other ready-made, possible source of realism was provided by philosophy. After all, an understanding, based on first-principles, of the real world was the very thing the natural philosopher was after. But how to tap that source? And what might make it attractive for anyone located outside the realm of philosophy to do so? Here the legacy of the medieval period was decisive in many ways—never had mathematical science sunken so low in comparison.

• *The chasm between the mathematician and the philosopher.* Due to the almost complete submersion of the Alexandrian legacy under the undisputed reign of

[6] Letter to Herwart von Hohenburg of 26 March 1598, 'Ego vero sic censeo, cum Astronomi, sacerdotes dei altissimi ex parte libri Naturae simus ...'; Kepler 1938–, 13, p. 193.

scholasticism, the mathematical humanists had to fight an uphill struggle to regain some recognition when, after the fall of Byzantium, the Alexandrian legacy finally had a chance to enter the European scene. The mathematical humanists fought for both prestige and income against the entrenched philosophers, whose near-monopoly during the medieval period had put them in secure possession of the university system then freshly established. That the chasm was acutely felt to exist appears from a range of 'in praise of mathematics' speeches by means of which men like Regiomontanus contrasted the certainty of knowledge gained the mathematical way to the perennial intellectual rivalry, hence less than full certainty, that beset the domain of the natural philosopher. Previously the making of this point against the philosophers had been confined to the sceptics, who, unlike the mathematicians, had no positive alternative to offer. But only in a few cases did such advertising of the virtues of mathematical science lead to a transgression of the boundaries separating mathematics and philosophy—for instance no trace of an infusion with philosophical realism is to be found in Regiomontanus' own work. With a few other, later men, however, such traces are clearly detectable, though in somewhat peculiar ways. The first in time was Copernicus.

• *Copernicus' cosmological views.* The reader may well have felt that throughout our survey the contribution made by Copernicus to the Scientific Revolution has been almost ludicrously downplayed. Indeed, neither the publication of Copernicus' *De revolutionibus* in 1543, nor the overwhelming portion of its reception over the half-century that ensued is being taken here as the onset of the Scientific Revolution. *De revolutionibus* and the bulk of its reception history fits far better with our picture of the dominant sixteenth-century mode of recovery-with-some-enrichment of the Greek legacy in mathematical science. It is more enlightening by far to regard Copernicus as Ptolemy's last and greatest heir, who throughout books II–VI of *De revolutionibus* carried the hoary art of 'saving the phenomena' to new heights by means of his heliocentric, Aristarchus-inspired hypothesis. It is only in retrospect that this change can be seen to have been instrumental in setting afoot the upheaval in the tradition to which Copernicus himself belonged in almost every respect. And that outcome is largely due to the unambiguously realist interpretation of Copernicus' heliocentrism given to it by two exceptionally perceptive men, Galileo and Kepler. It is true, however, that these men, whose exceptional perceptiveness we are here seeking to explain, did find one ready source of realism in the introductory book I of *De revolutionibus*. In adducing a motley selection of *ad hoc* natural-philosophical arguments in the context of a simplified version of his system, Copernicus made the most of the realist claim laid down there that the earth really and truly circles the sun in a year and really and truly rotates around its own axis every twenty-four hours. That claim was at bottom irreconcilable with what he went on to do over the full remainder of his book. There he settled down for the real business of working in all required planetary details, which he did using time-honoured fictional gadgets honed by Ptolemy. Whence, then, Copernicus' claim for the reality of heliocentrism if he did not even seek to uphold it in Books II–VI of *De revolutionibus* where he carried out his principal job as a mathematical scientist, and if he failed (as he was himself more than a little aware) to produce convincing arguments for it in Book I? A truly satisfactory answer to this question has not so far been forthcoming, even though the tentative consideration reported above that such snatching of a piece of the philosophers' gown might have brought some rewards in terms of enhanced social prestige is not perhaps to be thrown out of court without a further hearing.

Far less controversial is the empirical observation that, over the period 1543–1600, when the fictional schema of books II–VI was as widely applauded and applied by the common run of mathematical scientists as Book I went ignored by them, no more than some ten scholars were prepared to take Copernicus' realist claim at all seriously. More than that, only two of these ten did so, not only without any remaining ambiguity, but also from a profound awareness that the inner tension in Copernicus' heliocentrism (with its broadly realist claim incongruously joined to a 'saving the phenomena' mode of operation) could, and ought, to be resolved, not by glossing it over but by transforming it. Kepler, then, transformed Copernicus' heliocentrism-in-Ptolemaic-fashion into what he called with justice a 'New Astronomy, Based Upon Causes, or Celestial Physics'. And Galileo went on actually to shatter the Aristotelian world-view disrupted only potentially by Copernicus, using for the purpose this novel, mathematical conception of motion he developed during his Padua years. So our net conclusion on this score appears to be that one more factor propitious to their realism resided in the highly incongruous and ambivalent precedent set by Copernicus in his Book I.

• *Kepler's hybrid philosophy of nature.* The occasional appeals to elements of natural philosophy one encounters with Galileo were of a rather eclectic and inconsistent kind and fail to reveal any fixed allegiance, whether to Platonic cosmology, to Aristotelian methodology, or even to atomist speculation, all of which appear in his mature work. With him philosophy came by and large *after* his newly realist mode of mathematical science, so that the former yielded none but *ad hoc* justifications for the realism he had come to instil in the latter.

With Kepler all this was quite different. Kepler was the first (and, but for Newton, the last) creator of a uniquely hybrid philosophy of nature. Like all prior philosophies of nature it operated on first-principles—in his case, the geometric, archetypal ratios God had, at the creation, worked into His created world at various strategic spots. In it, too, one finds an ongoing search for fitting empirical evidence. But the fundamental difference with the four Greek philosophies of nature is that these were essentially *closed* systems, in that their respective first-principles were constructs of reason alone, whereas Kepler from the start adopted a characteristically *open* stance toward the two issues basic to his own brand of philosophy. These were, what exactly those archetypal ratios would, on investigation, prove to be, and where and how God appeared actually to have placed them in nature. Not *a priori* reason alone, but reason freely developed *a priori*, yet both expanded and held in check *a posteriori* by empirical evidence, ought to serve as the arbiter in the search for a fully satisfactory account of the world at large. Kepler displayed an openness, unprecedented in natural philosophy, in putting mathematical modes of operation of Alexandrian origin to quite novel ends. Time and again he sought to anchor his hybrid mathematical-science-cum-philosophy in empirical reality, refusing to regard it as enduringly settled for as long as potentially countervailing empirical evidence had not been brought into line with the hybrid philosophy or the hybrid philosophy with the evidence. It is, therefore, his openness that in the last resort accounts for the peculiar mode of realism of his mathematical science. Indeed without this openness it would not have been a science at all but (for all its thoroughgoing Alexandrian inspiration) just an idiosyncratic piece of mathematical philosophy. Kepler's openness was clearly reinforced in its turn by the peculiar manner in which he had worked God into his first-principles, together with a profoundly un-Greek respect for the facts of nature as God-given and therefore to be accepted in all humility. He shared this respect with many other men of mostly Protestant allegiance at that and later times. It came to the fore most decisively in his celebrated rejection of an otherwise satisfactory theory of his own in view of a seemingly tiny, apparent discrepancy with Tycho's observational data.

 • *The Jesuits and 'mixed mathematics'.* In bringing the budding universe of mathematical precision (in the sense defined above) into the world, Galileo and Kepler were unique. Nothing of the kind was being done by anyone else at the time. But this is not to deny that some tentative infusion of natural-philosophical tenets with a certain amount of mathematical rigour was taking place as well during the decades around 1600, nor that Galileo in particular was well-acquainted with this current of thought and activity among the mathematical fringe of the Jesuit order. Here, however, natural philosophy definitely came first, and the main point made by men like Clavius in their struggle to have more mathematics taken up in the standard set of Jesuit courses was the need for a further elaboration of Aristotle's category of 'mixed mathematics'. The perceived similarity of such efforts to what Kepler and Galileo were doing was sufficient to make both men enjoy the support of mathematical Jesuits like Guldin or Clavius, respectively (until, for essentially personal reasons, the order turned against Galileo). To us the difference far outweighs the similarity, thanks to our hindsight derived in part from a host of contemporary misunderstandings. Indeed it may be doubted whether this fourth propitious factor contributed more than a little—mostly by enhancing somewhat Galileo's realist bent—to the transformation of the Alexandrian legacy. The same factor, however, appears to have been overall more propitious in view of quite another, almost contemporary feat of transformation, the one that took place over the 1610s–1640s in natural philosophy.

9. THE SECOND TRANSFORMATION AND ITS SOMEWHAT OBSCURE ORIGINS

When in 1618 Isaac Beeckman and young René Descartes (fresh from the school desks of a Jesuit college) happened to meet they congratulated each other with their rare, shared capacity to 'join mathematics with physics'.[7] What Descartes then meant by that expression becomes clear from the musical treatise he composed for his new

---

[7] Beeckman 1939–1953, 1, p. 244: 'Hic Picto cum multis Jesuitis aliisque studiosis virisque doctis versatus est. Dicit tamen se nunquam neminem reperisse, praeter me, qui hoc modo, quo ego gaudeo, studendi utatur accurateque cum Mathematica Physicam jungat. Neque etiam ego, praeter illum, nemini locutus sum hujusmodi studii'.

friend. It offered a fairly traditional kind of 'mixed mathematics' in which the basic data of musical consonance were subjected to somewhat more rigorously geometric treatment than usual. To Beeckman the expression meant something else. Much less mathematically gifted than Descartes, by 1618 he was already far advanced in his ongoing construction of a natural philosophy of a partly novel, at times somewhat quantitative kind. Having settled on atomism as the most appealing philosophy of nature, he went ahead to enrich it with a decisively new element. This involved a conception of motion quite comparable to what, unbeknownst to him, Galileo kept in store unpublished until the appearance, in the 1630s, of his *Dialogo* and his *Discorsi*. Ancient atomism had dealt in assumed particles moving in some unspecified way through the void. In Beeckman's brand of what may perhaps best be called 'kinetic corpuscularianism', the way assumed particles actually move became a good deal more specific. Recall, for instance, that Democritus and Epicurus had explained sound by invoking the emission, by some source, of sound particles which on arrival at the ear produce the sensation of sound heard. Beeckman now went on to explain consonant sound by invoking the emission, by the vibrating string or pipe, of sound particles of specifically different sizes and speeds such as to produce, once arrived at the ear, the sensation of consonant sound heard. Beeckman joined his general principle of the preservation of motion once a body has acquired it, to appeals to a variety of velocities and directions assumed for a variety of differently sized and shaped particles so as to explain a variety of natural phenomena and effects. This strategy turned him into the pioneer of the mode of natural philosophy that, in quickly outstripping all its rivals, was to dominate a good part of innovative, seventeenth-century nature-knowledge. Beeckman's manifest inability to organise into a publishable treatise his disjointed notes on a disparate range of topics opened the gateway for Descartes, his more gifted disciple. Arriving at essentially the same outcome—a doctrine of kinetic corpuscularianism—along a rather different path of development, Descartes gave systematic expression to this new natural philosophy during the 1630s in his unpublished 'Le monde' (*The World*) and then, in 1644, in his immensely influential *Principia philosophiae* (*Principles of [Natural] Philosophy*).

Whence this transformation? Four men brought it about by respectively blending ancient resources, creative gifts, and personal encounters in somewhat different fashions. They were, of course, in chronological order: Beeckman, Descartes, Gassendi and Hobbes. A glance at their motives reveals little commonality beyond (1) dissatisfaction with the dominant, Aristotelian mode of natural philosophy, and (2) a felt need to stick to the mode of knowledge that traditionally went with the Athenian approach: that of first-principles directed at grasping the world in its totality and shored up by means of apparently well-fitting fragments of empirical reality. Under such circumstances, the appeal exerted by atomism is not too hard to fathom. Rather, the question at issue is why did these men move beyond just reviving atomism, as did several of their contemporaries, to enrich it with their new doctrine of motion? And the crucial subsidiary question, therefore, is how did they arrive at that new doctrine of motion in the first place?

Note carefully that this new, broadly mathematical conception of motion was the only point of overlap—although, to be sure, a profoundly significant one—between the universe of mathematical precision on the one hand and kinetic corpuscularianism with its utterly different mode of attaining nature-knowledge on the other. This being so, was not the one just rooted in the other? In other words, was not the decisively novel feature used to transform ancient atomism into kinetic corpuscularianism simply transferred to the latter from outside, that is, from Galileo? With Hobbes this was undoubtedly the case; with Gassendi it is highly likely; with Descartes it is something of a wild guess; and with Beeckman, who never had a chance to learn about Galileo's pertinent work, it is out of the question.

Beeckman's documented originality on this score proves in its turn that Galilean principles such as the preservation and the relativity of motion could indeed be independently developed in a general context of natural philosophy. And what Beeckman could pull off is not of course to be held beyond the powers of Descartes. Still, the shrillness with which the latter once denied ever to have owed anything to Galileo, in tandem with the silence in which he always enveloped his wanderings around Italy in the early 1620s (when Galileo's views on motion began to spread by word of mouth) may appear significant to those familiar with Descartes' obsession with priority and his gift for prevarication in a similar case, that of Beeckman and the explanation of musical consonance. But, in the end, this is bound to remain a matter of historical speculation. Nor does a further search for factors propitious to the turn of these men toward a mathematical conception of motion seem to be as productive as, to some extent, our comparable search for Galileo's and Kepler's turn toward realism has appeared to be. Let us move on instead to our third transformation, that is, to how a mode of pursuit of nature-knowledge marked by coercive empiricism began to consolidate around 1600 into the onset of fact-finding experimental science.

10. THE THIRD TRANSFORMATION AND ITS COMPARATIVELY SMOOTH COURSE

In conformity with the fairly disjointed nature of the motley of activities we have categorised together as Europe's 'coercive empiricism', the process of consolidation into a still fairly motley collection of experiment-infused sciences took place over a rather wide front. Thus, Gilbert's experimental treatment of the enigmatic phenomena associated of old with amber and the lodestone gave rise to a more generalised study of electricity and magnetism. Van Helmont subjected many Paracelsian tenets to experimental tests. Harvey drew experimentally confirmed conclusions from work done by Vesalius and later, likewise observation-prone anatomists. Bacon proposed an orderly ascension of systematically compared, experimental investigations of phenomena like heat or sound, subsequently carried out by hosts of scholars inspired by his broad vision of a general reform of nature-knowledge. Note carefully that this transition from a control-oriented empiricism to more systematically undertaken series of experiments, while surely the most significant innovation from our modern perspective, did little to undo such other features characteristic of Europe's mode of coercive empiricism as its numerous magical overtones and its profoundly organicist conception of things generally.

Activity in nature continued to be regarded primarily as the varied manifestations of a variety of forces, of attraction and repulsion, some material some spiritual, and more or less hidden ('occult') to human understanding. It is in this context that we must understand Bacon's call for a general reform of knowledge directed toward the conquest of nature for the benefit of humanity at large. It was as much a striking summing-up of all that seemed most dynamic and forward-oriented in what had been achieved in this coercive-empiricist mode of approach over the last century or so, as it was a ringing declaration of a daringly new, experimental science to be erected upon that very foundation.

Considered in this light, our customary 'why' question almost resolves itself. In this one case among our three transformations, what we are watching here is not a fairly sudden break involving a radically novel element—*kinetic* corpuscularianism; mathematical *realism*. Rather we have here an altogether rather smooth transition in the course of which one element already incidentally present—progressive series of experiments deliberately set up to detect hidden properties of nature—received a further boost in the same general direction of its constitutive values. Nor is there any difficulty in finding out why this particular transformation took place in Europe, so much were typically European values embodied in the whole movement. As we have insisted before, a Galileo-like figure may by some little stretch of the historical imagination be conceived to have arisen in Islamic civilization, but nothing even remotely comparable to so typically European a figure as Francis Bacon can by any stretch of that imagination be held possible to have come forth elsewhere.

What remains to be examined on this score, then, is solely the *timing* of this relatively smooth process of consolidation. Is the bald fact of history that it coincided in time with the emergence of the universe of mathematical precision just that—a coincidence? However hard it is to believe this, rare indeed are the facts pointing in another direction. A case can certainly be made that Galileo must have learned something from his father's musical experiments. Nor is any doubt possible that both Kepler and Galileo made grateful usage of some of Gilbert's magnetic insights, in seeking respectively to solve some major puzzle. But this is hardly tantamount to ascribing their radically novel approach to natural phenomena generally to incidental influences like these. More than the one propitious factor already invoked among others cannot be made of it. Nor, conversely, is there anything that points toward the consolidation of Europe's coercive empiricism having benefited from the simultaneous transformation of the Alexandrian legacy so glaringly ignored by Bacon, in particular. It is true that within decades the two principally different modes of experiment involved in each (heuristic in the one case, directed at validation in the other) were almost to blend with many a subsequent thinker. Here Mersenne's plodding yet Galileo-inspired experimentalism provides a uniquely early case. It is further true that that distinction does not even hold fully in the case of Galileo himself, who carefully effaced his previous, tentatively experimental search for natural regularities and in his published work confidently declared them experimentally proven *a posteriori*. Still, where the timing of our third transformation is concerned, it is not really satisfactory to leave it at the observation that this was just one possible, next step to be taken in the inner dynamics that noticeably propelled the mode of control-oriented empiricism forward.

This left-over issue of simultaneity, then, is the point where we appear to have reached the outer limits of our original explanatory strategy. That strategy has been guided by a determination not to commit ourselves to any among the customary plethora of broad causes purportedly covering the Scientific Revolution taken as one undifferentiated whole, prior to setting up a search for *specific* explanations mapped as carefully as we could upon *specifically* distinguishable portions of that truly complex event. Only upon completion of that search for specifics have we now put ourselves in a suitable position to ask what causal gaps still remain.

11.EXPLANATIONS AND THEIR LIMITS

Two distinct kinds of gap seem readily to present themselves. The first rests in what may well strike us as a certain poverty in the range of propitious factors encountered along the way. We have been dealing here throughout with events of extraordinary significance for the course of human history: [1] the onset of the subjection of natural phenomena to mathematical analysis in the frame of an intricate structure of idealised abstraction and empirical/experimental confirmation; [2] the natural philosophy of particles in motion now brought to bear in a much more intricate way upon a far broadened range of empirical phenomena; [3] the routine setting-up of whole ranges of fact-finding experiments. Somehow our harvest of propitious factors looks a bit bleak in comparison. A certain lack of explanatory imagination on the part of the author may well be responsible for that. It may also be (as Pascal reminds us) that big events do not necessarily require equally big causes. Finally, it may serve to underscore the principal causal thesis here defended, that, although we can see what turned Europe into a readier place to realise potentialities hidden in the Greek legacy than earlier recipients, we are not dealing here with events bound to happen regardless. In each case the hidden possibility might conceivably have been realised before (in locally different fashion, to be sure), or later, or not at all. Indeed, the less numerous and the less weighty the causes we have found to help account for Europe's greater readiness in this regard, so much the more does this confirm that contingency forms a major term in the full historical equation.

All these considerations notwithstanding, one more causal gap on another plane can surely be identified, as in fact it has already. It is, simply, that while so far we have been concerned to adduce a range of specific factors helping to explain why *each* of our three transformations occurred in Europe, not elsewhere, we still have to address the question of how it is that *all three* occurred there, and at almost the same time, too. Does it not defy belief to ascribe their concurrence in time and space to sheer chance?

12.INDIVIDUALIST NOVELTY AT A PREMIUM

Of course it does. And we may find a clue in the overall drift of how things had been developing in Europe over the fifteenth and sixteenth centuries. Not a turning inward, but, instead, a wave of concentrated, outward-bound curiosity, the exploration of foreign lands and of the forces of untrammelled individuality is what marks the period known of old as that of the Renaissance. Just when the Golden Age of Islamic nature-knowledge was reaching its peak by the early 5th/11th century, a period of inner-directed contraction set in for the civilization at large, leading in its turn, not yet to an overall cessation of the pursuit of nature-knowledge to be sure, yet to a certain sapping of the will at the exact point where (as the European experience strongly suggests) radical transformation might well have been the next step. In Europe around 1600, to the contrary, the will toward drastic innovation was furthered by an overall climate in which daring novelty was at a certain premium. Galileo, in particular, made himself a public figure in the very land where the cultivation of the individual had gone to the farthest extremes yet reached, and the admiration his public performance widely earned him is there to show it. Galileo's 'rugged individualism' has been called by the economist Joseph A. Schumpeter 'the individualism of the rising capitalist class',[8] and although the capitalist mode of doing business is really less relevant here than the values Schumpeter held to underlie it, he was quite right in sensing Galileo to symbolise in particularly outspoken fashion a uniquely European set of increasingly ascendant values.

Once again we must be careful not to carry this line of thought beyond its proper limits. This-worldly activity, and an individualist-innovative attitude toward it, stood at a premium at the very time when the imaginative reception of the Greek legacy at a level previously attained in Islamic civilization had once again reached its peak. This may help us understand further how steps of radical transformation of that legacy could now be made as they had not been previously. But, it does not help with an important problem that flows from these very transformations of the early seventeenth century: What caused them to continue to grow and to overcome various forms of sometimes severe resistance and opposition? For, as had occurred in Islam, the new modes of pursuing nature-knowledge found themselves represented as going against the reigning faith and its deepest values. The mere fact that these big transformations had occurred in no way allows us to assume that such challenges could just be lightly dismissed. It is true that those among the pioneers against whom charges were hurled in the first place, Galileo and Descartes, stuck to their positions in a more self-confident manner than, in Islam, the somewhat half-hearted defenders of the Greek legacy had. But this is not tantamount to saying that the eventual triumph of the new modes against an onslaught undertaken in view of alleged sacrilege and other grave defects was a foregone conclusion. Rather, the survival on the longer term of our three novel modes of nature-knowledge still hung in the balance of their no doubt numerous assets and their possibly (for who could safely predict the outcome?) even weightier liabilities as considered from the perspective of the times themselves.

[8] Schumpeter 1950, p. 124.

13.CONCLUSION: FLESHING OUT THE STORY OF THE SCIENTIFIC REVOLUTION DURING THE SEVENTEENTH CENTURY

Limitations of space preclude further exploration of the problem of explaining the fate of the three novel modes of nature-knowledge in the later seventeenth century. This would take us well beyond the point we have now reached—roughly the 1640s. In addition, it would involve, crucially, the identification of the dynamics that carried forward these developments over the remainder of the century. My forthcoming book develops this theme, aiming to provide full-scale elucidation and empirical anchoring for that account, as well as for the earlier stages of the argument, such as sketched above in this paper. So, in conclusion, let me foreshadow how the analytic framework proceeds for the remainder of the seventeenth century.

During the 1640s and early 1650s, an already brewing crisis of legitimacy became more and more manifest, but thanks in part to certain big-world events (notably, consequences of the Peace of Westphalia), the next generation managed to survive that crisis and turn the tables, so as to preserve and also extend by big steps the achievement of the pioneers. Among the principal components of this story are (1) the operation of such feedback mechanisms as made possible sustained advance in the universe of mathematical precision, in particular; (2) the partial yet, as such, unprecedented breakdown of barriers between modes of nature-knowledge; (3) the concoction, in Britain particularly, of a Baconian ideology bent on celebrating the still largely imaginary utility of the new science as such, while giving religious sanction to it; (4) three further revolutionary transformations, one (4a) effected by Huygens and young Newton in attempted fusions between elements of the Galilean and the Cartesian legacies, one (4b) effected by Boyle, Hooke, and young Newton in attempted fusions between elements of the Baconian and the Cartesian/Gassendist legacies, one (4c) effected by mature Newton alone. With his achievement in *Principia* and *Opticks*, as we all know, the Scientific Revolution did not come to an end, nor (considered from the point of view of the history of *science*) has such an end come in sight ever since. Still, with his *Principia* and *Opticks* the Scientific Revolution did come to an end in the sense that a coherent episode in the history of human thought and action had now found its first point of culmination. In the compass of less than a century humanity had moved from its customary pursuit of nature-knowledge in a variety of ways jointly bound to evoke unending, inherently irresolvable dispute, toward ways and means to make one's assertions about nature's realities truly conclusive. That particular move has since proven to be right at the heart of the making of our modern world. If it does not deserve the epithet 'revolutionary', then nothing in history does.

REFERENCES

Beeckman, I. (1939–1953) *Journal tenu par Isaac Beeckman de 1604 à 1634*, 4 vols, ed. C. de Waard, The Hague: Nijhoff.
Cohen, H. F. (1994) *The Scientific Revolution. A Historiographical Inquiry*, Chicago: University of Chicago Press.

— (2001) 'Les raisons de la transformation: la specificité europeenné (trans. A. Barberousse) in *L'Europe des sciences. Constitution d'un espace scientifique*, eds M. Blay and E. Nicolaidis, Paris: Le Seuil, pp. 51–94.

— (2004a) 'A historical-analytical framework for the controversies over Galileo's conception of motion' in *The Reception of the Galilean Science of Motion in Seventeenth-Century Europe*, eds C. R. Palmerino and J. M. M. H. Thijssen, Dordrecht: Kluwer, pp. 83–97.

— (2004b) 'The paradigm shift to beat all paradigm shifts' in *Scholarly Environments. Centres of Learning and Institutional Contexts 1560–1960*, eds A. A. MacDonald & A. H. Huussen, Leuven, Paris, Dudley MA: Peeters, pp. 1–14.

Copernicus, N. (1543) *De revolutionibus orbium coelestium*, Nuremberg.

Kepler, J. (1938–) *Johannes Kepler Gesammelte Werke*, 20 vols, eds W. von Dyck and M. Caspar, Munich: C. H. Beck.

Koyré, A. (1966) 'Du monde de l'"à-peu-près" à l'univers de la précision' in *Etudes d'histoire de la pensée philosophique*, ed. R. Taton, Paris: Presses Universitaires de France, pp. 341–362.

Monardes, N. B. (1577) *Ioyfull nevves out of the newe founde worlde*, trans. J. Frampton, London.

Palmerino, C. R. and Thijssen, J. M. M. H. eds (2004) *The Reception of the Galilean Science of Motion in Seventeenth-Century Europe*, Dordrecht: Kluwer.

Ptolemy (1984) *Ptolemy's* Almagest, trans. G. J. Toomer, London: Duckworth.

Schumpeter, J. A. (1950) *Capitalism, Socialism and Democracy*, 3rd edn, New York: Harper: 1st edn 1942.

Taton, R. ed. (1966) *Etudes d'histoire de la pensée philosophique*, Paris: Presses Universitaires de France.

JOHN A. SCHUSTER

'WATERWORLD': DESCARTES' VORTICAL
CELESTIAL MECHANICS

*A Gambit in the Natural Philosophical Contest of the Early Seventeenth
Century*

1.INTRODUCTION—UNCOMMON VORTICES

Nearly fifty years ago, Thomas Kuhn, in his best selling and often reprinted, *The
Copernican Revolution*, said this of Descartes' vortex universe: the 'vision was
inspired'; the 'scope tremendous'; but 'the amount of critical thinking devoted to
any of its parts was negligibly small'.[1] Typically more pointedly and poetically,
Gaston Bachelard had in 1938 condemned Descartes' plenist universe, including the
vortex mechanics, as the 'metaphysics of the sponge', an exemplary 'pre-scientific'
monstrosity, in other words, the sub-scientific progeny of cancerous metaphor and
baroque ego projection.[2] Other more mundane brush offs could also be cited.

Of course, Descartes' vortices do not posses for us the straight, presentist
scientificity of Newtonian mechanics, but they have an internal density and complex
genealogy—in Descartes' life work, and later, as Aiton has shown.[3] They are
deserving of study if we are to understand the structure and dynamics of natural
knowing in the early modern period. We can display how the vortices were
intellectually constructed, and why. This I intend to do, concentrating on Descartes'
Le Monde, The World or a Treatise of Light, his first systematic statement of the
mechanical philosophy, finished in 1633 but unpublished in his lifetime. In saying
this I in no way wish to imply that I introduced Bachelard and Kuhn above as mere
straw men. These two historian/philosophers of science initially most influenced my
understanding of the dynamics of seventeenth- and eighteenth-century natural
philosophy. I have argued elsewhere that Kuhn and Bachelard indeed misunderstood
the nature of that natural philosophy and the contestations over it—taking it as the
necessary but pre-scientific backcloth to the temporally splayed crystallisation of a
heterogeneous set of new 'real' sciences. However, as I have also claimed, that is
less important than the fact that their speculations prompted more positive modelling

---

[1] Kuhn 1959, pp. 240, 242.

[2] Bachelard 1965, p. 79, 'La métaphysique del'espace chez Descartes est la métaphysique de l'éponge'.

[3] Aiton 1972.

P. R. Anstey and J. A. Schuster (eds.), The Science of Nature in the Seventeenth Century, 35-79.

by historians of early modern natural philosophy, its nature, dynamics and trajectory.[4]

This paper is a modest essay in that very problematic. It focuses on a small but essential corner of Descartes' natural philosophical project. It will attempt to show the natural philosophical seriousness of Descartes' vortex universe as an intellectually constructed object and as a strategic gambit. I shall try to place Descartes' earliest celestial mechanics in relation to his manoeuvring in the natural philosophical contestation of his time.[5] This will involve exposing some of its minute design and biographical trajectory, thereby also relating it to similar aspirations and strategies of contemporary actors. They, including Descartes, were attempting to displace Aristotelianism, install some version of Copernicanism, and create alternative hegemonic natural philosophical syntheses. For many, Descartes included, such projects battened upon the achievements and promise of what Scholastics termed the mixed mathematical sciences, but which some of our struggling innovators occasionally termed 'physico-mathematical' disciplines, in particular hydrostatics, optics and mechanics. We shall need to say more below about such natural philosophical play upon the subordinate mathematical disciplines.

The argument of the paper will unfold as follows. Section 4 contains the fulcrum of the argument, an extended intellectual reconstruction of the inner toils of the vortex mechanics of *Le Monde*. Sections 5, 6, and 7 step back in time to trace three key moments in the genealogy of the vortex mechanics in the early work of Descartes, starting in 1619, and focusing, perhaps surprisingly, on his activities in hydrostatics and physical and geometrical optics, and his relations, spanning a decade, with his mentor in corpuscular-mechanism, Isaac Beeckman. The genealogy helps make sense of the already exposed anatomy of the vortex mechanics. Section 8 returns to 1633 and canvasses one small example of the coherence and power of the vortex mechanics. Finally, in section 9 the vortex mechanics is inserted into the context of the natural philosophical contest of Descartes' generation, with particular reference to Beeckman and Kepler. It also unveils the motivation for use of the odd term 'Waterworld' in the title, an outcome prepared by the genealogical and anatomical dimensions of the argument. But before any of this occurs, there are two items of preparation. Section 2 will explicate the key notions of mixed mathematical science and 'physico-mathematics', whilst section 3 will introduce the 'dynamics' of Descartes, the doctrine of causation, dealing with motions and tendencies to motion, through which he intended to 'run' the machinery of his vortex world. The evolution of this dynamics will be glimpsed throughout the genealogical sections 5, 6 and 7, as well, being part of the larger story of how the vortex mechanics became conceptually possible and strategically necessary.

---

[4] Schuster and Watchirs 1990; Schuster and Taylor 1996; Schuster 2002.
[5] Space constraints prevent discussion here of Descartes' 'cosmological optics', his theory of light in the context of the vortex universe. However, the development of the cosmological optics went hand in hand with that of the vortex mechanics, a relation to be treated at length in a monograph in progress, dealing with the development of Descartes as a physico-mathematician 1618 to 1633. In the present paper, it will at least be made clear that the genealogy of the vortex mechanics is entangled with the development of Descartes' work in physical optics and theory of light. See below, sections 5 and 6.

2.MIXED MATHEMATICAL SCIENCES AND PHYSICO-MATHEMATICS

As noted in the Introduction, Descartes' vortex mechanics emerged within a natural philosophical agenda which, in very general terms, he shared with other key anti-Aristotelian natural philosophical innovators of his generation: to displace Aristotelianism, install some version of Copernicanism, and create alternative hegemonic natural philosophical syntheses. For these innovators, Descartes included, such projects were premised on the exploitation of the achievements and promise of the mixed mathematical sciences, in particular hydrostatics, optics and mechanics. Because the competition to develop the mixed mathematical sciences and exploit them in the contest for natural philosophical dominance plays such an important role in Descartes' case, as well as that of others, we need to consider briefly what was happening and what was at stake in this domain in the generation of Descartes.

The term 'mixed mathematics' belonged to Aristotelianism. It referred to a group of disciplines intermediate between natural philosophy and mathematics. A natural philosophical account of something was an explanation in terms of matter and cause, and for Aristotle, mathematics could not do that. This meant that the mixed mathematical sciences, such as optics, mechanics, astronomy or music theory, used mathematics not in an explanatory way, but merely to represent physical things and processes mathematically. So in geometrical optics, one used geometry, representing light as light rays—this might be useful but did not get at the underlying natural philosophical questions: 'the physical nature of light' and 'the causes of optical phenomena.'

The question of the relation between mixed mathematics, on the one hand, and the 'superior', explanatory, discipline of natural philosophy, on the other hand, became extremely vexed in the generations around 1600. Strict Aristotelians did not grant any natural philosophical relevance to the findings of the mixed mathematical sciences; more avant garde Aristotelians such as some Jesuits, wanted to start extracting some natural philosophical juice out of the ripe fruit of mixed mathematical research discoveries.

Descartes and his mentor Isaac Beeckman, and others as well, used an alternative, more provocative term, 'physico-mathematics', which was gaining some prominence at the time. It signalled a more radical approach to the natural philosophical legitimacy of the mixed mathematical fields. As we shall see, Descartes and Beeckman went even further: they did not mean mathematical-physical disciplines subordinate to natural philosophy, especially Aristotle's natural philosophy, but a new realm of corpuscular-mechanical natural philosophy, in which the old mixed mathematical fields are explained in corpuscular-mechanical terms and therefore are not subordinate to, but are proper domains of, the new natural philosophy. They were meant to become areas in which could occur true natural philosophical explanation in terms of matter (corpuscles) and cause (the motion, impact and arrangement of corpuscles). Conversely, it meant for Descartes and Beeckman that novel findings in mixed mathematical sciences directly bespoke new insights into the realm of corpuscular-mechanical explanation. All this may seem to us just so much late Scholastic intellectual quibbling, waiting be brushed away with the advent of quite modern mechanics and celestial mechanics just a bit later in the Scientific Revolution. This would be to miss the point: these matters were explosive and challenging issues for contemporaries, and these were the struggles through which some of them, Descartes and Kepler especially, paved the way for those very turns in the Scientific Revolution that the Whig and the populariser are so happy to applaud out of context.[6]

3.CARTESIAN DYNAMICS—THE CAUSAL REGISTER OF CORPUSCULAR-MECHANISM

In the period of the Scientific Revolution 'natural philosophy' as a generic term denoted that common field of endeavour within which particular schools and varieties of systems contended: not only various species of neo-Scholastic Aristotelianism in the universities, but also natural philosophies of neo-Platonist, Stoic and qualitative atomist bent, to which in the generation of Descartes, Beeckman, Mersenne and Gassendi, we can of course add the genus 'corpuscular-mechanist'. Now, in broad terms the scope of 'natural philosophising' involved the identification of what causes material bodies to behave in particular ways. This was understood to be the case whether, as in Aristotelianism, natural processes were explained primarily on the basis of causes identified with the nature or essence of the matter in question, or, as in neo-Platonic natural philosophies, brute matter was worked upon from the outside by various types of non-material causal agents. Theorising about matter and an associated 'causal register' was traditionally taken as constitutive of natural philosophy. Whatever disputes there might have been amongst Platonists, Aristotelians, Stoics, and atomists, there was consensus on what kind of theory provided the ultimate explanation of macroscopic physical phenomena, namely a theory of matter and causation.

Descartes was no exception to this and we may characterise his natural philosophy as concerned with the nature and 'mechanical' properties of microscopic corpuscles and a causal discourse, consisting of a theory of motion and impact, explicated in particular through key concepts of the 'force of motion' and 'tendencies to motion'. It is this causal register within Descartes' natural philosophical discourse which scholars increasingly term his 'dynamics'. Descartes' vortex theory (and his celestial optics as well) depended upon this dynamics. If we do not take his dynamics seriously, we cannot take the vortex theory seriously. Later in sections 5 and 6 we shall examine some aspects of the genealogy of the dynamics between 1619 and the late 1620s, leading to its initial systematisation in *Le Monde*. But for the moment, before examining the vortex theory, we need to survey the fundamentals of this dynamics.

---

[6] Readers of this volume may wish to compare the interpretation of Kepler and Descartes implicated here with they way they emerge in H. F. Cohen's interpretative essay on the causes of the Scientific Revolution above.

In Descartes' *Le Monde*, the behaviour of Descartes' micro-particles is governed by a carefully articulated theory of dynamics. Descartes' dynamics of micro-particles had nothing to do with the mathematical treatment of velocities, accelerations, masses and forces. Rather it was concerned with accounting for the motion, collision and tendency to motion of corpuscles. Descartes held that bodies in motion, or tending to motion, are characterised from moment to moment by the possession of two sorts of dynamical quantity: (1) the absolute quantity of the 'force of motion'—conserved in the universe according to *Le Monde's* first rule of nature; and (2) the directional modes of that quantity of force, the directional components along which the force or parts of the force act, introduced in *Le Monde's* third rule of nature.[7] These Descartes termed actions, tendencies, or most often determinations.[8] Such are the central tenets underlying Descartes' dynamics.

[7] The third rule of motion in *Le Monde* states: (Descartes 1996, hereafter cited as *AT*, XI, pp. 43–45: Descartes 1998, pp. 29–30, hereafter cited as *SG*) 'I shall add as a third rule that, when a body is moving, even if its motion most often takes place along a curved line and, as we said above, it can never make any movement that is not in some way circular, nevertheless each of its parts individually tends always to continue moving along a straight line. And so the action of these parts, that is the inclination they have to move, is different from their motion.[...leur action, c'est à dire l'inclination qu'elles ont à se mouvoir, est differente de leur mouvement]'. And, Descartes continues, 'This rule rests on the same foundation as the other two, and depends solely on God's conserving everything by a continuous action, and consequently on His conserving it not as it may have been some time earlier, but precisely as it is at the very instant He conserves it. So, of all motions, only motion in a straight line is entirely simple and has a nature which may be grasped wholly in an instant. For in order to conceive of such motion it is enough to think that a body is in the process of moving in a certain direction [en action pour se mouvoir vers un certain côté], and that this is the case at each determinable instant during the time it is moving'.
In the passages cited above, Descartes in his discussion of the third law defines 'action' as 'l'inclination à se mouvoir'. He then says that God conserves the body at each instant 'en action pour se mouvoir ver un certain côté'. This would seem to mean that at each instant God conserves both a unique direction of motion and a quantity of 'action' or force of motion. In other words the first law certifies God's instantaneous conservation of the absolute quantity of tendency to motion, the 'force of motion'. The third law specifies that as a matter of fact in conserving 'force of motion' or 'action', God always does this in an associated unique direction. The first law asserts what today one would call the scalar aspect of motion, the third law its necessarily conjoined vector manifestation. Just because he recognises that some rectilinear direction is in fact always annexed to a quantity of force of motion at each instant, Descartes often slips into abbreviating 'directional force of motion' by the terms 'action', 'tendency to motion' or 'inclination to motion', all now seen in context as synonyms for 'determination'.

[8] The understanding of determination used here develops work of Sabra 1967, pp. 118–121; Gabbey 1980; Mahoney 1973; Gaukroger 1995; Knudsen and Pedersen 1968; Prendergast 1975; and McLaughlin 2000.

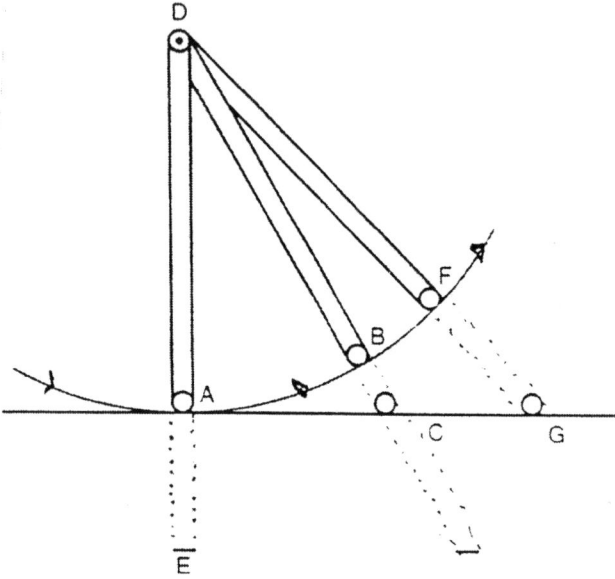

Figure 1. After Descartes, Le Monde, AT, *XI, p. 45 and p. 85.*

As corpuscles undergo instantaneous collisions with each other, their quantities of force of motion and determinations are adjusted according to certain universal laws of nature, rules of collision. Therefore Descartes' analysis focuses on instantaneous tendencies to motion, rather than finite translations in space and time. Indeed, Descartes offers a metaphysical account of translation which dissolves it into a series of inclinations to motion exercised in consecutive instants of time at consecutive points in space. Whilst the rudiments of this dynamics of instantaneously exerted forces and determinations dates back to Descartes' earliest work, as we shall shortly see, it was first systematically articulated in *Le Monde.*[9]

Descartes' exemplar in *Le Monde* for applying these concepts to celestial mechanics is the dynamics of a stone rotated in a sling [**Fig. 1**]. Descartes analyses the dynamical condition of the stone at the precise instant that it passes point A. The

[9] These rudiments appear in the so-called hydrostatic manuscript of 1619. See Schuster 1977, pp. 93–111; Gaukroger 1995, pp. 84–89; and Gaukroger and Schuster 2002. It should also be noted that *Le Monde* itself contains a reference to the text of the *Dioptrique* attributing the distinction between force of motion and directional force of motion to that earlier text; *AT*, XI, p. 9; cf. Descartes 1963, p. 321 n. 2. We shall see below that the key dynamical concepts probably did crystallise in Descartes' optical work of the 1620s, particularly his discovery of the law of refraction of light (cf. Schuster 2000).

instantaneously exerted force of motion of the stone is directed along the tangent AG. If the stone were released and no other hindrances affected its trajectory, it would move along ACG at a uniform speed reflective of the conservation of its quantity of force of motion.[10] However, the sling continuously constrains the privileged, principal determination of the stone and, acting over time, deflects its motion along the circle AF.[11] Descartes considers that the principal determination along AC can be divided into two components: one is a 'circular' determination along ABF; the other a centrifugal determination along AE. For present purposes, let us ignore the curious circular tendency. To discuss it would lead us further than we need to go into Descartes' manner of treating circular motion.[12] What Descartes is trying to do is decompose the principal determination into two components: one along AE completely opposed and hindered by the sling—so no actual centrifugal translation can occur—only a tendency to centrifugal motion; the other along the circle, which is as he says, 'that part of the tendency along AC which the sling does not hinder'.[13] Hence it manifests itself as actual translation. The choice of components of determination is dictated by the configuration of mechanical constraints on the system.

4.LOCKING AND EXTRUDING—PRINCIPLES OF THE VORTEX CELESTIAL MECHANICS

We turn now to a synthetic recounting of the vortex mechanics of *Le Monde*, the genealogy of which will be examined in following sections. It is important to note precisely what my interpretive strategy is and what is not, as well as how that strategy subserves the aims of this chapter, rather than some aims that might erroneously be attributed to my efforts here.

To be brutally frank, Descartes does not communicate well to the reader in the sections of *Le Monde* dedicated to the theory of vortices. The text, of course, is incomplete, unpolished and remained unpublished in his lifetime. Indeed, as we shall note in a couple of instances below, Descartes hardly helped his cause by his adoption of a commonsensical, *honnête homme* style. His appeal to commonly experienced analogies and observations—without explicating their limitations or precise modes of articulation to his underlying concepts and theories—tends to

---

[10] *Le Monde AT*, XI, pp. 45–46, 85.

[11] I have coined the interpretative concept of 'principal' determination to underscore this important concept, and differentiate this aspect of determination from the other determinations that can be attributed to the stone at that moment. I prefer this terminology to a perhaps too Whiggish concept of 'inertial' determination.

[12] *Le Monde, AT*, XI, p. 85. Descartes argues from the first and third laws of nature that at the instant of time the body is at point A, it tends in and of itself along the tangent AC. The circular tendency along AB is that part of the tangential tendency which is actively opposed by the physical constraint of the sling and hence gives rise to the centrifugal tendency to motion along AE. For the sake of Whiggish edification it can be noted that had Descartes dealt with the centrifugal constraint on the ball offered by the sling, instead of the circular tendency (which violates the first law in any case) he might have moved closer to Newton's subsequent analysis of circular motion.

[13] *Le Monde, AT*, XI, p. 85.

swamp and confuse his message. But, and this is the key point, Descartes arguably did possess a coherent and well thought out theory of vortices, of which the surviving text of *Le Monde* is a rather poor representation. It is, however, a representation that can lead the hermeneut to that underlying theory, provided three conditions of reading and analysis are fulfilled: [1] one must attend constructively to the likely trajectory of Descartes' work and struggle in natural philosophy and the mixed mathematical sciences in the decade or so leading up to the composition of *Le Monde*; [2] one must probe behind the breezy style of presentation and appeal to easy if somewhat misleading analogies in *Le Monde*, and interpret the text charitably in the search for deep and coherent theorising, consistent with and evolving out of the material studied in [1]; finally [3] one must be willing to use the much more systematically and coherently developed explication of vortex mechanics in the *Principles of Philosophy* as an heuristic guide to what Descartes might possibly have been entertaining in *Le Monde* (without falling into a vulgar retrospective Whiggism). In many ways, therefore, this reading of *Le Monde* for a strong and complex underlying theorisation rebounds upon our sense of the text itself, perhaps lending it the colouration of a more private, even solipsistic, document, somewhat akin to those sets of working notes and drafts that scaffold our own public utterances without ever seeing the light of day.

Now, the aim of such a reading is not to conclude that *Le Monde* 'really' teaches such a coherent theory of vortices which later seventeenth-century readers, and modern historians of science have, through some cognitive shortcoming, 'failed' to see. Nor is the aim to blame Descartes for failing to express what he had so systematically conceptualised. Such points are irrelevant in regard to my aims here. I aim, rather, to try to capture, via such a reconstruction, what arguably was the state of theorising that Descartes had reached about vortices at the end of almost fifteen years of work in physico-mathematics—a theory that lurks below the surface of *Le Monde*, but is recoverable from it. We shall see that Descartes' underlying theory was subtle and complex, reflecting upon and exploiting a sequence of technical achievements in physico-mathematics as well as his own lived experience as an increasingly mature, and competitive, player in the struggle to forge a new natural philosophy embodying Copernican realism. In the latter sense, Descartes' underlying conceptualisation was one instance of a more widely pursued problematic, worth studying as part of a mapping of other natural philosophical initiatives and aspirations of similar kind and intended scope—as in the work of Beeckman, Kepler, Gassendi, and Mersenne.

To all this one further and more pragmatic condition has to be added, as far as the account of the vortex mechanics offered here is concerned. My presentation will be synthetic and declarative, there being no space here to offer the more analytical and textual critical account of how Descartes' theory has been teased out of the text; and exactly how textual juxtapositions and interpretations, as well as judicious appeals to the *Principles*, can be used to clarify his analogies, reorganise his diffuse and confusing order of presentation, and explicate certain half articulated points and claims.[14] I shall, however, at various points indicate in footnotes the degree and type of interpretative work/reconstruction involved in presenting particular concepts and representations. Amongst the concepts and representations I shall use: [1] some arguably derive quite literally from the text of *Le Monde*; [2] some arguably express Descartes' theoretical intentions in ways he did not quite accomplish in the text; [3] some systematise or clarify concepts confusedly presented in *Le Monde* (but often better expressed later in the *Principia*) in a charitable attempt to elicit a coherent theory; [4] some are novel, my own interpretive inventions, advanced again in a charitable attempt to elicit a coherent theory from Descartes' text. Arguably, they could have been constructed by Descartes himself or a contemporary, but to my knowledge never were; [5] some representations and concepts correct misleading implications of some of Descartes' analogies in the interest of charitably supporting our vision of his underlying theory, and separating off misleading but understandable implications that have been or could be read into his surface analogies.[15] With all these caveats, let us now begin the explication.

---

[14] Indeed in oral presentations of this paper at seminars and conferences I have used, not unsuccessfully, the following conceit in synthetically presenting the vortex theory: that this is a pro-Cartesian university lecture in Cartesian natural philosophy circa 1660, assuming fairly widespread consensual acceptance of vortex mechanics. This allows the further conceit that the new diagrams and concepts I use below to explicate the vortex mechanics have actually become recognised parts of a Cartesian Scholastic tradition within a generation of his death. Perhaps if the remainder of this section is read in that spirit, the key points about the theory will come through, provided one remembers above all that I am not suggesting this was for anybody the explicit, publicly acknowledged version of vortices, but rather that this is very close, on a charitable reading, to Descartes' own best understanding of his vortex theory, as it related to his course of work and context of natural philosophical struggle up to the early 1630s.

[15] A more textual critical approach to teasing the underlying theory out the literal sense of *Le Monde* was begun in Schuster 1977 and will be fully explored in my monograph on Descartes as a physico-mathematician. Amongst the inadequately or misleadingly expressed analogies and claims that— revised, criticised and explicated—will find their place the synthetic presentation of the theory below are [1] the appeal to the behaviour of a large heavy boat compared to random flotsam in the confluence of two parallel rivers; [2] Descartes' mode of setting out the notion of a 'balance' of forces holding a planet in its orbit; and, [3] the articulation of the key concept of 'massiveness' or 'solidity' of an orbiting body.

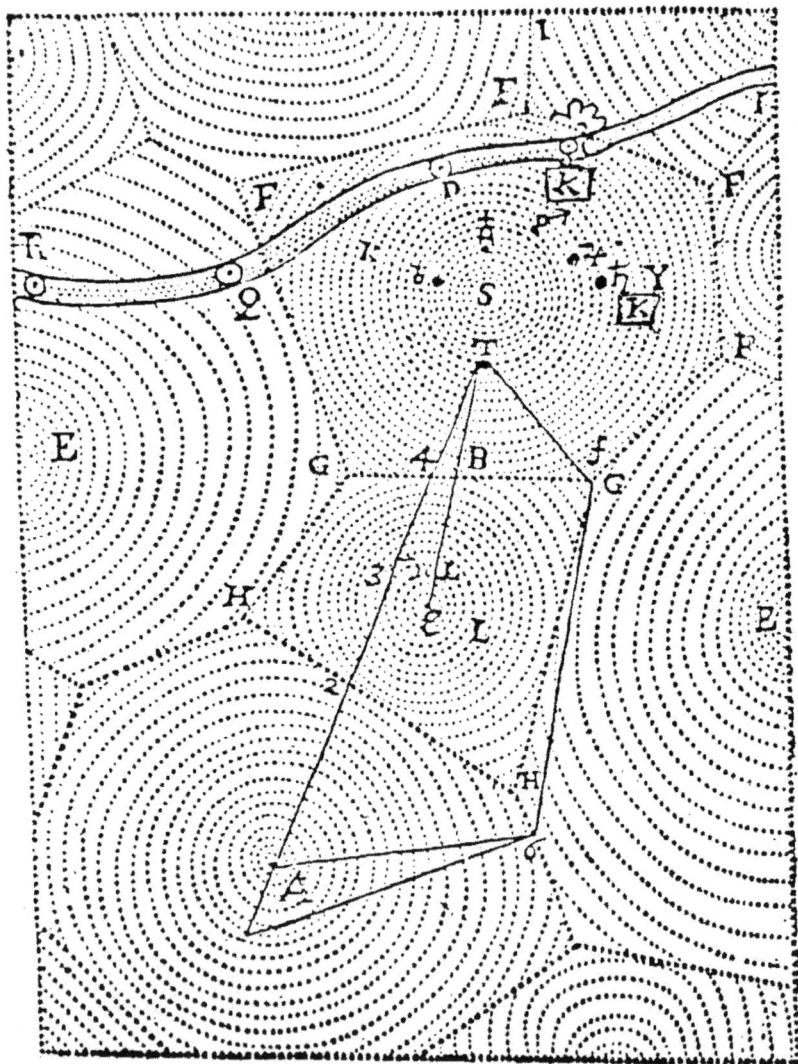

Figure 2. Descartes, Le Monde, AT, *XI, p. 54.*

People often take the celestial mechanics on its most superficial level, as if it was just a historical holding action waiting for Newton [**Fig. 2**]. Descartes imagined whirlpools or vortices of second element, rotating around their respective central stars to sweep along their planets like boats in a strong current.[16] In fact the swishing along of the planets in the vortex was the least of his concerns. He thought that the mere existence of a whirlpool of second element accounted for the orbital movement. What interested him was why the planets maintain relatively stable celestial distances and different distances; and why comets do what he imagined them to do, that is, continually oscillate between vortices, spiralling in toward the central star of one vortex, up to a specific, theoretically given radial distance, and then spiralling out again into a neighbouring vortex, up to a similar theoretically given minimum radial distance from its central star, and so on.[17]

The overall condition for stability of the vortex is that there be a uniform and continuous increase in the centrifugal tendency of the particles making up the vortex as one goes away from the centre.[18] Now, according to Descartes' dynamics, discussed above, centrifugal tendency is proportional to the force of motion in the tangent direction, and force of motion is measured by quantity of matter times speed, or more technically, quantity of matter and the instantaneously exercised principal tendency to motion. Descartes wants to specify how the size and speed of the particles of the vortex vary with distance from the centre. He does this twice and we shall need to attend closely to both moments in his exposition.

First Descartes describes the speed/size distribution of the particles making up the vortex in the earliest stages of vortex formation, prior to the production of his three types of stable particle, or elements, and hence prior to the formation of the sun, which, of course, is made up entirely of the highly agitated particles of the 'first element'—a critical moment in the theory as we shall shortly see. So, Descartes tells us that in this first, very early stage, as the vortex settled out of the original chaos, the larger corpuscles were, of course, harder to move, so there was a tendency for the smaller ones to acquire higher speeds more easily. Accordingly, in these early stages, the size of particles decreased and their speed increased from the centre out.[19] But the speed of the particles increased proportionately faster, so that force of motion increased continuously. In **Fig. 3** we see Descartes' first declared distribution of size and speed of the particles making up the vortex in the period before the formation of the three elements and the emergence of a star in the centre of the vortex: force of motion constantly rises, as does speed, while size decreases proportionately less than speed.[20]

---

[16] In fact in the key analogy used by Descartes, in a strong river current boats behave like comets, and it is light flotsam that behaves on analogy to planets. So untutored intuition misleads as to Descartes' own preferred analogy (and hence misses the theoretical points he will be elucidating through the analogy).

[17] Additionally, as we shall see, he was also interested in relating a theory of local terrestrial gravity to his vortex celestial mechanics—a nice trick, since on earth bodies of third element subjected to the local vortex fall down; but in the heavens, bodies of third element, subjected to the stellar vortex, find specific and stable orbital distances. Descartes thought there was a unified conceptual explication of these indubitable phenomena and he prided himself on designing it.

[18] Let us call this the 'force-stability principle'. Strictly speaking, however, more is involved in Descartes' full conception of the orbital stability of the particles, or planets, orbiting at a given radial distance. Descartes' articulated version of the force-stability principle will be developed below, note 32.

[19] *AT*, XI, pp. 50–51.

[20] Note in relation to this figure, as well as figures 4 and 5 below that they of course do not exist in *Le Monde* and are interpretative tools of my own design, used to picture the relationships Descartes sets

Force of motion: ▬▬▬▬

Speed: ▬ ▬ ▬ ▬

Size: ••••••••••••••••••

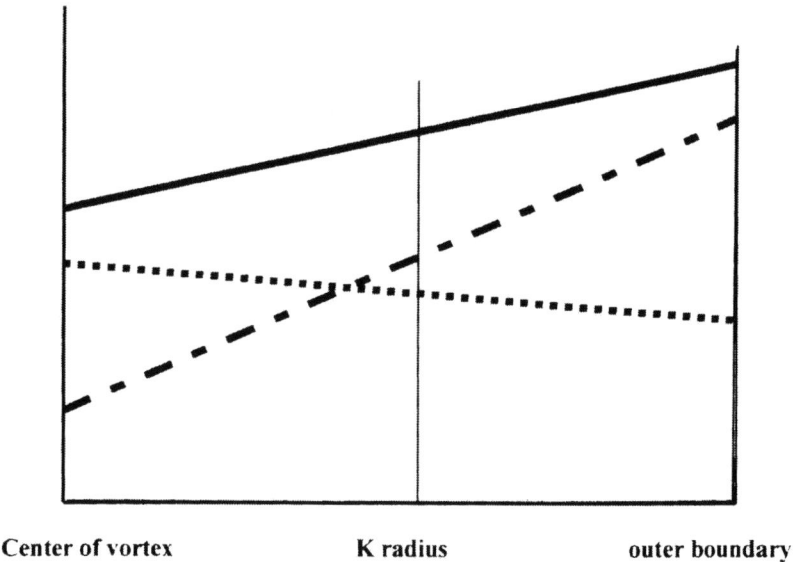

Center of vortex　　　　　　**K radius**　　　　　　**outer boundary**

Figure 3. Size, speed and force of motion distribution of particles of 2ⁿᵈ element, prior to existence of central star.

Descartes' second description of the speed/size distribution of the particles making up the vortex applies to the period after the formation of the three elements.[21] Descartes explains that as the vortex rotates in its first stage, the particles

---

out verbally. Additionally, it should be remembered that Descartes had no way of assigning empirically meaningful dimensions to the sizes and speeds of the *boules*. Nor would it have occurred to him to insist on any specific relationship for the variation of size and speed with distance. He limited his discussion to notions of proportionately greater or lesser increase or decrease of variables, which the figures then represent.

[21] Descartes adduces the elements at this stage in *Le Monde* in Chapter 8 (*AT*, XI, pp. 51–55), but he has already adumbrated their properties at the end of chapter 4. And, in chapter 5 he writes in more detail that, 'I conceive of the first, which one can call the element of fire, as the most subtle and penetrating

collide with one another, breaking off their rough angles and points. These cosmic scrapings form the first matter. Much of the first matter is forced to the centre of the vortex while the remainder fills the interstices left between the particles of the vortex. The latter particles, smoothed and polished by this process, become the spherical *boules* of the second element. Grosser particles of third matter are assumed to have existed all along. The first matter at the centre of the vortex is highly agitated and forms 'perfectly liquid and subtle round bodies', that is, stars, including the sun at the centre of our vortex.[22] It is the sun's presence in the centre of the vortex that alters the first distribution of size and speed of particles in the vortex. This is absolutely crucial to the final theory, for the star's disturbing effect on the original size/speed distribution produces a second, quite different stable distribution of size and speed of the vortex particles, and it is this second distribution that allows the planets to maintain stable orbits.

The sun is made up of the most agitated particles of first element; their agitation communicates extra motion to parts of the vortex near the surface of the sun; that is to those spheres of second element in the vortex lying near the sun. This increment of agitation decreases with distance from the sun's surface and vanishes to nothing at a certain radius, labelled by Descartes in **Fig. 2** as K.[23] In **Fig. 4** we represent Descartes' conception of the solar disturbance and its decrease with distance up to radius K.

---

fluid there is in the world... I imagine its parts to be much smaller and to move much faster than any of those other bodies. Or rather, in order not to be forced to imagine any void in nature, I do not attribute to this first element parts having any determinate size or shape; but I am persuaded that the impetuosity of their motion is sufficient to cause it to be divided, in every way and in every sense, by collision with other bodies, and that its parts change shape at every moment to accommodate themselves to the shape of the places they enter... As for the second, which one can take to be the element of air, I conceive of it also as a very subtle fluid in comparison with the first; but in comparison with the first there is need to attribute some size and shape to each of its parts and to imagine them as just about all round and joined together like gains of sand or dust. Thus, they cannot arrange themselves so well, nor press against one another, that there do not always remain around them many small intervals, into which it is much easier for the first element to slide in order to fill them. And so I am persuaded that this second element cannot be so pure anywhere in the world that there is not always some little matter of the first with it. Beyond these two elements, I accept only a third, to wit, that of earth. Its parts I judge to be as much larger and to move as much less swiftly in comparison with those of the second as those of the second in comparison with those of the first. Indeed, I believe it is enough to conceive of it as one or more large masses, of which the parts have very little or no motion that might cause them to change position with respect to one another' (*AT*, XI, pp. 24–26; Mahoney 1979, hereafter cited as *MSM*, pp. 37–39).

[22] *AT*, XI, p. 53, *MSM*, p. 85.

[23] Descartes insists that a central star can agitate the surrounding particles of second matter of its vortex: 'These spherical bodies] incessantly turning much faster than, and in the same direction as, the parts of the second element surrounding them, have the force to increase the agitation of those parts to which they are closest and even (in moving from the center toward the circumference) to push the parts in all directions, just as they push one another' (*AT*, XI, p. 53, *MSM*, p. 85).

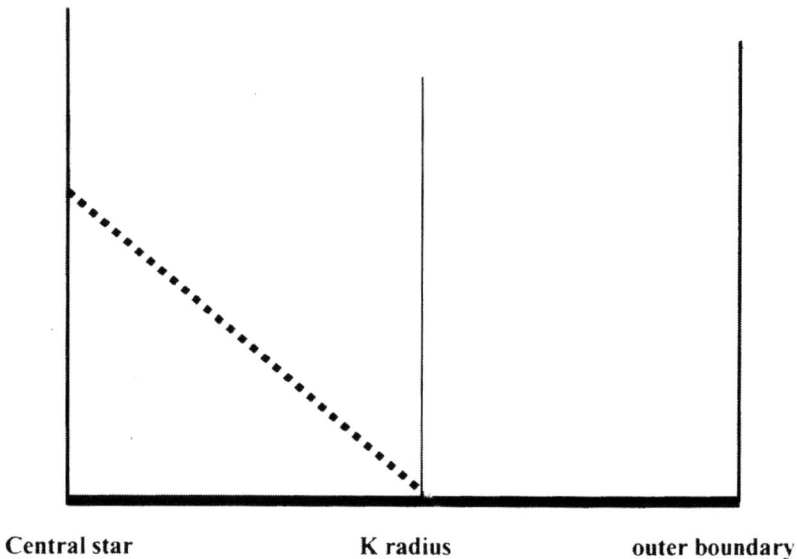

Central star   **K radius**   **outer boundary**

Figure 4. Agitation due to existence of central star.

The solar effect alters the original size and speed distribution of the spheres of second element in the vortex, below the K layer.[24] We now have greater corpuscular speeds close to the sun than in the pre-sun situation. But the force-stability principle, of course, still holds, so the overall size/speed distribution must change, below the K layer.[25] Descartes description of this situation is represented in **Fig. 5**.

---

[24] The special radial locus at distance K is present in Descartes' own discussion. Here for expository purposes I introduce the term 'K layer' not used by Descartes. Note as well that the existence and location of the K layer are caused by the existence and action of the sun.

[25] Descartes' final distribution of the size and speed of the particles of the second element is as follows: *AT*, XI, pp. 54–56; *MSM*, pp. 87–91 (**Fig. 2**): 'Imagine... that the parts of the second element toward F, or toward G, are more agitated than those toward K, or toward L, so that their speed decreases little by little [as one goes] from the outside circumference of each heaven [vortex] to a certain place (such as, for example, to the sphere KK about the sun, and to the sphere LL about the star) and then increases little by little from there to the centers of the heavens because of the agitation of the stars that are found there... As for the size of each of the parts of the second element, one can imagine that it is equal among all those between the outside circumference FGGF of the heaven and the circle KK, or even that the highest among them are a bit smaller than the lowest (provided that one does not suppose the difference of their sizes to be proportionately greater than that of their speeds). By contrast, however, one must imagine that, from circle K to the sun, it is the lowest parts that are the smallest, and even that the difference of their sizes is proportionately greater than (or at least proportionately as great as) that of their speeds. Otherwise, since those lowest parts are the strongest (due to their agitation), they would go out to occupy the place of the highest'.

Force of motion: ━━━━━

Speed: ▬ ▬ ▬ ▬

Size: ·················

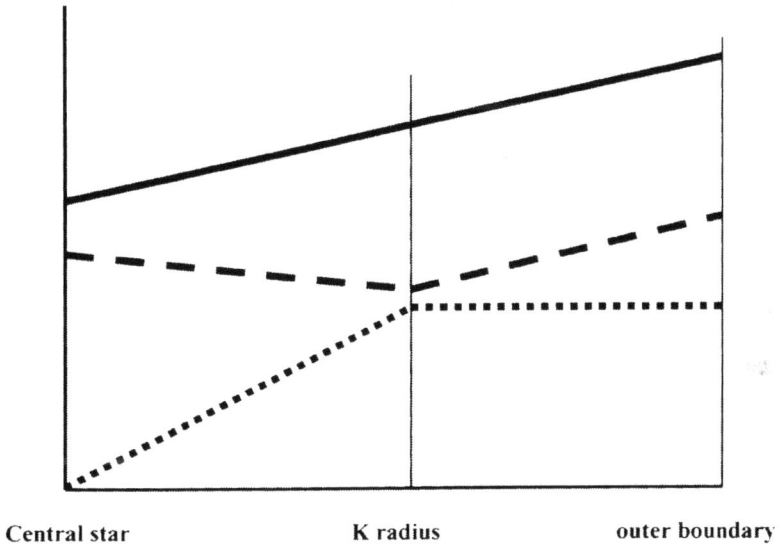

| Central star | K radius | outer boundary |

Figure 5. Size, speed and force of motion distribution of particles of 2nd element, in a solar vortex.

In the solar vortex as one moves away from the sun the agitation (speed) of the *boules* decreases, reaching a minimum at the distance K (where Descartes will locate the planet Saturn). From K outward to the boundary of the vortex the agitation increases again. The size of the *boules* increases from a minimum near the sun to K; and from K outward the size remains constant or perhaps diminishes a little. From the sun to K the size of the *boules* of second element increases proportionately more than their speed decreases; from K outward the speed increases proportionately more than the size decreases. Thus we can draw a line of positive slope representing the force of motion of the *boules* (agitation X size) at each distance from the sun.

Two points are crucial about Descartes' model, and they are particularly clear in our representations in **Fig. 3** and **Fig. 5**:

1. It is the action of the sun that transforms the distribution of **Fig. 3** into that of **Fig. 5**. The presence of the sun not only shifts the distribution of agitation, but it also as a consequence induces a change in the relative size distribution of the particles. This is due to the theoretical requirement that when the speed curve shifts, the size distribution must change accordingly so that the force condition on the stability of the vortex is maintained.

2. The K radius is the critical distance. It marks the locus beyond which the sun's added effect vanishes. Beyond K we have the *old*, stable pattern of size/speed distribution; below K we have a *new*, stable pattern of size/speed distribution—we still have force of motion increasing continuously with radius, but that comes about because size increases more quickly than speed decreases. This new distribution permits the observed celestial motions to occur. In effect it turns the vortex into a special kind of machine—a machine that *locks* planets into their appropriate orbits below K and that *extrudes* them from inappropriate orbital distances.[26] All this occurs in what, to Descartes, seems a straightforward mechanical fashion.[27]

Descartes' approach focuses on the centrifugal tendency of planets and of surrounding particles of second element in the vortex. Remember that, according to Descartes' dynamics and his sling exemplar, as a body or corpuscle moves on a curve, it has a certain force of motion along the tangent at any moment in its translation. Because it is constrained to move along a curved path, part of its tangential tendency manifests itself as a centrifugal tendency to recede along the normal to the path at that point on the curve. So, all bodies moving along a curve generate a centrifugal tendency to motion proportional to their size, quantity of matter, and instantaneously manifested tangential force of motion.

The key question in Cartesian celestial mechanics now becomes this: when and why is centrifugal tendency actualised as centrifugal motion, and when and why does that not happen? In the vortex, what plays the role of the sling, constraining the planet into a curved path and thus generating centrifugal tendency on its part? Well, it is of course the neighbouring, superjacent particles of second element that do this job—they surround and penetrate the pores of every piece of third matter making up a planet.

Why then do planets maintain orbits and why at different distances—all within radius K—from the sun? This depends on the amount of resistance the superjacent second element can put up, and that is dependent upon how much second element

---

[26] It is crucial to notice this moment in Descartes' theorising—it is the fact that a star, made of first element, happens to inhabit the centre of each vortex that transforms every vortex into an orbit-locking mechanism. This is Descartes' version of the Keplerian emphasis (compared to Copernicus himself) on the physical-causal role of the sun in orbital mechanics. Interestingly, and crucially, the central location and physical behaviour of each vortex's star, are also essential to Descartes' theory of light in the cosmic setting—again it is the central star that completes the theoretical picture explaining the phenomena of light in the vortex universe.

[27] The reconstruction that follows here skims over all the complexities of textual interpretation mooted above at the beginning of this section, including some hopefully non-Whiggish appeals to clarifications in the utterances of the *Principles* eleven years later.

can surround and *envelop* the parts of the planet, as what we may term a 'surface envelope'—a term of hermeneutical art that greatly helps our explication of *Le Monde* and the *Principles*.[28] The more matter of second element in this envelope, the more resistance the envelope will present to its being shoved aside by the planet's tendency to recede from the sun. **Fig. 6** is a schematic representation of this notion: a simple ball of third element in circular motion is surrounded by a smaller and a larger envelope of corpuscles of second element.

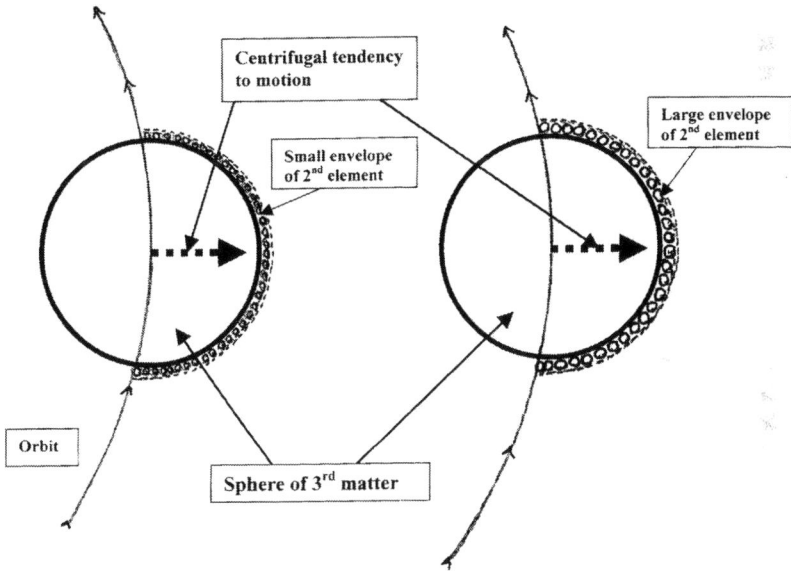

Figure 6. 'Surface envelopes' of 2nd element. Small envelope left. Large envelope right. Envelope sizes are a function of volume to surface area ratios of spheres of 2nd element.

What determines how large a surface envelope is relative to a given planet? Well, obviously, the size distribution of the second element with distance from the sun. Descartes recognised that the size of a surface envelope is dependent upon the volume to surface ratio of the spheres of second element. That ratio is function of their radii. The greater the radius of a sphere, the greater the V/S ratio.[29] Imagine a

---

[28] My notion of 'surface envelope' is a good example of a term of interpretative art belonging to my hermeneutical categories 2, 4 and 5, discussed earlier in this section.

[29] The second element, recall, is quite small compared to the pieces of third element, something Descartes goes out of his way to claim, in first describing the elements, as we saw above in note 21: 'Its parts [third element] I judge to be as much larger and to move as much less swiftly in comparison with

ball of third element in circular motion surrounded by an envelope of second element. As long as the spheres of second element are so small compared to the piece of first element that we do not reach the point at which only a few spheres of second element suffice to 'cover' its surface, we can get a great variation in overall, aggregate size of the envelope, hence its quantity of matter, and hence its resistance to being moved out of place by the centrifugal tendency of the piece of third matter.

Next recall the size distribution of the second element [**Fig. 5**]. We can turn this into a curve of V/S ratios, which in turn indicates the magnitude of the surface envelopes made out of the second element at different distances as related to a given piece of third matter [**Fig. 7**].

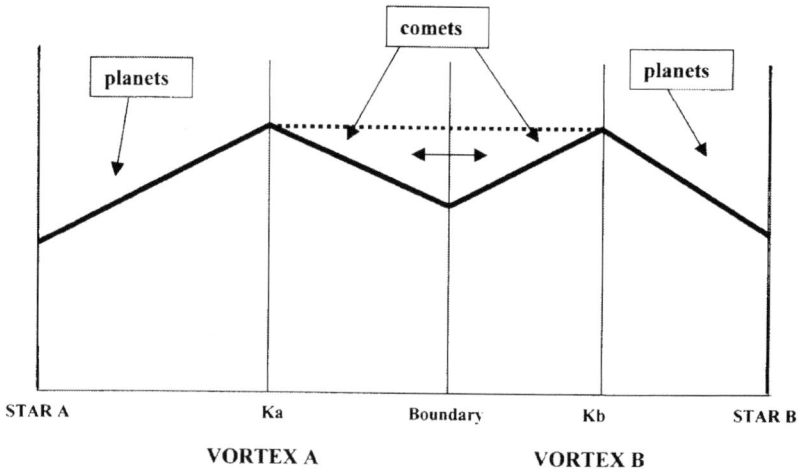

Figure 7. 'Resistance Curve': derived from V/S ratios of spheres of 2nd element.

The K layer marks an inflection point. From there outward, the spheres of second element get smaller not larger, and hence, surface envelopes made out of them are progressively less capable of resisting a centrifugally tending piece of third matter.

The bottom line is this: planets will always be locked into the vortex at a radius below the K layer. If you like, and Descartes speaks this way obscurely in *Le Monde*, more clearly in *Principles*, a planet will drift outward due to actualised centrifugal tendency, until it reaches a layer of the vortex where the spheres of second element have a V/S ratio sufficient to make the surface envelope they form resist any further centrifugal translation by the planet. The planet is locked in somewhere along the V/S curve of the spheres of second element. In his discussion of this part of the theory Descartes spoke of the 'massiveness' or 'solidity' of a

those of the second as those of the second in comparison with those of the first'. We are about to see one important reason why he has done this.

planet, meaning its aggregate volume to surface ratio.[30] This locking occurs at a radial distance from the sun at which the centrifugal tendency of the planet, a function of its massiveness, is exactly balanced by the resistance to centrifugal translation offered by the surface envelope in play at that location in the vortex. The greater a planet's V/S ratio or massiveness, the more distant that planet's orbit will be from the sun.[31]

Imagine a planet, hypothetically finding itself not in its proper orbital place, literally too high up in the vortex given its degree of massiveness. It will not be able to develop sufficient centrifugal tendency and will be extruded downward by subjacent spheres of second element. It will stop 'falling' when a balance is realised on the one hand between the centrifugal force of the subjacent second element at that radius in the vortex and the resistance offered by the planet (owing to its degree of massiveness), and, on the other hand, its own centrifugal tendency, conferred by its massiveness balanced by the resistance of the superjacent surface envelope at that layer in the vortex.[32]

Let us stop for a moment here and note that Descartes has constructed his mechanical heavens in such a way that mechanically efficacious stars are absolutely essential to the functioning of the celestial machine. If stars were inert, or if the

---

[30] In *Le Monde* Descartes did this somewhat confusedly, improving his explication of massiveness and its role considerably in the *Principles*. I am reconstructing the underlying model in *Le Monde*, using a crisp hermeneutics of 'solidity' as aggregate volume to surface ratio and meshing that concept with my analysis of the size/speed distribution of the *boules* in the vortex. By using the graphical representations of these ideas, mediated by my interpretive construct of 'surface envelopes', the resulting decoding of the underlying model emerges. Note that in this process of reading, the verbal descriptions of the size/speed ratios comes directly from the text, as does the concept of solidity. These are clarified and amplified graphically. The 'least Cartesian' notion used in this interpretation is that of 'surface envelopes', but even it has textual warrant in the overall direction of the theory, and in Descartes' various descriptions of the centrifugal tendency of planets (and comets) and the resistances they encounter at various levels of the vortex.

[31] This articulates the simple notion of centrifugal tendency as a function of size (quantity of matter) and force of motion only. In this mature application of the dynamics to a 'real' fluid vortex, it is clear that centrifugal tendency is a function of size, force of motion and 'solidity' (or massiveness), the latter taken in relation to the solidity of the relevant, resisting surface envelope.

[32] The condition for a piece of third matter to be in stable orbit in the vortex can thus be expressed as $F^m_b < R_{mu}$ and $F^m_{ml} < R_b$ Where F^m_b means Force of motion of the orbiting body; R_{mu} means resistance of superjacent layer of boules (upper medium) to being extruded downward by body; F^m_{ml} means Force of motion of subjacent layer of boules (lower medium) and R_b means resistance of orbiting body to being extruded downward by subjacent layer of boules. All these terms need to be taken in their full explication including the concepts of massiveness, surface envelopes, and the size/speed distribution of *boules* in the vortex. Note that the formula also expresses the conditions for a ball of second element to be in stable orbital motion as part of the total vortex, if we take F^m_b to mean the force of motion of the orbiting sphere, and R_b to denote the resistance of the orbiting sphere of second element to being extruded downward by the subjacent layer of spheres. This, then, would conduce to a fuller understanding of what we above termed the 'force-stability' principle for constitution of the vortex.

It must be reiterated that the systematic conclusions reached here constitute a charitable reading of the relevant passages in *Le Monde*, supplemented carefully by the somewhat more clear and cogent presentation in the *Principles*. There is no scope in this short paper for an explication of the construction of my reading, which will be reserved for a more copious discussion within the scope of a book length treatment of 'Descartes, physico-mathematician', dealing with all the matters touched upon in this and later sections of this paper, and other related topics as well.

second element filled the centres of the vortices, then two sets of consequences would follow: first, light would be propagated only along radial lines from the axis of revolution of the vortex (a matter we cannot pursue in the present paper); and, more importantly for our present concern, the resultant distribution of size and shape of the *boules* would not be proper for the existence of planets in stable orbits.

Descartes' theory of comets now follows with a kind of mechanistic inevitability. We already know that according to this theory of celestial mechanics, the more distant a planet is from the sun, the greater its V/S ratio or massiveness. Now what if a planet is very massive, and it has the centrifugal tendency sufficient to overcome even the most highly resistant surface envelopes formed by second element at or near the K level? Well then, the object will pass by actualised centrifugal tendency beyond the K level, beyond the hump in the resistance curve. Beyond K it will meet second element with decreasing V/S ratios, and less resistance, so that this object will move right on out of the vortex and stream into a neighbouring one. The locking mechanism fails for these extremely 'solid' or 'massive' 'planets'.

When such an object of great 'solidity' is flung into the neighbouring vortex, it meets increasing resistance to its centripetal trajectory—as we can see by looking at the curve in **Fig. 7**. The object picks up increments of orbital speed, until it starts to generate centrifugal tendency again, and again overcomes all obstacles—reading the curve in **Fig. 7** backward—and gets flung back out of that second vortex. These, of course, are Cartesian comets, planets of high massiveness that oscillate between vortices, never penetrating any lower than the K level—trapped on our representation in **Fig. 7** in the resistance depression between K levels of adjoining vortices.[33]

To summarise, then, each vortex is a locking and extrusion device. Its corpuscular make-up, size and speed distribution, given Descartes' theory of planet/comet make up, entails that planets are locked into orbits of differing radii. Comets are objects extruded from vortex to vortex, first 'falling' into a vortex and then being extruded out. The existence and make up and mechanical behaviour of the central stars are crucial, not to the existence of vortices, but to the existence of planet locking/comet extruding vortices. Otherwise extrusion would be the universal rule. Multiple vortices are necessary, as each vortex is set in a container made of contiguous vortices, exerting a kind of centripetal backwash at its boundary.

We now explore three genealogical steps in Descartes' trajectory to the vortex celestial mechanics of *Le Monde*. As foreshadowed in section 1, this exploration starts in 1619, and focuses on Descartes' activities in hydrostatics and physical and geometrical optics, and his relations, spanning a decade, with his mentor in

---

[33] There is of course much more to say about this theory of comets. It first of all makes some concrete empirical predictions, which could have stood unrefuted for at least a generation after 1633; to wit, comets do not come closer to stars than a layer K; they are 'more massive' than planets, they move in spiral paths oscillating out of and into solar systems. In addition, in dealing with the phenomena of comets' tails, Descartes had to attribute odd optical properties to the K layer as part of his overall theory of cosmological optics—raising thereby issues quite telling about the origin and import of his theorising, but beyond the scope of the present essay. See Schuster 1977. The matter will taken up in more detail in my monograph on Descartes as a physico-mathematician.

corpuscular-mechanism, Isaac Beeckman. The genealogy will illuminate a great deal about the structure of the vortex mechanics as we have just decoded it.

5.GENEALOGY PART A: 1619—FROM HYDROSTATICS TO DYNAMICS; FROM MIXED MATHEMATICS TO CORPUSCULAR-MECHANICAL NATURAL PHILOSOPHY[34]

In November 1618, Descartes met Isaac Beeckman. For two months they worked together on problems in natural philosophy, mechanics, theory of music, mathematics, and, hydrostatics. Descartes served a second natural philosophical apprenticeship with Beeckman. The scholastic vision purveyed during his schooldays at La Flèche was overlaid with an incipient corpuscular mechanism, derived from Beeckman, but about to take on a uniquely 'Cartesian' character, even at this early date. Sometime during this period Beeckman set Descartes four problems in hydrostatics culled from the work of Stevin. Beeckman was probably curious about how Descartes would explain Stevin's fundamental but strange result, the hydrostatic paradox. This was to be an exercise in their self-proclaimed style of 'physico-mathematics', briefly mentioned above in section 2, the agenda of which demanded that macroscopic phenomena be explained through reduction to corpuscular-mechanical models. Beeckman's questioning Descartes about Stevin's 'paradoxical' hydrostatical findings arguably sits squarely within this practice. This in general was what Beeckman and Descartes were envisioning when in 1618 they congratulated themselves on being virtually the only 'physico-mathematici' in Europe.[35] What they meant was that only they unified the mathematical study of nature with the search for true corpuscular-mechanical causes.[36] Beeckman wanted to see what his new friend and fellow 'physico-mathematicus' could do about reducing Stevin's work to corpuscular mechanical terms, thereby fundamentally explaining it.

In his *Elements of Hydrostatics* 1586, Stevin demonstrated that a fluid can exert a total pressure on the bottom of its container many times greater than its weight. In particular, he showed that a fluid filling two vessels of equal base area and height exerts the same total pressure on the base, irrespective of the shape of the vessel and hence, paradoxically, independently of the amount of fluid contained in the vessel. Stevin's argument proceeds entirely on the macroscopic level of gross weights and volumes. The rigour of the proof depends upon the maintenance of static equilibrium, understood in terms of Archimedes' hydrostatics.

Stevin proves that the weight of a fluid upon the horizontal bottom of its container is equal to the weight of the fluid contained in a volume given by the area

---

[34] Material in this section closely follows the argument of Gaukroger and Schuster 2002.

[35] *AT*, X, p. 52.

[36] *AT*, X, p. 52. In this regard Beeckman was to note in 1628 that his own work was deeper than that of Bacon on the one hand and Stevin on the other just for this very reason; Beeckman 1939–1953, III, pp. 51–52, 'Crediderim enim Verulamium [Francis Bacon] in mathesi cum physica conjugenda non satis exercitatum fuisse; Simon Stevin vero meo judico nimis addictus fuit mathematicae ac rarius physicam ei adjunxit'.

of the bottom and the height of the fluid measured by a normal from the bottom to the upper surface.[37] He employs a *reductio ad absurdum* argument: ABCD is a container filled with water [**Fig. 8**]. GE and HF are normals dropped from the surface AB to the bottom DC, dividing the water into three portions, 1 [AGED], 2 [GHFE] and 3 [HBCF].

Stevin has to prove that on the bottom EF there rests a weight equal to the weight of the water of the prism 2. If there rests on the bottom EF more weight than that of the water 2, this will have to be due to the water beside it, that is water 1 and 3. But then, there will also rest on the bottom DE more weight than that of the water 1; and on the bottom FC also more weight than that of the water 3; and consequently on the entire bottom DC there will rest more weight than that of the whole water ABCD, which would be absurd. The same argument applies to the case of a weight of water less than 2 weighing upon bottom EF.[38]

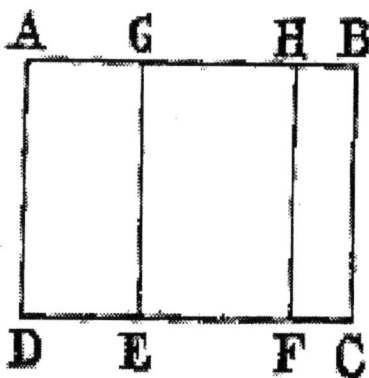

Figure 8. *Stevin,* Elements of Hydrostatics *(1586) in* Principal Works of Simon Stevin, *I, p. 415.*

Stevin then ingeniously argued that portions of the water can be notionally solidified, replaced by a solid of the same density as water. This permits the construction of irregularly shaped volumes of water, such as IKFELM, to which, paradoxically, the theorem can still be applied. [**Fig. 9**].[39]

---

[37] Simon Stevin, *De Beghinselen des Waterwichts*, Leiden, 1586 in Stevin 1955–1966, I, p. 415.

[38] *Ibid.*, I, p. 415.

[39] *Ibid.*, I, p. 417. 'Let there again be put in the water ABCD a solid body, or several solid bodies of equal specific gravity to the water. I take this to be done in such a way that the only water left is that enclosed by IKFELM. This being so, these bodies do not weight or lighten the base EF any more than the water first did. Therefore we still say, according to the proposition, that against the bottom EF there rests a weight equal to the gravity of the water having the same volume as the prism whose base is EF and whose height is the vertical GE, from the plane AB through the water's upper surface MI to the base EF'.

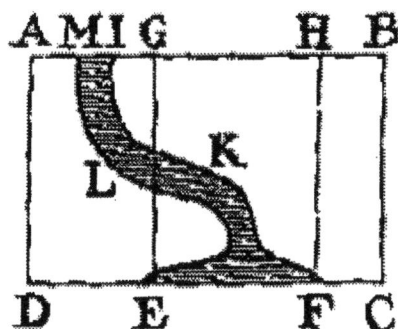

Figure 9. Stevin, Elements of Hydrostatics *(1586) in* Principal Works of Simon Stevin, *I, p. 417.*

That is, on bottom EF there actually rests a weight equal to that of a volume of water whose bottom is EF and whose height is GE. Stevin then applies these findings to the sides of containing vessels.

Descartes' response to Beeckman's request that he explain Stevin's results is given in a report preserved in Beeckman's famous *Diary*, which Gaukroger and I have termed 'the hydrostatic manuscript'.[40] It involves an attempt to 'improve' upon Stevin's work; that is, to provide a deep natural philosophical explanation for his results, based on an incipient corpuscular-mechanical ontology. The hydrostatic manuscript amounts to an 'ultrasound scan' of Descartes' embryonic corpuscular-mechanical agenda, disclosing its fine, foetal anatomy. But in order to understand the hydrostatics manuscript and its place in the genealogy of Descartes' vortex celestial mechanics, we need first briefly to consider the work of Beeckman, his then thirty-year-old mentor.

Beeckman was one of the very first individuals in Europe to pursue consistently the idea of a micro-mechanical approach to natural philosophy. He conceived of a redescription of all natural phenomena in terms of the shape, size, configuration and motion of corpuscles, and he insisted that the causal register of this account, that is, the principles of all natural change, had to be derived from the transdiction of the presumed mechanical principles of macro-phenomena, in particular the behaviour of the simple machines. Beeckman offered on a first-hand basis an approach to natural philosophy which was not available to Descartes from any other contemporary source.

---

[40] The text, *Aquae comprimentis in vase ratio reddita à D. Des Cartes* which derives from Beeckman's diary, is given in *AT*, X, pp. 67–74, as the first part of the *Physico-Mathematica*. See also the related manuscript in the *Cogitationes Privatæ*, *AT*, X, p. 228, introduced with, 'Petijt è Stevino Isaacus Middelburgensis *quomodo aqua gravitet in fundo vasis* b…'.

Beeckman held a fundamentally atomistic view of nature. His atoms possess only the geometrical-mechanical properties of size, shape, and impenetrability (being absolutely hard, incompressible and non-elastic). We may, with slight Whiggish licence, say that Beeckman conceptualised motion as follows: motion is, in effect, conceived as a simple state of bodies, rather than, in Aristotelian terms, as an end-directed process which they undergo. Unlike previous advocates of atomism and prior to any of the great mechanists of the later seventeenth century, Beeckman sought to explain the behaviour of his atoms by applying to them a causal discourse modelled on the principles of mechanics.

By 1613 or 1614 Beeckman had formulated a concept of inertia holding for both rectilinear and curved motions. Combining his principle of inertia with his atomic ontology, Beeckman was led to conclude that only corpuscular collision and transfer of motion can account for the initiation of motion of resting bodies or alternation of motion of moving ones. What he needed was rules of corpuscular collision. Since his atoms are perfectly hard, he formulated rules applicable to what we would term perfect inelastic collisions. He measured the quantity of motion of corpuscles by taking the product of their quantity of matter and their speed. Significantly, Beeckman linked his measure of motion to a dynamic interpretation of the behaviour of the balance beam. He evaluated the effective force of a body on a balance beam by taking the product of its weight and the speed of its real or potential displacement, measured by the arc length swept out in unit times during real or imaginable motions of the beam. Beeckman was able to build up a set of rules of impact by combining certain intuitively symmetrical cases of collisions with the dictates of the inertial principle and an implicit concept of the conservation of the directional quantity of motion in a system. His treatment of symmetrical cases of collision and his notion of the conservation of motion owed their form and their putative legitimacy to the model of the balance beam, interpreted in a dynamic rather than static fashion.[41] Indeed, Beeckman consistently demanded a dynamical

---

[41]Beeckman's rules fall into two broad categories: (1) cases in which one body is actually at rest prior to collision; and (2) cases which are notionally reduced to category (1). The concept of inertia and the stipulation that only external impacts can change the state of motion of a body provide the keys to interpreting instances of the first category. The resting body is a cause of the change of speed of the impacting body and it brings about this effect by absorbing some of the quantity of motion of the moving body. Beeckman invokes an implicit principle of the directional conservation of quantity of motion to control the actual transfer of motion. In each case the two bodies are conceived to move off together after collision at a speed calculated by distributing the quantity of motion of the impinging body over the combined masses of the two bodies. For example, in the simplest case, in which one body strikes an identical body at rest, '... each body will be moved twice as slowly as the first body was moved... since the same impetus must sustain twice as much matter as before, they must proceed twice as slowly'. And he adds, analogising the situation to the mechanics of the simple machines, '... it is observed in all machines that a double weight raised by the same force which previously raised a single weight, ascends twice as slowly' (Beeckman 1939–1953, I, pp. 265–266). Instances of the second category of collision are assessed in relation to the fundamental case of collision of equal speeds in opposite directions (*ibid.*, p. 266). Being perfectly hard and hence lacking the capacity to deform and rebound, the two atoms annul each other's motion, leaving no efficacious residue to be redistributed to cause subsequent motion. This symmetrical case, which was also generalised to cases of equal and opposite quantities of motion arising from unequal bodies moving with compensating reciprocally proportional speeds, derives from a dynamical interpretation of the equilibrium conditions of the simple machines. Instances in which the quantities of motion of the bodies are not

approach to statics, the theory of simple machines and mechanics in general, including hydrostatics.

It is important to realise that Beeckman's tactic here was following the classic model set forth in that sixteenth-century best seller, the pseudo Aristotelian text *Mechanical Problems* or *Mechanica*, which took a dynamical approach to the problems of statics and the behaviour of the simple machines. That is, in the *Mechanica* one views equilibrium conditions on a lever or simple machine as a balance of forces, where force is defined as weight times speed. The basic principle behind this comes from Aristotle: the same force will move two bodies of different weights but it will move the heavier body more slowly, so that the velocities of the two bodies are inversely proportional to their weights. In the case of the lever, when these are suspended from the ends of a lever, we have two forces acting in contrary directions, and each body moves in an arc with a force proportional to its weight times the length of the arm from which it is suspended. The one with the greater product will descend in a circular arc, but if the products are equal, they will remain in equilibrium. So, the *Mechanica* makes statics simply a limiting case of a general dynamical theory of motion, a theory that is driven by Aristotelian dynamics, above all by the principle of the proportionality of weight and velocity.

It is significant that some earlier anti-Aristotelian mathematicians and natural philosophers, such as Tartaglia, Benedetti and the young Galileo, had tried to exploit the dynamical interpretation of the balance beam and simple machines in the *Mechanica* against Aristotelianism, despite the Aristotelian underpinning of that text. Beeckman was working in this general vein but being more radical about it, by placing at the basis of his corpuscular-mechanism rules of collision founded on the *Mechanica*-type—dynamical—approach to the simple machines. We shall now see that in the 'hydrostatics manuscript' Descartes was to take an even more surprisingly radical turn in the search for a causal doctrine, a dynamics, through which to 'run' corpuscular-mechanical explanations.

Descartes takes as given the following conditions [**Fig. 10**]: A, B, C, and D are four vessels with equal areas at their bases, equal height and of equal weight when empty. B, C and D are filled to their tops. A is filled with water equal to the amount it takes to fill B.

---

equal are handled by annulling as much motion of the larger and/or faster moving body as the smaller and/or slower body possesses (*ibid.*). This in effect reduces the smaller and/or slower body to rest. The outcome of the collision is then calculated by distributing the remaining unannulled motion of the larger and/or swifter body over the combined mass of the two bodies (*ibid.*). It is obvious that Beeckman viewed this case through a two-fold reference to the simple machines; for he first extracts as much motion as can conduce to the equilibrium condition for symmetrical cases, and then he invokes the principle cited just above in this note to determine the final outcome.

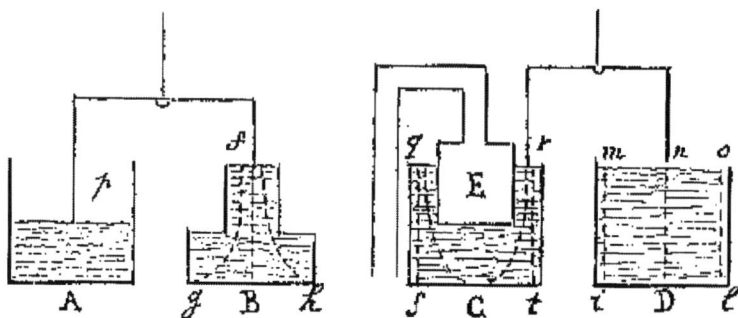

Figure 10. Descartes, Aquae comprimentis in vase ratio reddita à D. Des Cartes, AT, *X, p. 69.*

In the key problem Descartes proposes to show that, 'the water in vessel B will weigh equally upon its base as the water in D upon its base'—Stevin's paradoxical hydrostatical result.[42] While Stevin's approach is geometrical, Descartes' analysis and explanation are based on an attempt to reduce the phenomenon to micro-mechanical terms. First, Descartes tells us that of the various ways in which bodies may 'weigh-down' [*gravitare*], only two need be discussed: the weight of water on the bottom of a vessel which contains it, and the weight of the vessel and the water it contains.[43] By the weight of the water on the bottom of the vessel he does not intend the gross weight of the quantity of water measured by weighing the filled vessel and subtracting the weight of the container itself. He means instead the total force of the water on the bottom arising from the sum of the pressures exerted by the water on each unit area of the bottom. Next, and crucially, the term 'to weigh down' is explicated as 'the force of motion by which a body is impelled in the first instant of its motion'. Descartes insists that this force of motion is not the same as the force of motion which 'bears the body downward' during the actual course of its fall.[44] Finally, Descartes insists that we attend to both the 'speed' and the 'quantity of the body', since both factors contribute to the measure of the 'weight' or force of motion exerted in the first instant of fall.

---

[42] *AT*, X, pp. 68–69; '... the water in base B will weigh equally upon the base of the vase as does the water in D upon its base, and consequently each will weigh more heavily upon their bases than the water in A upon its base, and equally as much as the water in C upon its base'. This is the second of the four puzzles posed in the text, the others are: '(First), the vase A along with the water it contains will weigh as much as vase B with the water it contains. ... Third, vase D and its water together weigh neither more nor less than C and its water together, into which *embolus* E has been fixed. Fourth, vase C and its water together will weigh more than B and its water. Yesterday I was deceived on this point'.

[43] *AT*, X, p. 68.

[44] *AT*, X, p. 68. In the *Cogitationes Privatæ* (*AT*, X, p. 228) the inclination to motion is described as being evaluated 'in ultimo instanti ante motum'.

These three suppositions mark the first, embryonic appearance of some fundamental notions of Cartesian dynamics. Weight or heaviness reduces to the mechanical force exerted by a particle in its tendency to motion of descent. Weight is no longer an essential quality of bodies, but is jointly determined by the size of the body and its tendency to motion as conditioned by a given configuration of neighbouring bodies.

Descartes next solves the problem of accounting for the hydrostatic paradox. But, whereas Stevin offered an Archimedean argument from macroscopic conditions of equilibrium, Descartes manufactures a curious exercise in *ad hoc* micromechanical reductionism. He proposes to demonstrate the proposition by showing that the force on each 'point' or part of the bottoms of the basins B and D is equal, so that the total force is equal over the two equal areas.[45] He does this by claiming that each 'point' on the bottom of B is, as it were, serviced by a unique line of 'tendency to motion' propagated by contact pressure from a point (particle) on the surface of the water through the intervening particles [See **Fig. 10**].

He takes points g, B, h in the base of B, and points i, D, l in the base of D. He claims that all these points are pressed by an equal force, because they are each pressed by 'imaginable lines of water of the same length'. That is, the same vertical component of descent. He says,

> ... line fg is not to be reckoned longer than fB or [any] other line. It doesn't press point
> g in respect to the parts by which it is curved and longer, but only in respect to those
> parts by which it tends downward, in which respect it is equal to all the others.[46]

This rather strange material requires some unpacking. Assuming the points on the bottoms are indeed served by unique lines of tendency transmitted from points on the surface; then, in so far as we are only concerned with the tendency to descend, we may compare the lines of tendency in respect to their vertical 'components'. What, then, about the mapping of the lines of tendency? Descartes is saying that when the upper and lower surfaces of the water are similar, equal and posed one directly above the other, then unique normal lines of tendency will be mapped from each point on the surface to a corresponding point directly below on the bottom. But, when these conditions do not hold, then some other unstated rules of mapping come into play.

So, in the present case the area of the surface at f in the basin B apparently is one-third that of the bottom, hence each point or part on f must be taken to service three points or parts of the bottom. The problem, of course, is that no rules for mapping are, or can be, given. Descartes does not justify the three-fold mapping from f. He merely slips it into the discussion as an 'example' and then proceeds to argue that *given the mapping*, f can indeed provide a three-fold force to g, B and h. He proceeds to show by a syllogism, no less, that point f presses g, B, h with a force equal to that by which m, n, o press the other three i, D, l.

---

[45] Descartes consistently fails to distinguish between 'points' and finite parts. But he does tend to assimilate 'points' to the finite spaces occupied by atoms or corpuscles. Throughout we shall assume that Descartes intended his points to be finite and did not want his 'proofs' to succumb to the paradoxes of the infinitesimal.

[46] *AT*, X, p. 70.

1. Heavy bodies press with an equal force all neighbouring bodies, by the removal of which the heavy body would be allowed to occupy a lower position with equal ease.
2. If the three points g, B, h could be expelled, point f alone would occupy a lower position with as equal a facility as would the three points m, n, o, if the three other points i, D, l were expelled.
3. Therefore, point f alone presses the three points simultaneously with a force equal to that by which the three discrete points press the other three i, D, l. And so, the force by which point f alone presses the lower [points] is equal to the force of the points m, n, o taken together.[47]

In sum, there is a two-fold displacement away from what one might consider the original terms of the problem: [1] Descartes assumes an ad hoc mapping; and [2] invokes a hypothetical voiding and consequent motion.[48] The proof of this 'example' is then taken as a general demonstration without any indication as to how the procedure is to be generalised to all the points or parts in the surfaces.

Strange as all this may appear to us today, Descartes himself was quite pleased. He continued to use descendants of these concepts for the rest of his career. Here we have the key concept of instantaneous tendency to motion. Descartes' later mechanistic optics and natural philosophy will depend on the analysis of instantaneous tendencies to motion, rather than finite translations. Often Descartes will consider multiple tendencies to motion which a body possesses at any given instant, depending on its mechanical circumstances. There is evidence that even in 1619 Descartes was considering trying to systematise this set of new dynamical concepts to apply to corpuscular explanations, as he speaks in the 'hydrostatics manuscript' and surrounding correspondence of a treatise of 'Mechanics' he is planning to write.[49]

This 1619 performance of Descartes was quite portentous, and it is worth pausing to crystallise precisely what was going on in his work at this stage. First of all, it is obvious that Descartes certainly was not denying the rigour or correctness of Stevin's strictly mathematical, Archimedean account. What he was after was proper explanation, meaning explanation in terms of natural philosophy. Stevin's treatment of the hydrostatic paradox fell within the domain of mixed mathematics. The account Descartes substitutes for it falls within the domain of natural philosophy: the concern is to identify what causes material bodies to behave in the way they do. Fluids are physical entities made up, on Descartes' account, of microscopic corpuscles. Their behaviour determines the macroscopic behaviour of the fluid. So, we need to understand the physical behaviour of the constituent corpuscles, if we are

---

[47] *AT*, X, pp. 70–71.

[48] There actually is a third displacement away from the original terms of the problem: Notice that Descartes implicitly solidifies parts of the fluid not involved in the first two steps. That is, in working out the hypothetical case of descent, Descartes imagines away the rest of the fluid, *qua* fluid. It is in effect hypothetically solidified, so that its behaviour does not complicate the postulated mechanical relations between f and g, B and h. This sort of tactic, along with the first two, plays a key role later in his theory of light in the cosmic setting of vortices in *Le Monde* and the *Principles*.

[49] *AT*, X, p. 72 and in correspondence with Beeckman early in 1619, *AT*, X, pp. 159, 162. For more discussion see Gaukroger and Schuster 2002.

to understand the behaviour of the fluid. Descartes therefore is saying that hydrostatics is no longer merely a discipline of mixed mathematics in the Aristotelian sense; rather it is an application of, and indeed illustration of, corpuscular-mechanical natural philosophy. This is deeply anti-Aristotelian, for it unambiguously bids to shift hydrostatics from the realm of mixed mathematics into the realm of natural philosophy.

Described in this manner, the 'hydrostatic manuscript' sounds quite in tune with a general notion of 'physico-mathematics' that he shared with Beeckman. There was, however, more going on in the manuscript, bespeaking some subtle differences between what Descartes was actually doing, and their otherwise shared general sense of what it meant to do physico-mathematics. As mentioned earlier, attempts to wring anti-Aristotelian natural philosophical conclusions out of the mixed mathematical sciences were hardly new. Most such attempts—as in the work of the young Galileo or in Beeckman himself—depended on taking a dynamical approach to statics and the simple machines, following the lead of the pseudo-Aristotelian *Mechanical Questions*. But, in the hydrostatics manuscript, the young Descartes does not proceed via the *Mechanical Questions'* dynamical account of the lever. Rather he plays the Archimedes/Stevin card—he starts from a mathematically rigorous hydrostatics of all things, which he *then* fleshes out in terms of the micro-corpuscularian model he learned from Beeckman.[50] The young Descartes' hyper-radical program was this: he wanted to reduce Stevin's hydrostatics to an embryonic corpuscular mechanism in which discourse concerning causes or 'forces', elicited on the basis of that hydrostatics, would provide the basis for unifying the mathematical sciences. This difference, and the dynamic of research and concept formation that it unleashed, was going to play out in his optical work in the 1620s, and crystallise in the program of *Le Monde*.

6.GENEALOGY PART B: 1627 THE LAWS OF LIGHT AND THE LAWS OF NATURE[51]

The next step in our genealogy of the vortex celestial mechanics of *Le Monde* involves perhaps the most important, successful and fruitful physico-mathematical

---

[50] We should note here a point that makes Descartes' moves all the more interesting: The Archimedean account, exploited by Stevin, comes without any dynamical, or more broadly speaking natural philosophical commitments. In the hands of Stevin, statics and hydrostatics, are hardly mixed sciences at all, since they really do make no physical or dynamical claims. Stevin was an arch Archimedean, and champion of the practical mathematical arts over against natural philosophical verbal wranglings. He pursued an ultra Archimedean program. So, he rejected the *Mechanica*, denying that the arcs through which bodies would move if they ceased to be in equilibrium have any bearing on the problem of the lever: You cannot deduce equilibrium conditions from the supposition that motion has or would occur—that is absurd, since if motion occurs the forces are not in equilibrium. This led Stevin to his famous reasoned denial that the study of motion, i.e. natural philosophy, could ever be pursued in a rigorous mathematical manner. How extremely interesting it is, then, that Descartes seemed able to make natural philosophical capital, indeed innovative natural philosophical capital, by recourse not to the *Mechanica* but to the purely statical, purely mathematical, equilibrium science of Stevin. See Gaukroger and Schuster 2002, pp. 540, 545–549.

[51] For full details on claims in this section see Schuster 2000.

research Descartes ever attempted—his work in geometrical and physical optics in the 1620s. These endeavours climaxed with his discovery of the law of refraction of light around 1627; but, they did not end there, despite widely held views amongst scholars about this period in his life. Descartes immediately began to think about possible mechanical rationales or explanations for the law, and these attempts were intimately connected with a process by which he crystallised his emerging concepts of dynamics directly out of a 'physico-mathematical' 'reading' of his geometrical optical results. In short, his optical researches marked the high point of his work as a physico-mathematician transforming the 'old' mixed mathematical sciences and co-opting the results into a mechanistic natural philosophy: On the one hand, his results confirmed his 1619 agenda of developing a corpuscular ontology and a causal discourse, or dynamics, involving concepts of force and directional 'determination' of motions or tendencies to motion. On the other hand, his results concretely advanced and shaped his concepts of light as an instantaneously transmitted mechanical tendency to motion, as well as the precise principles of his dynamics.

Around 1620 Descartes explored Kepler's speculations about the refraction of light, using his newly acquired physico-mathematical style of 'reading' geometrical diagrams representing phenomena for their underlying message about the causal principles at work. I have demonstrated elsewhere that Descartes, in one of his physico-mathematics fragments dating from around 1620, attempted to appropriate in physico-mathematical fashion a particularly telling geometrical optical diagram he found in Kepler's *Paralipomena ad vitellionem* (1604).[52]

---

[52] See Schuster 2000. The optical fragment of Descartes appears at *AT*, X, pp. 242–243.

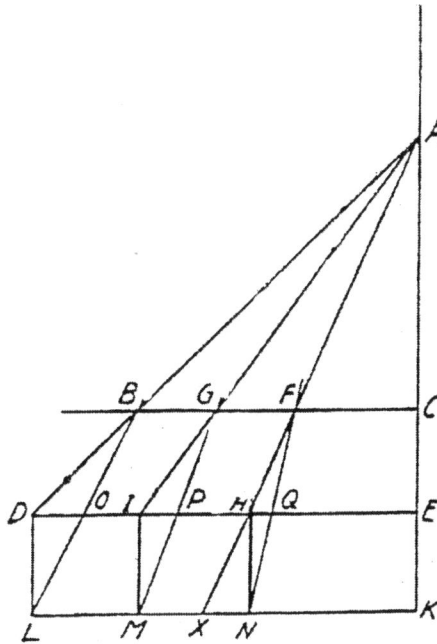

Figure 11. Kepler's diagram representing possible role in refraction of light of density of refracting medium and obliquity of incident ray. Kepler, Ad Vitellionem Paralipomena *(1604), in* Gesammelte Werke, *II, p. 85.*

In **Fig. 11** Kepler depicted the possibility that the density of the refracting medium and the inclination of the incident ray both exercise a geometrically representable physical effect on the refracted ray. Kepler takes AG incident upon a basin of water. The density of water is said to be twice that of air, so Kepler lowers the bottom of the basin DE to LK so that the new basin contains 'as much matter in the rarer form of air as the old basin contained in the doubly dense form of water'. Kepler then extends AG to I and drops a normal from I to LK. Connecting M and G gives the refracted ray GM. Its construction involves the obliquity of incidence and densities of media. Kepler then rejects this construction on empirical grounds.[53] But,

---

[53] Kepler, *Ad Vitellionem Paralipomena*, in Kepler 1938–, II, pp. 81–86. My analysis, Schuster 2000, pp. 279–285 shows how this passage provides the source for Descartes' speculation, which he further linked to two other passages in Kepler's optics, Kepler 1938–, II, pp. 89–90, 107.

the young Descartes, fresh from his physico-mathematical foray into hydrostatics, played physico-mathematical games with it.

What Descartes, physico-mathematicus, saw here was the *physical-causal* speculation that denser media bend light toward the normal, *and* the *physico-mathematical* notion that you can represent this geometrically in order to construct refracted rays. Let us recall that opticians, Kepler included, treated light rays in these situations in terms of normal and parallel components. Descartes' manuscript fragment on this indicates that he saw the lower medium as acting to increase the normal component of the force of the ray in a fixed ratio. So, Descartes was reading Kepler the way he had read Stevin: as a physico-mathematician. That is, he was attempting to elicit some mathematicised physical theory from a compelling geometrical diagram for refraction presented by Kepler.[54] It must be noted, however, that this physico-mathematical exercise actually hindered Descartes' eventual attainment to the law of refraction of light, because on the likely concomitant assumption that the parallel component of the force of the incident ray remained constant, Descartes' 1620 speculation yields a law of tangents, rather than the law of sines (or in fact cosecants) which he achieved later.[55] Nevertheless, when he did succeed six or seven years later, his physico-mathematical style again came into play, with portentous results.

It was only in 1626/7 that Descartes, in collaboration with Claude Mydorge, discovered of the law of refraction. This discovery took place entirely within the confines of traditional geometrical optics, without the benefit of dynamical or corpuscular-mechanical theorising. My detailed reconstruction, published elsewhere,[56] involves Descartes and Mydorge having done in practice, or merely on paper, what we know Harriot earlier had done to construct the law of refraction— that is use the traditional image locating rule in order to map the image locations of point sources taken on the submerged circumference of a disk refractometer **[Fig. 12]**.[57]

---

[54] Schuster 2000 pp. 281–282.
[55] Schuster 2000 p. 285 and note 69 thereto.
[56] Schuster 2000, pp. 272–277.
[57] Lohne 1959, pp. 116–117, 1963, p. 160. Gerd Buchdahl provided a particularly clear statement of the methodological role played by the image principle in Harriot's discovery of the law Buchdahl 1972, pp. 265–298 at p. 284. Willebrord Snel's initial construction of the law of refraction also followed the type of path indicated by the Lohne analysis. See Vollgraff 1913, 1936; de Waard 1935–1936.

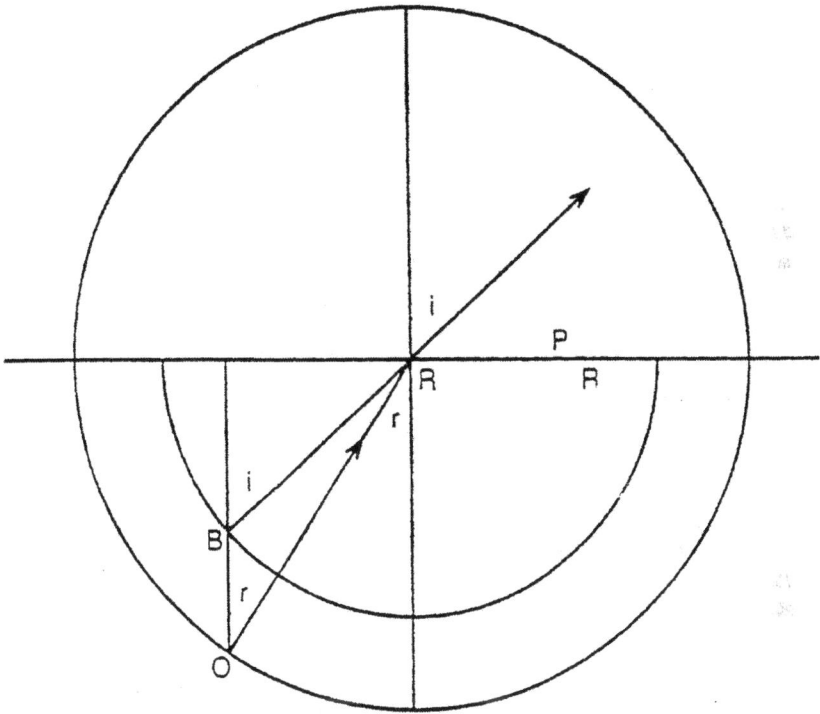

Figure 12. Harriot's key diagram: half submerged disk refractometer. Source points on lower circumference. Image points on smaller semi-circular locus. A Law of Cosecants results for refraction of light, Lohne 1963, p.160.

Even using Witelo's cooked data, one gets a smaller semi-circle,[58] and accordingly initially a law of cosecants, mathematically equivalent to the law of sines Descartes later published in the *Dioptrics*.[59] In a key letter describing the cosecant form of the law and a resulting theory of lenses, Mydorge later drew this diagram as a refraction predictor, by flipping the inner semi circle up above the interface as the locus of point sources for the incident light [**Fig. 13**].[60]

---

[58] Schuster 2000, p. 276, referring to confirmation of the work of Bossha 1908.

[59] For evidence on the movement from the original cosecant form of the law to the later sine form, based on Descartes' early work on lens theory, see Schuster 2000, pp. 274–275.

[60] On the important issue of the dating and content of Mydorge's letter containing this crucial diagram see Schuster 2000, pp. 272–275.

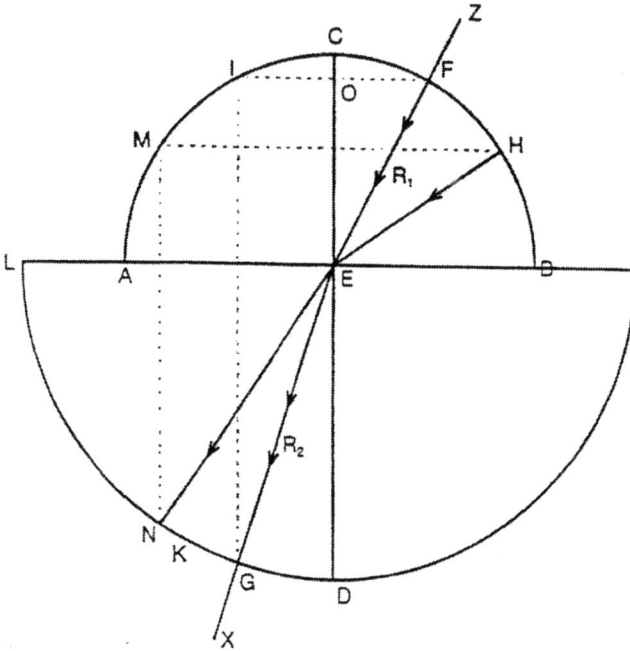

$$\frac{\cosec\ i}{\cosec\ r} = \frac{R1/OF}{R2/OI}$$

Figure 13. Mydorge's refraction prediction device, Mydorge to Mersenne in Mersenne 1932–1988, I, p. 405.

After the discovery of the law of refraction by these purely geometrical optical means, and with this sort of geometrical representation available to him, Descartes looked for better conceptions of the dynamics of light by which to explain the law. Unsurprisingly, these he found by doing physico-mathematics in the style of 1619: that is, he transcribed into dynamical terms some of the geometrical parameters embodied in this diagrammatic representation of the law [**Fig. 13** above]. The resulting dynamical principles concerning the mechanical nature of light were: [1] the absolute quantity of the force of the ray was increased or decreased in a fixed proportion; whilst, [2] the parallel component of the force of a light ray was unaffected by refraction. There is evidence dating from 1628 of Descartes using these concepts in an attempted analogical proof of the cosecant law of refraction, by appeal to the behaviour of a bent arm balance.[61] His proof of the law of refraction in sine form in the *Dioptrics,* published in 1637 but inscribed sometime between 1629 and 1633, also deploys precisely these dynamical concepts, applied to light as an instantaneously transmitted tendency to motion. But, as I have shown elsewhere, to discern this one must peer below the surface of his superficially confusing tennis ball model for the motion, reflection and refraction of light.[62] There is no doubt, however, that these dynamical principles were constructed prior to the writing of *Le Monde*, since the text alludes to their existence in the as yet unpublished *Dioptrics*.[63] In sum, it may safely be concluded that the insights [1] and [2] above, abstracted from the original geometrical representation of the newly discovered cosecant form of the law, actually suggested the form of the two central tenets of his mature dynamics, described earlier in section 3, when he came to consider the need for them in the course of composing *Le Monde*.[64] After all, to Descartes the physico-mathematician what could have been more revealing of the underlying principles of the punctiform dynamics of micro-corpuscles than the basic laws of light—itself an instantaneously transmitted mechanical impulse?

Descartes' dynamics, the causal register for talking about micro corpuscles, had eventuated from work in the mixed mathematical sciences—that was interesting and precisely in the agenda of physico-mathematics as conceived (and practised) in 1619. Even more interesting was that he focused on results and phenomena in which, paradoxically no motion of bodies took place at all—in hydrostatics, and in the exemplary refracting of instantaneously transmitted light rays. In these 'statical' exemplars, or 'phénoméno-techniques' Descartes found crisp, clean messages about the underlying dynamics of the corpuscular world and indeed about its laws. With these findings we are almost back at the text of *Le Monde*. But there was one final critical encounter in Descartes' trajectory of the 1620s from the early physico-mathematics to the dynamics and vortex celestial mechanics of *Le Monde*. Again, an encounter with Beeckman crystallised and shaped ensuing events.

---

[61] Schuster 2000, pp. 290–295.

[62] Schuster 2000, pp. 261–272.

[63] See above note 9. In discussing the distinction between the force of motion and its directional determinations, Descartes appeals to an already existing text on *Dioptrics*, *AT*, XI, p. 9.

[64] Schuster 2000, pp. 302–303. The two principles read out of the optical diagram suggest that one may treat absolute quantities of force of motion (or force of tendency to motion) separately from their directional modes, or determinations. The diagram, read in this fashion, tells Descartes that a light ray is refracted due to the facts that [1] a change is affected in the absolute quantity of the force of motion (here force of tendency to motion) which is a constant for the two media in question, but that [2] the component of its determination of tendency to motion parallel to the refracting surface is unaffected by the refraction. Later the first rule of nature in *Le Monde* will subsume [1] and the third rule of nature will subsume [2]. The results of the optical research directly parallel the two key dynamical concepts of Descartes as discussed above in section 3.

7.GENEALOGY PART C: 1629 COSMIC BALANCING ACTS

In late 1628 after a gap of ten years, Descartes re-established contact with Beeckman.[65] He found Beeckman ploughing through the astronomical works of Kepler, seeking to evaluate instances in which Kepler had invoked immaterial celestial forces or powers. In each case Beeckman sought to re-write the 'mechanisms' into corpuscular-mechanical terminology. As far as Beeckman was concerned, the key issues in astronomy did not involve the traditional activities of observation or even Kepler's work on elliptical orbits. Rather, Beeckman saw in Copernican astronomy, especially as transformed by Kepler, a broad, hitherto neglected field for natural philosophical explication, in particular corpuscular mechanical explanation. Beeckman specifically identified his celestial mechanical speculations as *desiderata* for a *restitutio astronomiae*.[66]

Similar concerns lay behind Descartes' celestial mechanics in *Le Monde*. Descartes, like Beeckman, avoided technical issues in observational astronomy, concentrating on plausible mechanical accounts of the causes of the motions of the planets in the Copernican system. Descartes and Beeckman were engrossed by the radical attempt to indicate how the latest conceptions in their 'physico-mathematics' might be brought to bear in explaining in a general way the causes of the motions of the planets in the Copernican system—allowing of course for the differences in content and trajectory in their respective versions of physico-mathematics, evident since 1619 and certainly quite further developed in Descartes' case by the late 1620s, as we have seen. In addition both Descartes and Beeckman sought to support their respective celestial physics by trading upon the suggestion that their celestial physics also explained the nature of light and thus was partially confirmed by its broad explanatory sweep.

Indeed Beeckman's review of Kepler starts with a penetrating mechanistic critique of Kepler's theory of light: light is corporeal, consisting in a type of heat particle emitted by stars. Kepler's law of illumination is explained by the way streams of light corpuscles spatially diverge from each other with distance from a source—an outcome impossible and unintelligible, he claims, on Kepler's own theory of light as an immaterial emanation.[67] This is crucial, because Beeckman's varied celestial mechanical speculations all play upon the idea of opposed, corporeally mediated forces that vary in strength with distance from source, hence constituting particular loci of equilibrium for the orbital placement of objects.

Beeckman then addresses a theory of lunar orbital placement: the moon is held in its orbit by a balance of attractive and repulsive actions delivered respectively, by rays of the sun reflected by the earth, and rays of the earth itself. The efficacy of the earth rays decreases with distance more quickly than that of the solar rays. The solar

---

[65] Schuster 1977, pp. 508–509; Beeckman 1939–1953, III, p. 114 n. 3; Mersenne 1932–1988, II, pp. 222, 217–218, 233–244; *AT*, X, pp. 341–343; Beeckman 1939–1953, III, p. 103; Schuster 1977, pp. 507–520.

[66] Beeckman 1939–1953, III, p. 103. In the period July 1628 to June 1629 roughly twenty-one out of fifty-nine pages of Beeckman's journal deal with celestial mechanical and related matters. Material in this section is treated in more detail in Schuster 1977, pp. 507–520.

[67] Beeckman 1939–1953, III, p. 74.

rays are presumably Beeckmanian light rays—streams of corpuscles; the earth rays are rays of Beeckman's version of earth magnetism.[68] The 'attraction' and 'repulsion' attributed to these corporeal rays is unexplicated at the corpuscular level. (It is worth noting here, in passing, just how much better Descartes would later have judged his own constructions to be. In *Le Monde* he will have a plausible locking and extrusion mechanism deeply embedded in findings about hydrostatics, optics and a general dynamics of corpuscles.)

There were of course problems with the lunar theory,[69] which Beeckman detected (perhaps aided by his French friend), before he rushed onto a grander vision of the entire celestial mechanism. By substituting the fixed stars for the sun, and the sun for the reflecting earth, his moon theory could perhaps be applied to the entire solar system. Sometime between October 1628 and late January 1629 Beeckman boldly writes:

> [the same] thing can be said about all the planets (among which I also number the earth)… the light or corporeal virtue of the eighth sphere reflected by the sun draws the planets to the sun and the sun [itself] repels them. And thus each planet will be affected by each of the virtues according to its magnitude or rarity and therefore they will be located at different distances from the sun.[70]

This is indeed a striking speculation: the heavens are criss-crossed with the direct and reflected corporeal emanations of the fixed stars and the sun, leading the planets being located in the network of differential forces according to their 'magnitude' and 'rarity'.[71] However, Beeckman then noticed that the sun's own emanations were now repulsive in nature, and so he quickly reverted to a simpler picture of paired sun-planet interactions, based, as before, on a balance of forces—in this case planetary magnetic attraction, corporeally mediated, working against the mechanical repulsion arising from impact of solar heat and light corpuscles.[72]

Continuing to jot in his *Journal* as his speculations wandered, Beeckman shifts his ground again: he reverted to the fixed stars sending a flow of effluvia through the solar system. There are always more solar emanations immediately within the orbit of a given planet than immediately beyond it, thus fewer celestial emanations can make their way within the orbit and exert a back-pressure on the sunward side of the planet. Hence each planet suffers a pressure toward the sun arising from the incoming stellar rays which is to be balanced by the light/heat repulsion of solar

---

[68] *Ibid.*, pp. 74–75.
[69] One problem is that Beeckman realised that the unreflected rays of the sun would attract the moon to it; *ibid.*, p. 75.
[70] *Ibid.*, p. 100.
[71] *Ibid.* Note Beeckman's emphasis on the magnitude and rarity (/density) of a planet. Beeckman was always acutely interested in how the volume to surface ratios of bodies, especially corpuscles, affected their mechanical interactions. The similarity in this respect to Descartes' later celestial mechanics is obvious.
[72] *Ibid.*, p. 101. Beeckman also applies this approach to the earth-moon problem, in which case he sees the earth as emitting both repulsive 'heat' and 'light' corpuscles and attractive magnetic 'virtue'. It is clear that he entertains a corporeal theory of magnetism, however, cf. *ibid.*, p. 102. For Beeckman's corpuscular-mechanical theory of magnetism see also Beeckman 1939–1953, I, pp. 36, 101–102, 309; II, pp. 119–120, 229, 339; III, pp. 17, 76.

emanations.[73] This indeed was a mechanical picture of orbital equilibria of causes which was to be supplied in a much more elegant fashion by Descartes' vortices.

By mid-1629 Beeckman had not achieved a settled view and in typical fashion he unceremoniously dropped the matter. Beeckman's work just pre-dates and overlaps the period of renewed contact with Descartes in late 1628. Arguably, the interest Descartes evidenced after 1628 in the problem of celestial mechanics, as well as his mode of approach to it, grew from his acquaintance with Beeckman's speculation. Descartes would have been all the more confident in his union in *Le Monde* of a theory of light with celestial mechanics, if he recognised the advance in comprehensiveness, coherence and mechanical rigour achieved in his work as compared with these wranglings of Beeckman.

Had Descartes been quite a bit more charitable and magnanimous to his erstwhile mentor, he might just have acknowledged what is clear to us—the underlying spirit and structure of the argument of the celestial mechanics of *Le Monde* harks back to the notions behind Beeckman's shifting speculations of 1628–1629. That was not Descartes' style, as is well known.[74] One can imagine him much more readily agreeing with an uncompromising technical judgment which we can now offer, recalling our discussion of the vortex mechanics in section 4. Having been spurred by Beeckman's highly interesting but inconclusive foray into a unified theory of light and celestial optics, Descartes was in a position to try to succeed where Beeckman was floundering, and he approached this by in effect cashing in the intellectual profit of his physico-mathematical endeavours since 1619. Instead of Beeckman's wandering and inconclusive jottings, Descartes elaborated his model of a celestial vortical locking and extrusion machine. He based himself on his principles of dynamics (the emergence of which was initiated in his hydrostatics of 1619 and articulated in his optical work of the 1620s); his theory of centrifugal tendency to motion; a theory of the make up of the stars and their surrounding vortices; and his notion of the massiveness of planets and comets.

A mechanistic theory of light as instantaneously transmitted tendency to motion could be fitted to this cosmic setting, providing the ultimate basis for the optical work and discoveries, and fulfilling the *de facto* challenge issued by Beeckman to render in corpuscular-mechanical and properly physico-mathematical terms the problematic of Kepler. *Le Monde* challenges Beeckman back by saying in effect: 'here is a physico-mathematical explanation of light in cosmic setting and of celestial "physics"; causes are not multiplied; the same concepts of dynamics, applied to the nature of stars and vortices, explain everything!'.

---

[73] Beeckman 1939–1953, III, p. 103. As Beeckman continued he also speculated about countervailing forces arising from impact of corpuscular emanations to explain, amongst other things, the eccentricity of orbits and precession of the equinoxes; *ibid.*, pp. 102, 108.

[74] Van Berkel 2000, Schuster 1977, pp. 530–533.

8.THE UNIFIED THEORY OF 'WEIGHT', ORBITING, AND EXTRUSION— THE CASE OF TERRESTRIAL BODIES

Having looked at the chief genealogical moments in its genesis, we have now arrived back at the vortex mechanics of *Le Monde*. Unfortunately, the scope of this essay precludes our now surveying the full range of detailed issues that fall into place, given this genealogy. Chief amongst these would be Descartes' theory of light in its cosmological setting, that is, in the context of his universe of vortices, each centred on a light giving star—a topic equally grounded in the genealogy and most enlightening about the status and aims of the vortex theory. Other topics would follow: the fall of heavy (third matter) bodies near the surfaces of planets; the motion of planetary satellites; and, for planets possessing both oceans and a moon, the nature and causes of the resulting tides. All of these articulations of the vortex mechanics will be addressed in my forthcoming monograph on Descartes, 'physico-mathematicus'.

For the moment, we must limit ourselves to a brief discussion of Descartes' vortex theory of the fall of heavy bodies near the surface of the earth (or any planet). This theory, like the other articulations of the vortex mechanics listed above, contains quite a few conceptual and empirical problems, which are often emphasised in a kind of Whiggish perspective eager to move on to discuss Newtonian theory. Of course, every theory, indeed even Newton's, has its limits, its strengths, weaknesses, pointed difficulties, lacunae and, perhaps, contradictions. Descartes' bold, internally complex, and strategically thought out vortex mechanics is hardly an exception to this. But, in the spirit of this chapter, I choose here to analyse the little appreciated coherences and strength of Descartes' account of local fall, as a token of my larger attempt to wring as much systemic cogency, and contextual and developmental 'reason' out of the vortex theory and its several articulations.

As we know, Descartes articulated his vortex theory by claiming that smaller, local vortices form around the planets orbiting stars. Using the case of the earth, Descartes attributes to its local vortex the explanation of the motion of the moon, the diurnal motion of the earth, the local fall of 'heavy' bodies, and, conjointly with the moon's motion, and the existence of oceans on the earth, the tides.[75] In very general terms the local fall of heavy bodies follows on this theory as a case of extrusion downward in the planetary vortex of bodies possessing less centrifugal tendency than the surrounding matter of the local vortex.

> Weight is the force ... that makes all parts of the earth tend toward its center, each more or less according as it is more or less large and solid. That force consists the fact that, since the parts of the small heaven surrounding it turn much faster than its parts about its center, they also tend to move away with more force from its center and consequently to push the parts of the earth back toward its center.[76]

The analogy is to planets in the solar vortex which, located at the 'wrong' orbital distance, 'fall' (indeed spiral) downward toward the star until they pick up enough

---

[75] *AT*, XI, pp. 64–83.
[76] *AT*, XI, pp. 72–73, 34; *SG*, p. 47; *MSM*, p. 123.

centrifugal tendency to stabilise in an orbit—at a distance determined by the 'solidity' of the planet, and the speed, size distribution of balls of second element in the vortex, as we have explored in section 4.

Descartes' account, as is well known, meets an immediate large objection, as obvious to contemporaries as to us: because the local vortex spins on an axis coincident with that of the earth, fall on or near the earth should be in direction normal to the axis of rotation, not radially toward the centre of the earth. Moreover, why do not falling bodies sweep laterally across the surface of the earth, spiralling downward, rather than apparently falling in straight lines normal to the local surface of the earth? Finally, to add to these commonly adduced puzzles, we are in a position to add a third one, grounded in our genealogy of the vortex theory: how in detail could Descartes consistently explain or reduce Stevin and Archimedes' rigorous, geometrical, and macro-descriptive hydrostatics to a full vortex theory of fall and weight? In 1619, let us recall, he had been inspired by such an hydrostatics, but had only asserted a piecemeal corpuscular-mechanical explanation limited to assertions about particles in the water, rather than a full vortex theory of weight, specific weight, the behaviour of air, water and circulating vortex particles of second element. Descartes had things to say about the first two issues, but never assayed the third, much more profound one. Be that as it may, I wish here to emphasise the virtues of the theory, from Descartes' point of view, rather than explore further its deficiencies and its (arguable) anomalies.

In *Le Monde* Descartes furnishes an initial clue about what he thought was most striking about his theory of fall. It is precisely the fact that the theory of fall, on his view, *is completely consistent with his vortex theory of planets and comets at the level of basic explanatory machinery*. The very first issue Descartes discusses after his explanation of fall is the following likely misunderstanding on the part of the reader:

> You may find some difficulty in this, in light of my just saying that the most massive and most solid bodies (such as I have supposed those of the comets to be) tend to move outward toward the circumferences of the heavens, and that only those that are less massive and solid are pushed back toward their centers. For it should follow that only the less solid parts of the earth could be pushed back toward its center and that the others should move away from it.[77]

Descartes is directing us to his key concept of 'solidity' and to the fundamental theory of speed/size distribution of balls of second element in the star-centred vortex. He continues,

> But note that, when I said that the most solid and most massive bodies tended to move away from the center of any heaven, I supposed that they were already previously moving with the same agitation as the matter of that heaven. For it is certain that, if they have not yet begun to move, or if they are moving less fast than is required to follow the course of this matter, they must at first be pushed by it toward the center about which it is turning. Indeed it is certain that, to the extent that they are larger and more solid, they will be pushed with more force and speed. Nevertheless, if they are solid and massive enough to compose comets, this does not hinder them from tending to move shortly thereafter toward the exterior circumferences of the heavens, in as much as the agitation

[77] *AT*, XI, p. 73, *MSM*, p. 125, *SG*, p. 47.

they have acquired in descending toward any one of the heavens' centers will most certainly give them the force to pass beyond and to ascend again toward its circumference.[78]

Using precisely the conceptual terms of the larger vortex theory Descartes is distinguishing between: (a) the potential orbital distance of a body in a vortex as determined by its volume-to-surface ratio or solidity, and (b) the amount of force of motion the body possesses at any given time and radial place in the vortex. Now, although he is more precise in his expression about this later in the *Principles*, even in *Le Monde* it is clear that comets or pieces of terrestrial matter have definite solidities, that ultimately determine their placement in a vortex, but *only on condition* that they have gradually acquired circulatory motion and have begun to translate centrifugally. A stone initially sharing only in the diurnal rotation will be forced down toward the centre of the earth, and that is what we habitually observe. But, Descartes is also saying that if the earth's vortex were large enough and if the stone were released from a sufficiently great distance from the centre, it might acquire sufficient circulatory speed in the course of its descent to begin to rise as a result of the centrifugal tendency thus gained. It would rise through the vortex until it reached a level at which the resistance to centrifugal motion of the *boules* balanced its own centrifugal tendency. And such an object of terrestrial matter does exist in stable orbit in the terrestrial vortex—the moon.

The greater the solidity of a body, whether in a solar or planetary vortex, the more difficult it will be for the surrounding second matter to impart motion to it. Thus, upon being released, a body of great solidity will yield more readily to centripetal extrusion than one of lesser solidity. In local fall near the earth, heavy bodies generally do not attain sufficient centrifugal tendency to begin their ascent. Viewing stellar and planetary vortices under a unified theory of vortex mechanics, we can formulate the following theorem, reflecting the essentials of Cartesian celestial mechanics—*comets, planets in the wrong orbits and heavy bodies released near any planet's surface are all doing the same thing for the same reasons.*

Finally, as an explanatory conceit, let us imagine Descartes himself, brought back to discourse with us, commenting upon this unified theory, as well as other competing theories, including his post-mortem acquaintance with Newton's work. Perhaps such a revived, typically self-regarding and feisty Descartes might lecture us as follows:

> I know all of you are, so to speak, in love with Newton—he's like you, or so you think. Well, for me, he is like Kepler, brilliant but ontologically unsound. Here is Newton's leading question—orderly procedure starts with the right question: 'What single immaterial causal agency explains the motions of the planets, comets, satellites, the fall of bodies on earth, as well as the tides?'. Very elegant, is it not? And to be sure, nobody ever posed that precise question: not Aristotle obviously; certainly not Copernicus, not even Kepler—he multiplied such unintelligible immaterial causal agencies, rather than look for one elegant one.
>
> Very well, my question, the methodologically appropriate one, was: 'What unique and certain set of dynamical principles applied to the vortex motion of corpuscles explains the motions of the planets, comets, satellites, the fall of bodies on earth, as well as the

---

[78] *AT*, XI, pp. 73–74, *MSM*, p. 125, *SG*, p. 47.

tides?'. You have seen my dynamics and general vortex theory and can work out for
yourselves why they constitute a unified general theory of the key phenomena in
question. Newton pursued the same problematic. He had the benefit of my example. He
grasped the aim or the problematic, but faltered badly on the issues of causation and
ontology.[79]

9.CONCLUSION: 'WATERWORLD'—A CRAFTY BUT POORLY EXPRESSED GAMBIT IN THE NATURAL PHILOSOPHICAL AGON

In 1619 Descartes had begun to develop his conceptions of force and tendency to
motion in a hydrostatical context: by 1633, having been crystallised in his profound
work in optics, they sat at the centre of the corpuscular-mechanical 'hydro-
dynamics' that ran 'the world', or as my friend and occasional collaborator, Stephen
Gaukroger, incisively dubbed it, Descartes' 'Waterworld'.[80] The ambitious but
embryonic physico-mathematical project of 1619 had borne some hefty dividends.
Descartes, physico-mathematicus, was building a novel corpuscular-mechanical
natural philosophy that would entrain new, non-Aristotelian relations between
natural philosophising and the mathematically based physical disciplines. Indeed,
once one grasps the underlying conceptual framework of *Le Monde*, and the
genealogy of that framework, one sees that *Le Monde* was a work deeply
symptomatic of a contemporary problematic in natural philosophy shared by certain
bold, mathematically oriented anti-Aristotelian innovators, regardless of their own
ontological differences. The vortex celestial mechanics was not just a fanciful and
amusing advertisement for Copernican realism in infinite universe mode, nor was it
just a representation of Copernicanism inside a proffered, alternative system of
natural philosophy. Descartes' 'Waterworld' was in fact a post-Keplerian play for
hegemony in the field of natural philosophising, in its particularly overheated and
contested early seventeenth-century state.

Le Monde, as Descartes would have seen it, was built in part on the basis of a
concatenation of achievements in natural philosophising key chunks of the mixed
mathematical sciences. He had come to terms with, competed with, and, in his view,
surpassed Beeckman's natural philosophical strivings, themselves partially shaped in
the shadow of Kepler. In the mixed mathematical sciences, Kepler's own master
strokes had been the elliptical orbit of mars, and the laws of planetary motion in
general. Descartes' competing jewels, in his view at least, were his corpuscular-
mechanical reduction of hydrostatics, and his solution of the ancient and prestigious

[79] I am not advocating here history as mere literature or entertainment. Rather I believe that Descartes had
intentions and conceptual structures reconstructable on the basis of textual and contextual evidence.
My conceit is meant to motivate and focus proper historical scholarship on *Le Monde* and related
texts, not to displace those texts or dissolve disciplined historical inquiry into more or less amusing
creative writing. What 'Descartes' says here is also arguably a good heuristic guide to what to look
for in post-Newtonian Cartesians.
[80] The conceit arose out of Gaukroger's reflection on Gaukroger 2000 as well as issues arising in the
composition of our joint study, Gaukroger and Schuster 2002. I have accordingly entitled the present
chapter, as well as previous conference and seminar presentations of this argument, 'Waterworld', in
homage to Gaukroger's striking and amusing term.

refraction problem, and he too had a celestial mechanics, which followed Beeckman's critique of Keplerian spiritual neo-Platonic nonsense, but which outplayed Beeckman by being based on a coherent dynamics of corpuscles, itself the product of the same course of physico-mathematical research.

Consider this short list of the characteristics of Descartes' *Le Monde* program: Descartes was articulating Copernicus' claims; he was displaying what he thought was best dynamical practice, best causal discourse practice, to explain planetary motion and the dynamical role of stars; he was associating in the same problematic local gravity, the behaviour of satellites, orbital motions of planets, cometary motions, the nature and causal role of the sun, or of any star, in all this and in a theory of light in cosmic setting. Now, on each of these points, there are notable parallels to the enterprise of Kepler, allowing for complete difference of natural philosophical content (but not of aim). The problematic is the same in both cases: what Descartes unifies as *explananda* by virtue of his dynamics of vortices,[81] including the key role of stars within vortices, Kepler unifies by a theory of a set of hierarchically arranged causal forces, similar to each other in respect of their immaterial nature, and law-like, mathematical functioning. Both natural philosophers attempted a unified set of explanations under the aegis of a new, alternative natural philosophy prominently advertising highly anti-scholastic dynamical registers, or causal doctrines. In other words, and in conclusion, the vortices were serious business, and, as Aiton brilliantly showed, they remained serious business amongst a small committed crew of serious celestial mechanicians, such as Huygens and Leibniz, well into the eighteenth century, in competition with the Newtonian view.[82]

REFERENCES

Aiton, E. J. (1972) *The Vortex Theory of Planetary Motions*, London: Macdonald.
Bachelard, G. (1965) *La formation de l'esprit scientifique*, 4th edn Paris: Vrin: 1st edn 1938.

[81] Admittedly somewhat different types of vortices in detail—star-centric and planet-centric.

[82] The rigorously contextual approach of this paper in regard to understanding the vortex mechanics and its genesis should not be taken to signal a denial of larger, long term, diachronic relevances of this inquiry or its findings. One important diachronic dimension immediately presents itself to the technical and internalist historian of classical mechanics: the natural philosophical contestation carried out by Descartes and Kepler was pursued with special attention to the subsumption of astronomy, i.e., Copernican astronomy, variously interpreted, and to its problem of celestial causation, in particular the function of stars. The nature of one's dynamics, the causal doctrine at the heart of one's system of natural philosophy, was thus focalised, and this drove both to contribute claims woven by later players in unintended and unforeseeable ways into what we recognise as the process of emergence of classical mechanics. Similarly, we should note the role of optical inquiries, in natural philosophical contexts, in the shaping the later crystallisation of classical mechanics, a matter hinted at in this paper and related work, and currently under serious study by Russell Smith of University of Leeds (personal communication). It would seem, as Stephen Gaukroger has expressed to me in discussion of themes of this and related work, that the long term genealogy of classical mechanics should be written, at least in part, in terms of the concatenation of unintended conceptual windfalls bequeathed to the emerging discipline by this and other nodes in the natural philosophical turbulence of the early- and mid-seventeenth century.

Beeckman, I. (1939–1953) *Journal tenu par Isaac Beeckman de 1604 à 1634*, 4 vols, ed. C. de Waard, The Hague: M. Nijhoff.

Bossha, J. (1908) 'Annexe note', *Archives Neerlandaises des Sciences Exactes et Naturelles*, ser 2 t. 13, pp. xii–xiv.

Buchdahl, G. (1972) 'Methodological aspects of Kepler's theory of refraction', *Studies in History and Philosophy of Science*, 3, pp. 265–298.

Descartes, R. (1963) *Oeuvres philosophiques de Descartes*, Tome I, ed. F. Alquié, Paris: Garnier Frères.

— (1996) *Oeuvres de Descartes*, eds C. Adam and P. Tannery, 11 vols, Paris: J. Vrin.

— (1998) *Descartes, The World and Other Writings*, ed. and trans. S. W. Gaukroger, Cambridge: Cambridge University Press.

de Waard, C. (1935–1936) 'Le manuscrit perdu de Snellius sur la refraction', *Janus*, 39–40, pp. 51–73.

Gabbey, A. (1980) 'Force and inertia in the seventeenth century: Descartes and Newton', in *Descartes: Philosophy, Mathematics and Physics*, ed. S. W. Gaukroger, Sussex: Harvester, pp. 230–320.

Garber, D. ed. (2002) *La Rivoluzione Scientifica*, Rome: Istituto della Enciclopedia Italiana.

Gaukroger, S. W. ed. (1980) *Descartes: Philosophy, Mathematics and Physics*, Sussex: Harvester.

— (1995) *Descartes: An Intellectual Biography*, Oxford: Oxford University Press.

— (2000) 'The foundational role of hydrostatics and statics in Descartes' natural philosophy' in *Descartes' Natural Philosophy*, eds S. W. Gaukroger, J. A. Schuster and J. Sutton, London: Routledge, pp. 60–80.

Gaukroger S. W. and Schuster, J. A. (2002) 'The hydrostatic paradox and the origins of Cartesian dynamics', *Studies in History and Philosophy of Science*, 33, pp. 535–572.

Gaukroger, S. W., Schuster, J. A. and Sutton, J. eds (2000) *Descartes' Natural Philosophy*, London: Routledge.

Kepler, J. (1938–) *Johannes Kepler Gesammelte Werke*, 20 vols, eds W. von Dyck and M. Caspar, Munich: C. H. Beck.

Knudsen, O. and Pedersen, K. M. (1968) 'The link between "determination" and conservation of motion in Descartes' dynamics', *Centaurus*, 13, pp. 183–186.

Kuhn, T. S. (1959) *The Copernican Revolution*, New York: Vintage: 1st edn 1957.

LeGrand, H. E. ed. (1990) *Experimental Inquiries: Historical, Philosophical and Social Studies of Experiment*, Dordrecht: Reidel

Lohne, J. (1959) 'Thomas Harriot (1560–1621) the Tycho Brahe of optics', *Centaurus*, 6, pp. 113–121.

— (1963) 'Zur Geschichte des Brechungsgesetzes', *Sudhoffs Archiv*, 47, pp. 152–172.

McLaughlin, P. (2000) 'Force determination and impact' in *Descartes' Natural Philosophy*, eds S. W. Gaukroger, J. A. Schuster and J. Sutton, London: Routledge, pp. 81–112.

Mahoney, M. (1973) *The Mathematical Career of Pierre de Fermat 1601–1665*, Princeton: Princeton University Press.

— trans. (1979) *René Descartes, Le Monde ou Traité de la Lumière*, New York: Abaris.

Mersenne, M. (1932–1988) *Correspondence du P. Marin Mersenne*, 17 vols, eds C. de Waard, R. Pintard, B. Rochot and A. Baelieu, Paris: Centre National de la Recherche Scientifique.

Prendergast, T. L. (1975) 'Motion, action and tendency in Descartes' physics', *Journal of the History of Philosophy*, 13, pp. 453–462.

Sabra, A. I. (1967) *Theories of Light from Descartes to Newton*, London: Oldbourne.

Schuster, J. A. (1977) *Descartes and the Scientific Revolution 1618–34: An Interpretation*, 2 vols, unpublished Ph.D. dissertation, Princeton University.

— (2000) 'Descartes' Opticien: the construction of the Law of Refraction and the manufacture of its physical rationales, 1618–29' in *Descartes' Natural Philosophy*, eds S. Gaukroger, J. A. Schuster and J. Sutton, London: Routledge, pp. 258–312.

— (2002) 'L'Aristotelismo e le sue Alternative' in *La Rivoluzione Scientifica*, ed. D. Garber, Rome: Istituto della Enciclopedia Italiana, pp. 337–357.

Schuster, J. A. and Taylor, A. B. H. (1996) 'Seized by the spirit of modern science', *Metascience*, 9, pp. 9–26.

Schuster, J. A. and Watchirs, G. (1990) 'Natural philosophy, experiment and discourse in the 18th century: beyond the Kuhn/Bachelard problematic' in *Experimental Inquiries: Historical, Philosophical and Social Studies of Experiment*, ed. H. E. LeGrand, Dordrecht: Reidel, pp. 1–48.

Stevin S. (1955–1966) *The Principal Works of Simon Stevin*, 5 vols, eds E. Cronie et al., Amsterdam: Swets & Zeitlinger.

Van Berkel, K. 'Descartes' debt to Beeckman: inspiration, cooperation, conflict' in *Descartes' Natural Philosophy*, eds S. Gaukroger, J. A. Schuster and J. Sutton, London: Routledge, pp. 46–59.

Vollgraff, J. A. (1913) 'Pierre de la Ramée (1515–1572) et Willebrord Snel van Royen (1580–1626)', *Janus*, 18, pp. 595–625.

— (1936) 'Snellius' notes on the reflection and refraction of rays', *Osiris*, 1, pp. 718–725.

PETER DEAR

CIRCULAR ARGUMENT

Descartes' Vortices and Their Crafting as Explanations of Gravity

1. INTRODUCTION

Vortices are well known to be an important conceptual component of Descartes' mechanical universe.[1] That analogies with everyday experience are an important feature of his writings about the physical world—tennis racquets and wine-vats, for instance—is equally familiar.[2] The present chapter considers how those two aspects of Descartes' philosophy relate to one another; how Descartes' vortices could function in a natural philosophical discourse as elements of specific physical explanations. Descartes was evidently always unable to give a full, formal mathematised treatment of circular motion, of the sort that Huygens or Newton would do a decade or two after the *Principles of Philosophy*. Instead, he made his arguments plausible in a different way, relying on appeals to experiential analogies that would persuade his readers to go along with him as his explanatory accounts of phenomena proceeded. In this sense, the kinds of everyday examples that Descartes liked to employ were never really simple analogies; instead, they always functioned as appeals to kinds of experience that he took to be directly relevant to the phenomena to be explained. Thus his analogised explanations less take the form of similitudes to already-known processes, and more involve implicit claims that they are true in the same sense that the invoked comparisons are already known to be true. Whether Descartes was always successful in doing this is another question; nonetheless, he did not present himself as merely a rhetorician who made convenient but non-demonstrative use of analogy and metaphor.

Vortices first appear in *The World* (*Le monde*), or in the work that produced *The World*, in the early 1630s. From the outset, their primary role was not so much to account for the orbital *motions* of planets around the sun, or of the moon around the earth, but more centrally to explain the apparent forces, or tendencies, of bodies. Descartes presents vortices as consequences of his fundamental characterisation of matter. At the outset, he says, speaking of the parts of matter making up all of space,

---

[1] See Schuster's chapter in this volume. For another recent discussion see Gaukroger 2002, pp. 150–160; also Schuster 1977, chap. 8; and the classic treatment in Aiton 1972.

[2] See, e.g., Galison 1984; Eastwood 1984.

P. R. Anstey and J. A. Schuster (eds.), The Science of Nature in the Seventeenth Century, 81-97.

since they were all just about equal and as equally divisible, they all had to form together into various circular motions. And yet, because we suppose that God initially moved them in different ways, we should not imagine that they all came together to turn around a single centre, but around many different ones, which we may imagine to be variously situated with respect to one another.[3]

This picture can then be applied to the 'small heaven' that contains the earth and its circling moon. It enables Descartes to explain the weight, or gravity, of heavy bodies as being due to their tendency towards the earth's centre, the centre of its vortex, the result of their balancing the contrary outward, centrifugal tendency of the surrounding and more rapidly-moving subtle matter. That centrifugal tendency had, in turn, already been explained as a consequence of Descartes' metaphysically justified law of rectilinear inertia. The earth's vortex also provided the basis of his explanation of the prized phenomenon of the tides.

There was a clear model for the use of a vortex to explain motion of bodies towards its centre, one that would have made Descartes' basic idea unremarkable. It would have been unremarkable because it appears in a canonical, in this case Aristotelian, text that was quite well-known by writers on mechanics in this period and that had received a number of lengthy printed commentaries. This was the work, now regarded as pseudo-Aristotelian but assumed to be authentic in the early seventeenth century, known as the *Mechanica*, or *Questions in Mechanics*. In this volume Helen Hattab discusses a number of features of this work that shed light on various of Descartes' procedures, but I want to focus on just one, which appears in the final chapter of the *Questions in Mechanics*:

> Why is it that an object which is carried round in whirling water is always eventually carried into the middle? Is it because the object has magnitude, so that it has position in two circles, one of its extremities revolving in a greater and the other in a lesser circle?[4]

An important diagram in Descartes' *The World* seems to call this very image to mind: it is the one that represents the moon orbiting the earth.

In this case, the greater and lesser circles about the centre correspond to the furthest and nearest parts from the earth of the moon's own vortex, and help to explain why the moon orbits the earth. But the same idea applied, of course, to Descartes' subsequent explanation of terrestrial *gravity*, where the objects made of lumpy third element are intruded towards the centre of the terrestrial vortex by the

---

[3] *The World*, Descartes 1996 (hereinafter *AT*), XI, p. 49; trans. in Descartes 1998, p. 33.
[4] Aristotle 1984, 2, p. 1317.

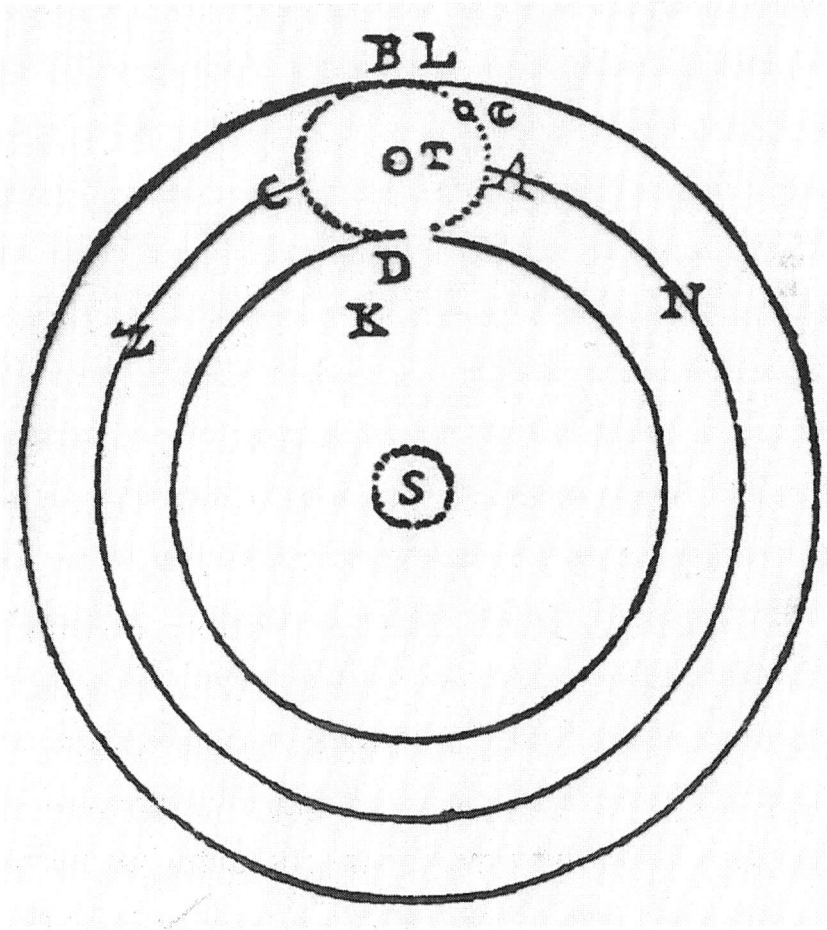

Figure 14. The moon orbiting the Earth, from The World, *AT, XI, p. 74.*

greater outward tendency of the more rapidly moving aether or second element whirling around the earth.

The above-quoted passage from the *Questions in Mechanics* commences from a given initial premise: 'Why is it that an object which is carried round in whirling water is always eventually carried into the middle?' This is, in its way, an authentically Aristotelian technique for setting up a problem, one similar to those found in another pseudo-Aristotelian work, the *Problemata*. Book XVIII of the *Problemata* in fact concerns quite similar subjects to those of the *Questions in Mechanics*, but its most notable feature overall is the text's style. The little chapters each begin with questions of the same form as the vortex query in the *Questions in Mechanics*: 'Why is it that those who are blind from birth do not become bald?',[5] or 'Why is the wax in the ears bitter?',[6] as well as 'Why are bubbles hemispherical?'[7] The point about these clearly rhetorical questions is that they take for granted that everyone will unproblematically agree with the premise on which they are based. So everyone knows that bubbles are hemispherical; everyone knows that earwax tastes bitter; and everyone knows that people blind from birth do not go bald. So, similarly, in the example from the *Questions in Mechanics*, everyone is taken to know that objects carried around in whirling water tend to move towards the whirlpool's centre.

Of course, it cannot be assumed that the apparent ideal reader of the *Questions in Mechanics* corresponds to that work's actual readers in early-seventeenth-century Latin Europe, that is, educated people like Descartes and his own readers. But there is in fact a good deal of evidence on this issue: several commentaries were written on the work in the sixteenth and early seventeenth centuries. These commentaries have been inventoried by Paul Lawrence Rose and Stillman Drake, who identified the importance and notoriety of the text in the decades around 1600.[8] As for Descartes' own circle, his friend Mersenne discussed aspects of the *Questions in Mechanics* at some length in a work of 1626, among other places, although not the specific issue of the vortex.[9] The *Questions in Mechanics*, therefore, is a text that someone like Descartes could be expected to have known, or at least to have known *of*. As for the style of making arguments found in the *Questions in Mechanics*: in the case of bodies swept around in a vortex, and the taken-for-granted character of that chapter's premise, there was nothing out of the way in that at all. Stating common experiential truths as the starting point in providing an explanation for them was standard Aristotelian practice, and perfectly usual in the commentary tradition on Aristotle's physical works right into the seventeenth century.[10] However, such statements were not always readily *accepted*—even when they were supposed to have been said by Aristotle himself.

---

[5] *Ibid.*, 2, p. 1507.

[6] *Ibid.*, 2, p. 1513.

[7] *Ibid.*, 2, p. 1421. See Blair 1999 for a study of the nature and career of the *Problemata*.

[8] Rose and Drake 1971; see also on these matters Laird 1986.

[9] *Synopsis mathematica*, 'Mechanicorum libri', pp. 146–168. See in general on Mersenne and the *Questions in Mechanics*, Dear 1988, esp. pp. 117–127.

[10] Dear 1995, chap. 1, sect. III; also *idem*. 1985.

2.CRITICISING 'ARISTOTLE'

One of the most widely known commentaries on the *Questions in Mechanics* was by Giovanni Battista Benedetti. Benedetti did not question the phenomenon of the vortical intrusion of solid objects, but he did criticise its proffered explanations. The pseudo-Aristotelian author had invoked the work's dominant theme of circles and their relative sizes:

> The greater circle, then, on account of its greater velocity, draws it [i.e. the object] round and thrusts it sideways into the lesser circle; but since the object has breadth, the lesser circle in its turn does the same thing and thrusts it into the next interior circle, until it reaches the centre.[11]

Benedetti says, instead, that the reason why 'objects found in whirlpools always come together toward the middle' is simply because vortices form inverted cones, the water rising up at the sides and leaving a depression in the middle; objects gather in the middle because the middle is also the lowest point—and the objects have 'weight and heaviness'.[12]

An example of an even more fundamentally critical response comes from the Jesuit mathematician Giuseppe Biancani, in his little work *Aristotelis loca mathematica* of 1615. This is a survey of mathematical matters in Aristotle's works, including a fairly substantial discussion of the *Questions in Mechanics*. But, rather like Mersenne in 1626, Biancani dispenses with a discussion of that work's final chapter. The difference is that he mentions it explicitly and explains why he will not discuss it. Biancani notes that his own text is concerned with mathematics in Aristotle, and the vortex discussion is, Biancani remarks, actually a matter of *physics*.[13] Presumably this is because, first, the discussion of vortices in the *Questions in Mechanics* concerns actual motion rather than potential motion, whereas most of the Aristotelian text concerns what had long since become standard sorts of problems in statics (such as why people have to lean forward in order to stand up from a seated position); and secondly, the vortex problem was (in Biancani's view) about physical causation, namely why objects will move towards the centre of the whirlpool. The Jesuits were, as William Wallace and others have argued, particularly strict on this point of physics versus mathematics.[14] But Biancani's objection went further: he says that, in any case, Aristotle's explanation was erroneous because its premise was false: objects do *not* get pushed to the centre when they are whirled around in a circle; instead, they are thrown outwards. And Biancani can say this because it's something that 'experience teaches' (*experientia enim docet*).[15]

The basis of Biancani's scepticism can be seen in Isaac Beeckman's journal from around the same time, in an entry from March 1618. Beeckman there stresses that in

---

[11] Aristotle 1984, 2, p. 1317.

[12] Giovanni Battista Benedetti, *Diversarum speculationum... liber*, 1585, p. [167], as trans. Drake and Drabkin 1969, p. 195.

[13] Blancanus 1615, p. 195.

[14] See, e.g., Wallace 1984, esp. pp. 136–148; see also, for further discussion and references, Jardine 1988.

[15] Blancanus 1615, p. 195.

an artificial whirlpool made by swirling a stick around in a vessel containing water, one observes (*videbis*) a tendency outwards from the centre of rotation, just as a stone on a horizontal spinning wheel gets flung outwards; this is seen in the way in which the water banks up around the outside as the centre tends to form a depression.[16]

However, Beeckman's presentation of the situation showed it as less straightforward than Biancani seems to have thought. Fittingly, Beeckman provides the most direct precedent for Descartes' ideas about vortices and gravity. Four years before the passage just considered, Descartes' erstwhile mentor had written on the subject in another entry in his journal. The brief discussion asks the question: 'Why are heavy bodies moved downwards?'. Beeckman's very first suggested answer states: 'Whether because things higher up are perpetually in motion, and the same thing happens to the earth as with stones tending towards the centre of an aqueous vortex?'.[17] He then goes on to provide an alternative answer, speculating about the existence of porosities in matter through which a tenuous, subtle substance could press down, pushing bodies before it.

> And because this descent of subtle parts generally penetrates, neither does the whole substance move forward on account of the larger porosities; [such bodies] are called *light*. The rest, which are of a more compact nature, are called *heavy* because that descent takes place more strongly with them: for, on account of the insufficient joining-together of those (albeit subtle) parts, it flies through [them].[18]

The discussion, while certainly quite sketchy and rather unclear, nonetheless indicates that Beeckman, in 1614, had already speculated about central elements of the mechanism that Descartes subsequently adopted for his own causal account of weight and fall: the vortex parallel and the idea of a subtle substance that could pass through the pores in ordinary matter.

Even while accepting that Beeckman was an exceptional individual whose ideas were not necessarily widely shared, we can see that the pseudo-Aristotelian vortex model was very much available for mechanical or quasi-mechanical application to matters of physics—despite there being no positive evidence that Beeckman himself had actually read the *Questions in Mechanics* until 1619.[19] The vortex conception of the pseudo-Aristotelian work, here now applied explicitly to heavy bodies and fall, was cultural property to which Descartes could readily make appeal (and, no doubt,

---

[16] Beeckman 1939–1953, I, p. 167.

[17] 'Cur gravia deorsum moventur? An quia superiora in perpetuò sunt motu idemque Terrae accidit quod lapidi ad medium vorticis aquarum tendenti?', *ibid.*, I, p. 25.

[18] *Ibid.* The entire passage reads, continuing directly from the previous note: 'Aut an tenuis est quidam defluxus subtilium corporum à superioribus partibus aequaliter circumcirca, qui obvia quaeque deprimit? Et quia hic defluxus est subtilium partium pleraque penetrat, nec tota substantia premit propter poros majusculos, eaque *levia* dicuntur. Reliqua, quae sunt compactioris naturae, *gravia* dicuntur quia iste defluxus fortius illis occurrit: propter compactionem enim parum istarum partium, licet subtilium, pervolat'. Beeckman discusses vortices at some length in *ibid.*, I, pp. 167–168.

[19] *Ibid.*, I, p. 318; see also van Berkel 1983, p. 223.

another piece of evidence favouring Beeckman's claim of Descartes' unacknowledged debt to him).[20]

3.CIRCULAR MOTION AND GRAVITY

How, then, does Descartes approach this issue? It appears to be more than likely that he knew about the problem in *Questions in Mechanics* concerning vortices pushing bodies to the centre, and hence its relative familiarity—but in any case, he still would have needed to deal with the practical question of how to present the issue in a persuasive fashion. In 1640, in one of his long letters to Mersenne that answered a miscellany of questions relating to his ideas, Descartes tried to explain his view of gravity as coming from what he called 'subtle matter, revolving very quickly around the earth, [which] chases terrestrial bodies towards the centre of its movement'. Mersenne had apparently requested a fuller explanation for the effect, the synthetic account found in *The World* having presumably been the basis of Descartes' earlier account to Mersenne. Descartes therefore provides a comparison, one that exhibits the same phenomenon as the one that he believes to be involved (although less evidently) in the case of falling bodies. He says of this account of gravity that it is

> just as you can see [*experimenter*] in making water revolve in some large vessel, and throwing into it small pieces of wood; you'll see that these little wood chips will gather together towards the middle of the circle made by the water, and will sustain themselves just like the earth does in the middle of the subtle matter…[21]

Another letter, this time to Regius the following year, referred to the same thing more briefly, when the two were still on good terms and Descartes wanted to make sure that Regius understood his ideas properly. Descartes wrote:

> The cause why bodies put into whirlpools are carried towards the centre I hold to be because the water itself, while it's moved circularly in a whirlpool, tends towards the outside, so therefore other bodies that do not have such rapid circular motion push themselves into the centre.[22]

A noteworthy feature of the way that Descartes explains himself both to Mersenne and to Regius is that he makes appeal *both* to available experience (in the case of Mersenne) *and* to an interpretive explanation (in the case of Regius).

In fact, not very long before, in another letter from 1639, Descartes had already attempted to explain himself to Mersenne on the same subject. This earlier account is presented as an explanation as well as, simultaneously, an experiential illustration.

---

[20] On the relationship between the two, including Beeckman's letter of 1629 complaining at Descartes, see van Berkel 2000, pp. 46–59; see also Schuster 1977, chap. 2.

[21] *AT*, III, pp. 134–135.

[22] *Ibid.*, III, pp. 445–446.

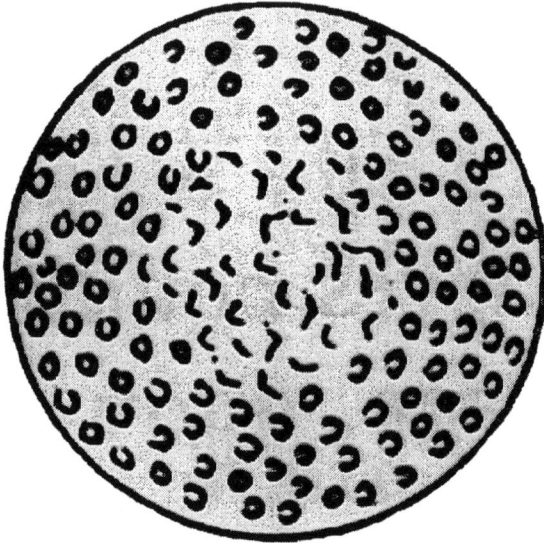

Figure 15. from Lettres de Descartes, *ed. C. Clerselier, II, 1659, p. 593.*

To understand how the subtle matter that revolves around the earth chases heavy bodies towards the centre, fill some round vessel with tiny lead shavings, and mix with this lead bits of wood or other material lighter than the lead, which are coarser than these shavings. Then, making the vessel revolve rapidly, you will show [*esprouverez*] that these little shavings will chase all these bits of wood, or other such material, towards the centre of the vessel, just as the subtle matter chases terrestrial bodies.[23]

For Descartes, this illustration—whether or not he expects Mersenne actually to try it, and whether or not he tried it himself—was supposed to help Mersenne understand Descartes' explanation of terrestrial gravity. In the unpublished *The World*, Descartes had eschewed this kind of illustration of his idea. Instead, he had simply explained why terrestrial bodies are heavy by direct reference to the relevant kinds of matter and their rapidity of motion, and used a simple Archimedean kind of principle to explain that when celestial matter is propelled outwards from the centre,

[23] *Ibid.*, II, pp. 593–594.

terrestrial matter must be correspondingly propelled inwards.[24] Although the corresponding account in the *Principles* is rather more elaborate,[25] it too avoids using such an illustration—perhaps, again, because it could not stand as a formal demonstration (although this was not a problem that deterred Descartes overmuch in other cases in the *Principles*).

The illustrations that Descartes gave in the letter to Mersenne are interesting because they seem to come close to narrated experimental accounts—although they both in fact take the common recipe-form of a set of instructions.[26] This contrasts with the way in which Descartes, Mersenne, and Beeckman all talk, in various places, about the behaviour of spinning bodies by making reference to a common children's toy, the top: children were always good stand-ins for what absolutely everyone could be expected to know already; no experiments were needed in such cases.[27] Descartes' efforts show that vortex motion was more difficult to render evident than he may have hoped.

The version of the explanation in the *Principles*, in 1644, adopts a rather different approach from that used in *The World*. The third part of the *Principles* is devoted to 'the visible world' (meaning the celestial universe) much as the third part of Isaac Newton's own *Principia* would later concern 'The System of the World', with a similar correlation of previously stated theoretical mechanical principles and theorems with observed phenomena. Descartes introduces the solar vortex that carries the planets around the sun not by deducing the necessary existence of such vortices from first principles, which was his procedure in *The World*, but by postulating it: 'For by that alone, and without any other devices, all their [i.e. the planets'] phenomena are very easily understood'.[28] There follows an analogy with straws carried around by an eddy in a river, and the pronouncement: 'Thus we can easily imagine that all the same things happen to the Planets; and this is all we need to explain all their remaining phenomena'.[29] Although Descartes attempts in the immediately neighbouring pages to vindicate himself of the suspicion that his planetary earth should properly be thought to move, contrary to recent Church rulings, he does not do so through a stress on the hypothetical nature of his arguments; the lack of dogmatism in his presentation of the vortical solar system must be understood, therefore, as part of a recognition of its methodological status, one inferior to that of, for example, his rules of motion and collision. Analogies with everyday experience play an irreducible role.[30]

---

[24] *Ibid.*, XI, p. 76.

[25] *Ibid.*, VIII, pp. 213–214 (Pt. 4, art. 23).

[26] On the 'recipe' format, see Dear 1995, chap. 2, section III.

[27] On tops: *The World, AT*, XI, p. 75; *Principles, AT* VIII, p. 212 (Pt. 4, art. 21); Marin Mersenne 1626, 'Mechanicorum libri', p. 150, citing Bernardino Baldi; Beeckman 1939–1953, I, p. 257; see also remarks regarding later observations of Beeckman's concerning children's games in van Berkel 1983, p. 185. On other uses of children's games see Dear 1995, pp. 147–149; I am grateful to Domenico Bertoloni Meli for informing me that the example quoted from Pierius on pp. 148–149 has its origin in Galen, *On the Natural Faculties*, Bk. 1, section 7.

[28] *AT*, VIII, p. 92, trans. in Descartes 1983, p. 96.

[29] *Ibid.*

[30] See below and n. 32.

In the fourth part of the *Principles*, Descartes tackles the phenomena found in the terrestrial world, and it is here that he deals with gravity. Section 23 explains 'How all parts of the Earth are driven downward by this heavenly matter, and thus become heavy'. The essence of the explanation, which is the same as that in *The World*, is contained in its first sentence:

> Next, it must be noted that the force which the individual parts of the heavenly matter have to recede from [the center of] the Earth cannot produce its effect unless, while those parts are ascending, they press down and drive below themselves some terrestrial parts into whose places they rise.[31]

All this is argued within the previously-established frame of hypotheticalism, Descartes using explanatory fruitfulness as the means for justifying his individual hypotheses.[32]

One of the central difficulties with vortex motion was that, although Descartes could take its statement from a standard, canonical source, the *Questions in Mechanics*, its analysis in his own mechanical terms was much less straightforward. The *Questions in Mechanics* itself took circular motion as a fundamental, primitive form of motion in order to explain mechanical problems, prototypically those concerning the lever, in which a great weight was moved by a small force. 'The original cause of all such phenomena is the circle',[33] said the pseudo-Aristotle. But despite that privileging of the circle, for the author of the *Questions in Mechanics* the circle was not a simple concept: 'the circle is made up of contraries', and it was therefore unsurprising that so many marvels should be owed to its properties.[34]

In an analogous way, Descartes privileged circular motion as a basic building block of his physical system of the world. But this ubiquitous vortex motion was not simultaneously a primitive *concept*; it was actually rather complicated when examined in the light of Descartes' fundamental laws of motion (see Schuster above). As Helen Hattab observes elsewhere in this volume, Descartes' ready use of circular motion in making a physics was likely a consequence of its familiarity due to the *Questions in Mechanics* and contemporary commentaries upon that text. Nowhere does Descartes claim that motion in a circle is a simple idea.[35] Given the complex character of the circle, and the even more complicated nature of vortical motion based on it, the additional step of explaining gravitational effects through consideration of vortex motion appears even more elaborate.

---

[31] *AT*, VIII, p. 213; trans. Descartes 1983, p. 191; the bracketed words represent an addition in the authorised French translation of 1647.

[32] See Clarke 1982, esp. pp. 148–155; Sakellariadis 1982. See also Daniel Garber 1993.

[33] Aristotle 1984, 2, p. 1299.

[34] *Ibid.*

[35] Cf. Westfall 1971, pp. 81–82, however, who notes that Descartes nonetheless at the same time tends to give circular motion a certain analytical priority.

4.HUYGENS' DEMONSTRATION

The most notable subsequent attempt to make the pseudo-Aristotelian/Cartesian model of vortex-behaviour plausible and relevant to understanding gravity was that developed by Christiaan Huygens in his mechanistic theory of gravity.[36] Originally presented to the Royal Academy of Sciences in Paris in 1669, it takes the general form of an adaptation of Descartes' own explanation of gravity.[37] Huygens accounts for the tendency of heavy bodies to move towards the earth's centre by imagining countless particles circling the earth's centre in all possible planes and in all directions. These particles tend to push outwards, centrifugally, against the constraints of the surrounding aetherial medium. Meanwhile, objects made of ordinary matter, which do not share this rapid circular motion, tend reciprocally to be pushed in towards the centre. That is the basic explanation for why such bodies are heavy and why they tend to fall. Huygens—the inventor, after all, of the term 'centrifugal force'—calculates the speed that his gravity-inducing particles would need to have in order to produce the requisite (empirically-measured) effect.[38]

A notable feature of Huygens' model is its level of elaboration and contrivance: it is a long way from everyday practical experience, and Huygens makes little attempt to compare it to analogical parallels. Instead, the conclusions are generated from highly suppositious premises that require all sorts of practical objections to be held in abeyance. When it produces the 'right answer' (which it would have to, since the answer is part of the problem itself), Huygens has succeeded, he hopes, in rendering gravity, as he says, 'intelligible'.[39] As with Cartesian-style explanations in general, this one really only showed the possibility of explaining its phenomenon mechanically, rather than demonstrating the necessary *truth* of that explanation.

But Huygens was not content with getting numbers that worked. He now also wanted to make his hypothetical cause of gravity more plausible by presenting a concrete physical illustration. But this was not to be one of the rather ambiguous analogies in which Descartes had dealt; rather than trying to make his claims more immediately acceptable by simply adducing an analogy that he hoped his audience might accept, Huygens devised a contraption to *show* the general feasibility of his explanation—or rather, to show the feasibility of Descartes' own explanatory strategy, of which Huygens' was a variant.

Thus, Huygens' discourse describes to the other members of the Academy an apparatus that involves a circular dish filled with water that contains some small lumps of wax very slightly denser than the water itself. The dish is placed in the centre of a rotating table-top, which is then spun around until the water is itself sharing the motion of the dish. During this procedure, the pieces of wax move out to

---

[36] On various later seventeenth-century pursuals of Descartes' idea, including Huygens and Rohault, see Aiton 1972, chap. 4.

[37] Christiaan Huygens 1888–1950, 19, pp. 631–640. See also, for an interpretation of methodological issues in Huygens' work, Elzinga 1980. For a valuable study of Huygens' relation to Descartes see Westman 1980.

[38] Huygens 1888–1950, 19, pp. 631–640.

[39] *Ibid.*, p. 631.

the rim of the dish, impelled by centrifugal force. However, if the dish's rotation is suddenly halted, the pieces of wax within it slow more rapidly than the still-revolving water, owing to their bumping on the bottom of the dish. Consequently, they are impelled inwards to gather at the centre of the dish's rotation.[40] In describing this contrived experience (presented in the form of a recipe, in the usual conventional style), Huygens says that 'it shows to our eye a certain image of gravity'.[41] He also talks about what Descartes had said on the question, citing remarks from the letters quoted above, from Clerselier's then recently-published edition of Descartes' correspondence. In particular, Huygens gently criticises Descartes' proposed demonstration involving the woodchips (see above, section III). Huygens says:

> if, as it appears, he means wood that floats on the water, there won't be any concentration [of the wood at the centre of the whirlpool]; but if he wishes that it sink to the bottom, it will truly be the same experience that I proposed a little earlier, and the wood will mass in the centre...[42]

Huygens' 1690 *Discourse on the Cause of Gravity* presented a diagram detailing a refinement of his demonstrative or illustrative apparatus.[43]

---

[40] *Ibid.*, p. 633; cf. also *ibid.*, p. 626 for an earlier MS version.
[41] *Ibid.*, p. 633.
[42] *Ibid.*, p. 634.
[43] *Ibid.*, 21, p. 452.

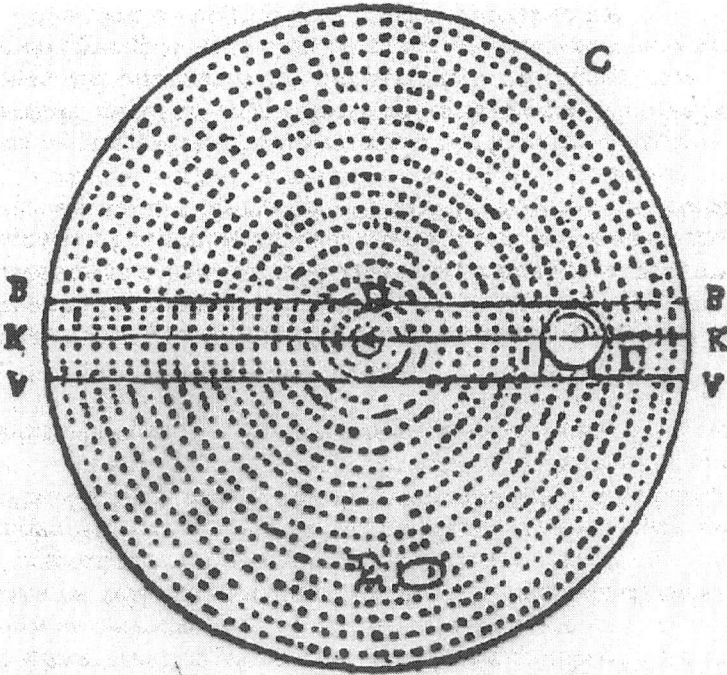

Figure 16. Huygens, Discourse on Gravity, Oeuvres complètes de Christiaan
Huygens, *21, p. 452.*

Here, instead of using free lumps of wax (or unusually heavy wood), a ball is
constrained to slide along stationary strings stretched across the diameter of the dish.
When the dish and its water are rotating, suddenly stopping the dish makes the
constrained ball slide between the string guides to the centre of the dish. Such an
arrangement clearly worked much more efficiently than allowing bodies in the water
to revolve, but having their speed of revolution retarded by their bumping against
the bottom of the dish.[44]

In either case, however, the inferential movement from an ocular demonstration
of the principle to a conclusion about the true cause of terrestrial gravitation was
large and insecure. It is therefore important to note the methodological point that
Huygens had made in his well-known *Treatise on Light*, which was written only a
few years after the 1669 presentation on gravity (although, like the *Discourse on the
Cause of Gravity*, not published until 1690).[45] At the outset of the *Treatise*, Huygens

---

[44] *Ibid.*, pp. 452–453. A version of this constrained variant also appears in the original 1669 presentation:
ibid., 19, pp. 632–633.
[45] *Ibid.*, 19, pp. 451–537; the English translation is from Huygens 1912.

justifies his presentation of a wave-theory of light not on the grounds that there is good evidence for supposing its truth, but on the grounds, again, that such mechanical explanations are especially intelligible and, in that sense, probable (*vraisemblable*).[46] The same general perspective evidently applied to his hypothesis about gravity: Huygens' table-top demonstration of its feasibility is really an illustration of its intelligibility; an aid to comprehension of the theory. As he commented in 1669, following his presentation of the spinning dish: 'Now, having found in nature an effect similar to that of gravity and of which the cause is known, it remains to see if one can suppose that something similar happens in regard to the earth...'.[47] In a sense, Huygens' spinning dish is an attempt to domesticate the phenomenon and the associated argument. It reminds us of studies concerning more recent work in physics that emphasise 'table-top' experiments and the 'mimesis' by apparatus of natural phenomena.[48]

A tempting way of making sense of such cases is to point to the ways in which these demonstrations create links to things with which people are familiar in everyday life. That seems a reasonable way of understanding Descartes' frequent technique of comparing literally unfamiliar micromechanical interactions with familiar observations drawn from daily life, to do with boats or wine-vats. Such an analogy might sometimes be of value even when it could not be articulated by analytical, causal explanation of the *familiar* thing with which the analogy is made (such an example appears in Huygens' original presentation of his hypothesis on gravity, which involves a ball of molten silver skittering around in a little dish).[49] In this sense, familiarity might breed intelligibility. However, Huygens' revolving-dish demonstration for gravity seems rather different from an appeal to simple familiarity: Huygens intended it as a *real* demonstration rather than an analogy between gravity and an already familiar phenomenon. The clustering of the pieces of wax at the centre of the dish was not something that could be taken as self-evident, as Biancani's doubts about the account of vortex-behaviour in the *Questions in Mechanics* illustrate.

Huygens' 'table-top experiment' acted as a kind of 'boundary object', in the general sense originally suggested by Star and Griesemer.[50] It was an object that sat between the world of Huygens' theory and another world of everyday objects familiar to his audience. In the theoretical world, the table-top model could be related point by point to terms in the theory—the speed of the water in the vortex mapping onto the speed of the particles circling the earth; the pieces of wax mapping onto bits of ponderable matter, and so on. In the familiar world, there was a different mapping: the water in the dish recalled the water in a wash-basin, and stirring it was like a whirlpool; the bits of wax were things generated by guttering candles, and so forth. Each mapping was self-sufficient, and only the fact that the two sets had a common mediating object—the demonstration apparatus itself—connected them at

---

[46] Huygens 1888–1950, 19, p. 459.

[47] *Ibid.*, p. 634.

[48] Galison and Assmus 1989; Schaffer 1995; cf. Dear 1985, pp. 159–161.

[49] Huygens 1888–1950, 19, p. 635.

[50] Star and Griesemer 1989; see also, for a critical development of the category, Fujimura 1992.

all. Huygens' demonstrative trick lay in getting his audience to accept that the contrived phenomenon of the dish had anything at all to do with the gravitational hypothesis.

5.ROHAULT AND A MECHANICAL 'NATURE'

Huygens' gravitational hypothesis found another iteration in Jacques Rohault's famous Parisian lectures, published in 1671 as the *Traité de physique*. Rohault presented a version of Huygens' recent vortex demonstration involving the bits of wax (the later version, involving the constraining strings, was, of course, unpublished until long afterwards).[51] Remarkably, Rohault portrays Huygens' account as an experiment that Huygens had performed, the outcome of which showed the true nature of gravity as the result of differential *levity*.

> He took an earthen Vessel which was white and round, about seven or eight Inches in Diameter, flat at the Bottom, and the Sides about three Inches high, and filled it with Water; then putting into it some beaten *Spanish Wax*, whose Weight made it sink to the Bottom, and whose red Colour made it very visible upon the white Bottom, he covered the Vessel with a Plate of very transparent Glass, and sealed up the Edge so that nothing could get out; having done this, he fastened the Vessel on an Engine or Pivett, so that he could turn it about and stop it at Pleasure. While the Vessel was turning round in this manner, the Wax Powder which was at the Bottom of the Vessel could not slip upon it so readily as the Water, but stuck a little to it, and therefore was more easily carried about... he then stopp'd the Motion of the Engine on a sudden, and the Vessel which was fixed to it consequently stopp'd also; whereupon the *Spanish Wax* grating against the Bottom, and its Particles being rugged, did not move so quick as the Water....[52]

Rohault's account is historicised in this English version, thus resembling Huygens' published version of 1690 (but unlike the version of 1669). However, the French original uses the present tense, in effect generalising Huygens' practices to 'what he usually does'.[53] Furthermore, Rohault's referring to Huygens in the third person, with Rohault himself now serving as an additional witness, tends if anything to add rhetorical weight to his account. Following this narrative, Rohault then explains how Huygens related the demonstration to the hypothesis concerning gravity:

> At this Instant of Time he shews us, that the Water resembles the Fluid Matter which surrounds the Earth, and the Powder of *Spanish Wax* resembles Pieces of the Earth which we see descend in the Air; for the Powder was then forced to approach to the Center of its Motion, being driven thither by the Particles of the Water which endeavoured to recede with greater Force than the Powder which gathered into a little round Body in the Center like the Earth.

Rohault's fairly authentic version of Huygens thus correlates the experimental model with the true nature of gravity by means of comparison and resemblance rather than by way of formal demonstration. But Rohault immediately sums up in a much less methodologically-restrained way, remarking dogmatically:

[51] Jacques Rohault 1723, 2, p. 94.
[52] *Ibid.*
[53] Cf. Aiton 1972, pp. 81–82, who presents some of the material here quoted in his own translation from the French.

By this Experiment we see clearly that Gravity is, properly speaking, nothing else but less Levity; and though it follows from hence, that the Bodies which descend have no Disposition in themselves to descend; yet this Motion ought however to be called Natural, because it is the Result of the established Order of Nature.[54]

Rohault's conclusion thus takes for granted the propriety of using a mechanistic ontology, in a way much less epistemologically-reserved than that of Huygens.[55]

The availability of the experience of the vortex as common knowledge, as a shared resource to be used in building, synthetically, explanations of more complex natural phenomena, has in Huygens' and (even more) in Rohault's secondary accounts disappeared altogether. What remained was an unproblematically-replicable *experiment*, a performance intended to reveal a previously unfamiliar aspect of nature. The 'circular argument' of the vortex has turned into a pragmatically-resolved experimenter's regress.[56]

REFERENCES

Aiton, E. J. (1972) *The Vortex Theory of Planetary Motions*, London: Macdonald.

Aristotle (1984) *The Complete Works of Aristotle: The Revised Oxford Translation*, ed. J. Barnes, 2 vols, Princeton: Princeton University Press.

Beeckman, I. (1939–1953) *Journal tenu par Isaac Beeckman de 1604 à 1634*, ed. C. de Waard, 4 vols, The Hague: M. Nijhoff.

Benedetti, G. B. (1585) *Diversarum speculationum... liber*, Turin.

Berkel, K. van (1983) *Isaac Beeckman (1588–1637) en de mechanisering van het wereldbeeld*, Amsterdam: Rodopi.

— (2000) 'Descartes' debt to Beeckman: inspiration, cooperation, conflict' in *Descartes' Natural Philosophy*, eds S. W. Gaukroger, J. A. Schuster and J. Sutton, London: Routledge, pp. 46–59.

Blair, A. (1999) 'The Problemata as a natural philosophical genre' in *Natural Particulars: Nature and the Disciplines in Renaissance Europe*, eds A. Grafton and N. G. Siraisi, Cambridge MA: MIT Press, pp. 171–204.

Blancanus, J. (1615) *Aristotelis loca mathematica*, Bologna.

Bos, H. J. M. *et al.* eds (1980) *Studies on Christiaan Huygens: Invited Papers from the Symposium on the Life and Work of Christiaan Huygens, Amsterdam, 22–25 August 1979*, Lisse: Swets & Zeitlinger.

Buchwald, J. Z. ed. (1995) *Scientific Practice: Theories and Stories of Doing Physics*, Chicago: University of Chicago Press.

Clarke, D. (1982) *Descartes' Philosophy of Science*, Manchester: Manchester University Press.

Collins, H. M. (1985) *Changing Order: Replication and Induction in Scientific Practice*, London: Sage.

Dear, P. (1985) '*Totius in verba*: rhetoric and authority in the early Royal Society', *Isis*, 76, pp. 145–161.

— (1988) *Mersenne and the Learning of the Schools*, Ithaca: Cornell University Press.

— (1995) *Discipline and Experience: The Mathematical Way in the Scientific Revolution*, Chicago: University of Chicago Press.

Descartes, R. (1983) *Principles of Philosophy*, trans. V. R. Miller and R. P. Miller, Dordrecht: D. Reidel.

(1996) *Oeuvres de Descartes*, eds C. Adam and P. Tannery, 11 vols, Paris: J. Vrin.

(1998) *The World and Other Writings*, trans. S. W. Gaukroger, Cambridge: Cambridge University Press.

Drake, S. and Drabkin, I. E., eds (1969) *Mechanics in Sixteenth-Century Italy: Selections from Tartaglia, Benedetti, Guido Ubaldo, and Galileo*, Madison: University of Wisconsin Press.

Eastwood, B. S. (1984) 'Descartes on refraction: scientific versus rhetorical method', *Isis*, 75, pp. 481–502.

---

[54] Rohault 1723, 2, p. 94.

[55] See in general McClaughlin 2000.

[56] The experimenter's regress (a version of the hermeneutic circle) is the central theme of Collins 1985.

Elzinga, A. (1980) 'Christiaan Huygens' theory of research', *Janus*, 67, pp. 281–300.

Fujimura, J. (1992) 'Crafting Science: standardized packages, boundary objects, and "translation"' in *Science as Practice and Culture*, ed. Andrew Pickering, Chicago: University of Chicago Press, pp. 168–211.

Galison, P. (1984) 'Descartes' comparisons: from the visible to the invisible', *Isis*, 75, pp. 311–326.

Galison, P. and Assmus, A. (1989) 'Artificial clouds, real particles' in *The Uses of Experiment*, eds D. Gooding, T. Pinch, and S. Schaffer, New York: Cambridge University Press, pp. 225–274.

Garber, D. (1993) 'Descartes and Experiment in the *Discourse* and *Essays*' in *Essays on the Philosophy and Science of René Descartes*, ed. S. Voss, Oxford: Oxford University Press, pp. 288–310.

Gaukroger, S. W. (2002) *Descartes' System of Natural Philosophy*, Cambridge: Cambridge University Press.

Gooding, D., Pinch, T. and Schaffer, S. eds (1989) *The Uses of Experiment*, New York: Cambridge University Press.

Grafton, A. and Siraisi, N. G. eds (1999) *Natural Particulars: Nature and the Disciplines in Renaissance Europe*, Cambridge MA: MIT Press.

Huygens, C. (1888–1950) *Oeuvres complètes de Christiaan Huygens*, 22 vols, The Hague: Nijhoff.

— (1912) *Treatise on Light*, trans. S. P. Thompson, London: Macmillan.

Jardine, N. (1988) 'Epistemology of the sciences' in *The Cambridge History of Renaissance Philosophy*, eds C. Schmitt, Q. Skinner, E. Kessler, and J. Kraye, Cambridge: Cambridge University Press, pp. 685–711.

Laird, W. R. (1986) 'The scope of Renaissance mechanics', *Osiris*, n.s., 2, pp. 43–68.

McClaughlin, T. (2000) 'Descartes, experiments, and a first generation Cartesian: Jacques Rohault', in *Descartes' Natural Philosophy*, eds S. W. Gaukroger, J. A. Schuster and J. Sutton, London: Routledge, pp. 330–46.

Mersenne, M. (1626) *Synopsis mathematica* , Paris.

Pickering, A. ed. (1992) *Science as Practice and Culture*, Chicago: University of Chicago Press.

Rohault, J. (1671) *Traité de physique*, Paris.

— (1723) *Rohault's System of Natural Philosophy Illustrated with Dr. Samuel Clarke's Notes*, 2 vols, London.

Rose, P. L. and Drake, S. (1971) 'The Pseudo-Aristotelian *Questions in Mechanics* in Renaissance Culture', *Studies in the Renaissance*, 18, pp. 65–104.

Sakellariadis, S. (1982) 'Descartes' Use of Empirical Data to Test Hypotheses', *Isis*, 73, pp. 68–76.

Schaffer, S. (1995) 'Where experiments end: tabletop trials in Victorian astronomy' in *Scientific Practice: Theories and Stories of Doing Physics*, ed. Jed Z. Buchwald, Chicago: University of Chicago Press, pp. 257–299.

Schmitt, C., Skinner, Q., Kessler, E. and Kraye J. eds (1988) *The Cambridge History of Renaissance Philosophy*, Cambridge: Cambridge University Press.

Schuster, J. A. (1977) 'Descartes and the Scientific Revolution 1618–1634: An Interpretation' (Ph.D., Princeton University).

Star, S. L. and Griesemer, J. R. (1989) '"Translations" and boundary objects: amateurs and professionals in Berkeley's Museum of Vertebrate Zoology, 1907–39', *Social Studies of Science*, 19, pp. 387–420.

Voss, S. ed. (1993) *Essays on the Philosophy and Science of René Descartes*, Oxford: Oxford University Press.

Wallace, W. A. (1984) *Galileo and His Sources: The Heritage of the Collegio Romano in Galileo's Science*, Princeton: Princeton University Press.

Westfall, R. S. (1971) *Force in Newton's Physics: The Science of Dynamics in the Seventeenth Century*, London: Macdonald.

Westman, R. S. (1980) 'Huygens and the problem of Cartesianism' in *Studies on Christiaan Huygens: Invited Papers from the Symposium on the Life and Work of Christiaan Huygens, Amsterdam, 22–25 August 1979*, eds H. J. M. Bos, *et al.*, Lisse: Swets & Zeitlinger, pp. 83–103.

HELEN HATTAB

FROM MECHANICS TO MECHANISM

The Quaestiones Mechanicae and Descartes' Physics

Concluding his account of the purely material properties of the universe in Part IV of the *Principia Philosophiae* René Descartes writes, 'Indeed up to this point I have described this earth and, what is more, the whole observable universe, like [*instar*] a machine, considering nothing except the shapes and motions in it'.[1] Descartes justifies this approach, claiming that it is much better to take what we perceive to happen in large bodies as a model for what occurs in imperceptible small bodies, than to invent 'extraordinary things which I am unable to know, having no resemblance to those which are sensed'.[2] Thus for Descartes, understanding and explaining natural phenomena requires transposing our knowledge of what constitutes and drives visible machines to the impenetrable realm of nature's ultimate constituents. What could be more different than Aristotle's organic conception of nature which explained change at the macroscopic level, in terms of a thing's matter and form, its first source of change, and the natural end towards which it strove? And yet Descartes boldly proclaims:

> But I would also like it to be noted that, in having tried here thus to explain the universal nature of material things, I have certainly not used any principle for this which was not admitted by Aristotle and all the other philosophers of all ages. Therefore, this philosophy is not new, but the oldest and most common of all. To be sure I have considered the shapes, motions and sizes of bodies and examined, according to the laws of Mechanics, confirmed by certain and everyday experiences, what must follow from the mutual concourse of these bodies.[3]

Coming from the author of the *Discours de la Méthode* and the *Meditationes de Prima Philosophia* this claim sounds rather suspect. But to a contemporary it would have been quite plausible, for the principles on which Descartes bases his physics bear more than a superficial resemblance to the principles of the ancient art of mechanics. Mechanics enjoyed a revival and a boost in status in the sixteenth century when its lineage was traced back to the Philosopher himself.[4] This paper

---

[1] *Principia*, IV, §188, Descartes 1996 (henceforth *AT*), VIIIA, p. 315. Translations are mine unless otherwise indicated.

[2] *Principia*, IV, §201, *AT*, VIIIA, p. 325.

[3] *Principia*, IV, §200, *AT*, VIIIA, p. 323.

[4] The *Quaestiones Mechanicae* were widely believed to be by Aristotle at the time. The work is now believed to date back to the Lyceum in the fourth or third century BC and while not by Aristotle, is of his school of thought; Rose and Drake 1971, p. 72.

P. R. Anstey and J. A. Schuster (eds.), The Science of Nature in the Seventeenth Century, 99-130.

traces developments within Renaissance mechanics in order to examine the significance of Descartes' overt positioning of his new natural philosophy in relation to an older tradition, and to explore the role that Aristotelian mechanics played in shaping key explanatory principles of Descartes' mechanical explanations of the heavens.

1.PRELIMINARIES AND BACKGROUND

The pseudo-Aristotelian *Quaestiones Mechanicae* was first translated into Latin by the humanist Vittore Fausto (1480–1551?) in 1517.[5] It is a grab bag of thirty-five questions ranging from explanations of simple mechanical devices such as the balance, lever, pulley and wedge, to the application of these devices in such diverse professions as seafaring and dentistry.[6] The unifying thread of these seemingly diverse questions are the marvellous properties of the circle, which is said to be 'The original cause of all such phenomena'.[7] The subject matter of mechanics is said to include, on the one hand, perplexing phenomena which occur according to nature but from hidden causes, and on the other hand, effects that are contrary to nature but for the benefit of humans. The mechanical art is defined as the skill that helps us overcome our perplexity in order to act against nature and produce useful results. The following century saw the publication of a series of commentaries on this work, as well as other texts that took up its subject matter in various ways. In these texts one finds a form of explanation which, while not in contradiction with Aristotelian physics, nevertheless offers an alternative—one based on geometrical principles rather than the four causes.

The first part of this paper explores the relationship between the disciplines of mechanics and natural philosophy through the eyes of Renaissance intellectuals who commented and elaborated on the subject matter of the *Quaestiones Mechanicae*. In recasting mechanics as a science between mathematics and physics as opposed to an art, these commentators brought to the foreground a form of explanation that combined geometrical principles with considerations regarding the physical causes of motion. While the development of a mathematical physics came later, and was the result of a complex set of factors that lie beyond the scope of this paper, the revaluation of mechanics prompted by the recovery of the *Quaestiones Mechanicae* and ensuing discussions of its scope and position among the other sciences can be seen as an early, unwitting step in that direction.

---

[5] Paul Lawrence Rose and Stillman Drake (1971) were the first to trace the reception of this work in Renaissance Europe.

[6] For example, the author not only explains how the balance works, but applies the principle of the lever derived from this to the following questions: 'Why is it that the rudder, being small and at the extreme end of the ship, has such great power that vessels of great burden can be moved by a small tiller and the strength of one man only gently exerted?' and 'How is it that doctors extract teeth more easily by applying the additional weight of a tooth-extractor than with the bare hand only?'; *Mechanics*, Aristotle 1984, 2, pp. 1304, 1310.

[7] Aristotle 1984, 2, p. 1299.

The pseudo-Aristotelian work inspired a variety of authors at this time, including humanists like Niccolò Leonico Tomeo (1456–1531) and Alessandro Piccolomini (1508–1579); self-taught mathematical practitioners like Niccolò Tartaglia (1499/1500–1557); the eclectic polymath, Girolamo Cardano (1501–1576); professional mathematicians such as Francesco Maurolyco (1494–1575) and Bernardino Baldi (1553–1617); mathematicians who were employed as engineers by the nobility, such as Giovanni Battista Benedetti (1530–1590); noblemen like Guidobaldo the Marquis of Monte (1545–1607); Professor of Medicine and Mathematics and student of Petrus Ramus, Henri de Monantheuil (1536–1606) and of course Galileo Galilei. Many of them commented on at least some of the mechanical questions and some consciously reflected on the nature and scope of mechanics.[8]

As we shall see, the mechanical arts were nothing new, but the recovery of ancient texts offering theoretical explanations of mechanical devices (most notably the *Quaestiones Mechanicae* attributed to Aristotle[9]) played an important role in the renegotiation of the boundaries between *ars* and *scientia*.[10] It would seem that, at the very least, the elevation of mechanics into the realm of theoretical knowledge opened up a new space for theoretical inquiry, thus making it possible for philosophers like Descartes to develop a mechanical philosophy of nature. However, the legitimation of mechanics as a science is in itself not sufficient to account for the incursion of its explanations of artificial devices into the realm of the natural. As a mixed mathematical science, mechanics could quite plausibly have been kept subordinate to and separate from physics—after all, it was the science of machines not of nature. Still, continuing to keep the two apart would have meant ignoring the two-part definition of mechanics in the *Quaestiones Mechanicae* and the author's subsequent application of the same geometrical forms of explanation he uses to explain mechanical devices to perplexing natural phenomena. In other words, the idea of applying the same kind of explanation that accounts for levers and balances to several phenomena of nature is already there in the pseudo-Aristotelian text. Why should it surprise us, therefore, that a century after its revival, philosophers like Descartes attempt to explain nature as a whole in this manner?

The second part of this paper will examine the mechanical questions dealing with natural phenomena so as to compare the explanations given by the author of the

---

[8] These authors and their contributions to mechanics are discussed in more detail in Rose 1975 and Laird 1986.

[9] The other ancient texts on mechanics that received attention at this time were Hero of Alexandria's *Pneumatica* and *Automata* and two works by Archimedes dealing with mechanical subjects. The works by Hero were taken up mainly by engineers as they did not deal with mechanical laws in a systematic way; Rose and Drake 1971, pp. 69–70.

[10] A concurrent development that also contributed to this was the rejection of the Aristotelian view that each natural body had a natural place towards which it moved, and the ensuing collapse of the distinction between natural and violent motions. The Copernican revolution called into question the strict division between the natural up and down motions of the four elements of matter, and violent or forced motions that overcame these natural tendencies. With this division undermined, the strict division between the motions of machines, which were supposedly 'against nature' and natural motions made less sense. For an overview of changing conceptions of motion during this period see Gabbey 1998.

Quaestiones Mechanicae and his Renaissance commentators to the mechanistic explanations Descartes employs in his accounts of the heavens found in *Le Monde* and the *Principia Philosophiae*. I do not thereby intend to imply that Aristotelian mechanics is the only or even the primary influence on Descartes' mechanism. There are certainly significant differences between the particular problems and explanations laid out in these texts and the details of Descartes' approach to problems and his development of a physico-mathematics.[11] I leave to others the task of articulating the full details of Descartes' mechanistic physics and tracking down the various ingredients that went into its making. What interests me as a philosopher is the general form of explanation introduced by Descartes, now commonly known as mechanism. I will argue that, in its generic form, Descartes' mechanism is consistent and continuous with the forms of explanation found in commentaries on the *Quaestiones Mechanicae*.

Either first-hand or indirectly, Descartes was familiar with some of the subject matter of Renaissance commentaries on mechanics.[12] He proposed a solution to question 24 of the *Quaestiones Mechanicae*[13] and, at the request of Constantijn Huygens, wrote a short treatise on the five simple machines that were the focus of Renaissance texts on mechanics.[14] Peter Dear shows that Marin Mersenne was

---

[11] Stephen Gaukroger and John Schuster argue that, 'Far from being something in the tradition of the Aristotelian "subordinate sciences", this "physico-mathematics" pursues a completely different route to the quantitative understanding of physical processes, attempting not to "mix" mathematics and physics, but to translate physical problems into the quantitatively characterisable behaviour of microscopic corpuscles making up material things (bodies and fluids), and then invoking a causal register of forces, tendencies, components and (later) 'determinations', which are completely different from Aristotelian "principles"'; Gaukroger and Schuster 2002, p. 549.

[12] Rose and Drake claim that by 1638 'the *Mechanica* was virtually dead', adding that the work did not arouse much interest in France even though Petrus Ramus lectured on it and his pupil, Henri de Monantheuil (1536–1606) published the Greek text with his own Latin translation and commentary in Paris, in 1599: Rose and Drake 1971, p. 68, pp. 99–100. However, this is contradicted by more recent scholarship which reveals that at least one of the Pseudo-Aristotelian questions, namely, question 24 known as the *Rota Aristotelis* paradox (the paradox of Aristotle's wheel) continued to be discussed throughout the seventeenth century (this is the very question Descartes solved at the behest of Marin Mersenne, although Descartes seems to treat it strictly as a mathematical puzzle). Not only was this question discussed by early modern philosophers, but it has been suggested that it occupied a central place in the development of the new physics. In particular, Carla Rita Palmerino has argued that Galileo's analysis of the *Rota Aristotelis*, found in his digression on the properties of the infinite in Book I of the *Two New Sciences*, has the 'clandestine purpose' of linking 'two of the most problematic hypotheses of the *Two New Sciences*: a) that matter is ultimately constituted by an infinite number of non-extended atoms, and b) that the total speed of a body is the sum of an infinite number of indivisibles of speed'; Palmerino 2001, p. 421. Palmerino further argues that in Pierre Gassendi's *Syntagma* 'the paradox serves above all the purpose of overcoming the greatest difficulty besetting Gassendi's physics, namely to bring the respective laws governing the microscopic *and* macroscopic levels of reality into agreement'; Palmerino 2001, p. 422. It is clear that the pseudo-Aristotelian paradox of the wheel was still a topic of debate in the late seventeenth century, as Robert Boyle's *A Defence* (published in 1662) in response to Francis Line includes a chapter by Robert Hooke which criticises Line's solution to the paradox and proposes an alternative solution. See Hooke's 'An Explication of the *Rota Aristotelica*' in Boyle 1999–2000, 3, pp. 89–93.

[13] de Gandt 1986, p. 394. Descartes criticises Galileo's solution to the problem in a letter to Mersenne, 11 October 1638, *AT*, II, p. 383.

[14] In his letter to Descartes of 1637, Huygens asks Descartes for a gift, such as the one he presented to Isaac Beeckman in the form of the *Compendium Musicae*. Specifically, he asks for three pages on the

heavily influenced by the pseudo-Aristotelian *Mechanica*, drawing on its principles when proving the existence of God in the *Quaestiones in Genesim* of 1623 and incorporating its dynamical explanation of the balance in other works. In the *Quaestiones* Mersenne responds to Baldi's criticisms of the Aristotelian mechanical explanations and also refers to the laws of Archimedes, Guidobaldo and Blancanus.[15] In his *Questions Theologiques* of 1634 Mersenne cites the four most recent commentaries on the *Mechanica*, namely those by Baldi, Blancanus, Monantholius and Giovanni de Guevara.[16] Since Descartes was in Paris during the time that Mersenne published the *Quaestiones in Genesim*, and remained a part of Mersenne's circle of correspondents after this, it is safe to assume that Descartes would have been familiar with the content of these commentaries through his discussions with Mersenne, if not by a first-hand reading.

Given this probable historical link between the *Mechanica* tradition and Descartes, it would not be surprising if there were some conceptual affinities between the explanations found in the *Quaestiones Mechanicae* and Descartes' mechanistic explanations. However, one cannot assume that the presence of a likely historical connection between mechanism and Renaissance mechanics points to a continuity in Descartes' understanding and use of mechanical explanations, and those found in the Aristotelian tradition. As Daniel Garber implies, since Descartes wrote separately on the traditional subject matter of mechanics, why should we assume that Descartes was appropriating elements of Aristotelian mechanics when he developed his physics?[17] In fact, the revival of Platonism, Pythagoreanism, ancient atomism, and contemporaneous developments in other mixed mathematical sciences, such as astronomy and optics, are equally plausible sources for the Cartesian mathematisation of nature and the search for hidden structures underlying our sensory perceptions, as is the more strictly mathematical Archimedean approach to the traditional problems of mechanics.[18] Nevertheless, one good reason to also explore the explanatory forms found in the *Quaestiones Mechanicae* and their relationship to Descartes' physics is that, setting aside his laws of motion and the rules of collision, there is nothing particularly mathematical about Descartes' mechanistic explanations. Rather he relies heavily on analogies to simple mechanical devices to infer similar mechanical principles at work in the *cosmos* at large. Descartes even admits to having been inspired by the study of machines in developing his mechanistic explanations of what goes on at the microscopic level of bodies:

'foundations of mechanics' and the 4 or 5 machines demonstrated from them (among which he counts the balance, lever and pulley). Huygens claims to be unsatisfied with what he read by Guidobaldo and Galileo (referring to Mersenne's translation of Galileo's treatise on mechanics) but is sure that Descartes can explain it much more clearly and concisely: *AT*, I, pp. 396–399. Descartes obliges Huygens on 5 October 1637 by sending him a brief account of the pulley, inclined plane, wedge, paddle-wheel or potter's wheel, screw, and lever; *AT*, I, pp. 431–447.

[15] Dear 1988, pp. 119–120.

[16] *Ibid.*, p. 126. Giovanni de Guevara's commentary was published in 1627.

[17] Garber 2002, p. 195.

[18] For example, in his chapter in this same volume, John Schuster argues that Descartes, under the influence of Isaac Beeckman, is attempting to give a corpuscular account of problems tackled by Simon Stevin and Johannes Kepler in hydrostatics and astronomy respectively.

And to this end, things made by art helped me quite a bit: for I recognise no distinction between them and natural bodies, except that the operations of things made by art are for the most part performed by instruments so large that they can easily be perceived by the senses: indeed this is necessary in order that they may be fabricated by men. By contrast, however, natural effects almost always depend on certain tools [*organis*] so minute that they escape all the senses. There really are no reasonings [*rationes*] in Mechanics which do not also pertain to Physics, of which it is a part or species: nor is it less natural for a clock composed of these or those wheels to tell the time, than it is for a tree originating from this or that seed to produce a certain kind of fruit. For this reason, just as when those who are trained in considering automata, whenever they know a certain machine and inspect some of its parts, easily conjecture from these parts in what way the others which they do not see are made, so from the sensible effects and parts of bodies I have attempted to investigate which are their causes and insensible particles.[19]

Given this explicit denial of a division between the objects studied by physics and mechanics, and given the bridge from explanations of machines to natural phenomena already suggested by the attempt of the *Quaestiones Mechanicae* to explain select natural occurrences, it is worthwhile to explore possible affinities between Aristotelian mechanical explanations of natural phenomena and Descartes' mechanistic physics, keeping in mind that one is thereby isolating and focusing on a single strand in what is obviously a very complex tapestry. In fact, this chapter is best read in conjunction with the chapters by Peter Dear and John Schuster found in this volume, as they follow somewhat different strands, linking sixteenth-century mechanics to Descartes' physics and cosmology.

2. THE RECLASSIFICATION OF MECHANICS

At first glance there appears to be no obvious conceptual connection between Cartesian mechanics and the ancient art of mechanics, as Descartes classifies mechanics not as an art but as a part or species of physics.[20] What seems to justify the use of mechanistic explanations in physics for him is that mechanics, as the explanation of machines, derives from and makes use of the very same principles as physics. This is a very different sense of mechanics than the pseudo-Aristotelian art of acting against nature to produce useful results. The strict Aristotelian division between artefact and natural body has disappeared for Descartes with the result that mechanics has been absorbed into physics. In what follows I will trace the steps leading up to this obliteration of the traditional boundary between natural philosophy and mechanics.

---

[19] *Principia*, IV, §203, *AT,* VIIIA, p. 326.

[20] Gabbey 1993 discusses the distinction between what we now call 'mechanics' (which we trace back to Descartes, Huygens, Newton, etc.) and Renaissance mechanics, which is the theoretical study of machines and argues that we miss much if we limit ourselves to studying the history of mechanics in our sense.

2.1From humble art to mixed mathematical science

It is not clear whether the *Quaestiones Mechanicae* were known in the Middle Ages,[21] but the classification Hugh of St Victor gives in the *Didascalicon* in the late 1120s, as well as most subsequent classifications, indicate that mechanical knowledge, which included fabric-making, armament, agriculture, hunting, medicine and theatrics, kept its lower status as an art in medieval times.[22] In the *Didascalicon* mechanical know-how falls under 'knowledge' and is distinguished from understanding (which includes theoretical and practical philosophy). Hugh of St Victor explains that this kind of knowledge is called mechanical or adulterate because 'it pursues merely human ends'.[23]

It was not until after the rediscovery of the *Quaestiones Mechanicae* in the Renaissance that mechanics was elevated into a theoretical science. As a science, it occupied an interesting position, straddling mathematics and natural philosophy. This point is already made in the pseudo-Aristotelian text when the author comments that the mechanical questions 'have something in common with both Mathematical and with Natural Speculations; for while Mathematics demonstrates *how* phenomena come to pass, Natural Science demonstrates *in what medium* they occur'.[24] Despite this suggestive observation, the author is clear that mechanics, as the skill that helps us overcome the difficulty and ensuing perplexity inherent in acting against nature, is an art.

In his *Opuscula* (1525), which contains a translation of the *Quaestiones Mechanicae* and the first rather brief commentary on it, Niccolò Leonico Tomeo closely follows the pseudo-Aristotelian text in classifying mechanics as an art.[25] He explains that it is an art because it accomplishes things that are contrary to nature for the sake of the utility of men. Unlike nature, which is simple and uniform, mechanics, by overcoming nature, can produce many different works that are useful to us, such as lifting stones, erecting foundations, and lifting beams and trunks to the

---

[21] According to Matthias Schramm (1967, p. 153), no Arabic or medieval Latin translation of the text is extant but Iordanus de Nemore's *De ratione ponderis* contains a set of propositions with a strong affinity to the Aristotelian mechanics, so the work may have been known in the Middle Ages.

[22] A notable exception is the twelfth-century translator and philosopher, Domenicus Gundissalinus (also known as Domenico Gundisalvo) who adapted al-Farabi's ideas, classifying the *scientia de ingeniis* under mathematics. 'His rationale for this switch is taken from Avicenna's *De anima* and Boethius' *De arithmetica*, and acknowledges that, properly speaking, mathematics considers the pure forms of things according to syllogistic demonstration: the science of 'devices' serves this study by providing the mechanical means by which 'pure forms' can be apprehended'; Ovitt 1983, p. 98.

[23] Hugh of St Victor 1961, Bk I, chap. 8, p. 55.

[24] Aristotle 1984, 2, p. 1299.

[25] Niccolò Leonico Tomeo (1456–1531) was an Italian philosopher and humanist who studied philosophy and theology at Venice and Padua. From 1497–1509 he was a professor of Philosophy at Padua and was one of the first there to comment on Greek texts of Aristotle. In addition to the *Opuscula*, he also wrote *Dialogi* (1524) and *De varia historia libri III* (1531). See Schmitt *et al.* 1988, p. 824. Leonico's translation went through many editions becoming the standard translation of the *Quaestiones Mechanicae*. Galileo owned the original edition of 1525 edition of the *Opuscula* which included Leonico's commentary; Rose and Drake 1971, p. 80. Mersenne also cites this commentary; Dear 1988, p. 120, n. 15.

roofs of buildings.[26] Leonico Tomeo goes on to explain that the Greeks applied the term 'Mechanics' to the part of the art of building that used machines to accomplish works and adds, 'There are, however, machines composed out of the conjunction of the material they contain, by which, through certain rotations and goings-round of orbs, great weights of things are moved, and also rise to [their] placements'.[27] In situating the mechanical questions relative to the sciences, Leonico Tomeo relates them to natural philosophy, on the one hand, since machines are made of matter which is natural, and to mathematics, on the other hand, because the explanations given as to how these machines work abstract from the matter. He relies on Aristotle's authority:

> In this place, the philosopher said that the mechanical questions were common to the contemplation of mathematics and of natural [things], [for] around this underlying natural matter indeed, all things are certainly made. They talk about iron levers (for example) and wooden or brass spheres: about heights and balances and things of this sort, which exist naturally without controversy, and have physical matter. Regarding the mode or force of working, they turn away to mathematical matters. They investigate circles, diameters and circumferences: and even the weights and measures that are granted to exist in natural matter, certainly seem to abstract from it and not undeservedly seem to lead them away from it, and to display and represent only the reasons of the forms.[28]

While mechanics is still an art for Leonico Tomeo, he situates its mode of explanation squarely under mathematics since mechanics abstracts from the nature of the matter.

By the middle of the sixteenth century the *Quaestiones Mechanicae* had been popularised as a result of the publication of Alessandro Piccolomini's Latin paraphrase in Rome in 1547. Like Leonico Tomeo, he was a philosopher and a humanist and taught moral philosophy at the University of Padua, where Leonico Tomeo had also taught philosophy earlier in the century.[29] One wonders why someone with no apparent background in mechanics or mathematics should undertake a paraphrase of the *Quaestiones Mechanicae*, but Piccolomini's prefatory remarks reveal what might have been at stake.

Piccolomini's classification of mechanics reveals a significant departure from Leonico Tomeo's, which, as we saw, closely followed the pseudo-Aristotelian text. Unlike Leonico Tomeo, Piccolomini is concerned to situate the mechanical questions within the intellectual and disciplinary landscape at large. Clearly signalling his allegiance to the Peripatetics, as opposed to the Stoics or Academics, he divides philosophy into the operative and contemplative parts, the difference

---

[26] 'For this very art is manifold, which performs many and varied things of use to the race of men: for nature being always simple and uniform, it always makes use of the same manner of movement'; Tomeo 1525, p. xiii.

[27] Tomeo 1525, pp. xiii–xiv.

[28] *Ibid.*, p. xiv.

[29] Alessandro Piccolomini was born in Sienna in 1508 and died there in 1579. He studied both philosophy and theology before becoming a lecturer at the University of Padua in 1539. Piccolomini published a commentary on Aristotle's *Meteorology* in 1540, and after publishing his paraphrase on mechanics, he went on to author two influential Italian compendia, one on natural and one on moral philosophy, as well as Italian translations of Aristotle's *Rhetoric* and *Poetics*; Schmitt *et al.* 1988, p. 824.

being that, even though speculation is involved in both, in the former the speculation 'goes out to work', whereas in the latter it 'is truly perfected or rests contented in itself'.[30] He goes on to divide the operative part into the active and the productive and situates the sellularian arts under the productive because it 'looks more to the useful work than the [work that is] good-in-itself'.[31] Contemplative philosophy is divided into the natural, the mathematical and the divine. Piccolomini gives standard definitions for these three disciplines and then further divides Mathematics into Arithmetic and Geometry, Arithmetic being the contemplation of number, and Geometry the examination of continuous magnitude.[32] Then Piccolomini diverges from Leonico Tomeo and classifies mechanics, not as an art, but as a science which belongs under Geometry.

> But one or the other of these [two] parts of mathematics [arithmetic and geometry] comprehends in return the other parts: not, however the sellularian arts (though some want to, they do not rightly set them under the tenth book of Euclid, in which book magnitude is to be had potentially), but indeed Arithmetic recommends [*vendicat*] Music to itself, and Geometry truly Stereometry, Perspectives, Cosmography, Astronomy, and Mechanics. But nevertheless all of these, even if they cannot be called pure or genuine [*syncerae*] mathematics, since they regard matter in a certain mode, it is still most convenient [to call them] Mathematical, rather than natural, which Aristotle himself declares of Astronomy in the divine [matters], moreover, [he declares it of] music and Perspective, in the second book on natural Principles.[33]

In order to classify mechanics as a science on a par with the other mixed mathematical sciences, Piccolomini distinguishes mechanics from the arts Hugh of St Victor called mechanical and renames the latter the 'sellularian arts'. Following Aristotle, Piccolomini then locates the mixed mathematical sciences under mathematics rather than natural philosophy, even though they share in both. He also offers a justification for this classification that is independent of Aristotle's authority.

> This is true, nonetheless, even if his declaration or his authority were not present, since what is studied by means of the instrument of mathematics, ought to be called mathematical. For in this manner, wherever the proposition follows a certain word of a subject, it is fitting that it be named from the same diction. Thus by its mode and force of demonstrating, any science whatsoever will be rightly named. Indeed when we say that a man is generated, or becomes white, we will audaciously pronounce propositions of this kind to be natural because just as a generation indicates a motion so [does] an eduction of white... By the same stipulation, in asserting that man could be divided *ad infinitum* we would construct a mathematical proposition. Wherefore, even sciences from those which are intermediate in demonstrations ought to be called [mathematical]. Since therefore, Perspectives, Astronomy, Music, and the faculties of such kinds, are studied by the mathematical instrument and (as I thus say) are intermediate, it is no wonder if they are rightly called mathematical.[34]

---

[30] Piccolomini 1565, A3v.

[31] *Ibid.*, A3v. 'Sellularius' derives from 'sella', which means 'seat' and refers to manual workers who did their work seated, i.e., craftsmen and artisans.

[32] Piccolomini follows Boethius in holding that geometry is posterior to and less noble than arithmetic.

[33] *Ibid.*, A4r–v.

[34] *Ibid.*, A4v–A5r.

Towards the end of his Preface, Piccolomini articulates the relationship of mechanics to the sellularian arts. Mechanics is a science that provides the causes and principles for the many sellularian arts. However, these arts are not rightly called mechanical, but rather should be called 'banausicae' (following the Greek) or 'humble'. Piccolomini includes the production of machines, both domestic and military, under these arts, and remarks that, insofar as mechanical principles are used to think up these machines, they may be called mechanical. He concludes:

> But since the mechanical faculties, however much touching on matter or motion, as for instance heavy and light things, quickness and slowness, are nevertheless studied by the mathematical way or mode, for this reason it must be judged that they are to be numbered among the mathematical. For although the mechanical instruments, or the very mechanisms themselves, are thought out towards some work, the Mechanic [*Mechanicus*], though a craftsman, for this reason he is a mechanic, which [is]: simply considering their causes and principles, he rests and stands in contemplation itself. [35]

Piccolomini also claims that the mathematical instrument is considered the most certain because it 'shows at the same time both that a thing is and wherefore it is',[36] but denies that mathematical demonstrations are the most excellent ones that Aristotle sought, instead tracing this view back to Proclus' *On Euclid's Elements*.

What seems to be at stake here is the status of mechanics as a science. Piccolomini in effect elevates mechanics from an art to a science, by arguing that it is the contemplation of the causes and principles behind machines that is properly called mechanics. The employment of these machines to produce useful effects is distinguished from the theory on which they are based, and this practical aspect is classified separately under the sellularian or humble arts. This stands in sharp contrast to the classification Hugh of St Victor gives in the *Didascalicon* composed in the late 1120s. The latter distinguishes knowledge from understanding (which includes theoretical and practical philosophy) and calls knowledge 'mechanical' or 'adulterate' because 'it pursues merely human ends'.[37] The belief that Aristotle himself wrote the *Quaestiones Mechanicae* no doubt contributed to the reclassification of mechanical knowledge under theoretical philosophy in the Renaissance. In fact, Piccolomini devotes part of his preface to presenting the philological evidence supporting his conclusion that this work is indeed by Aristotle.[38] Piccolomini furthermore highlights the view (quite common at this time) that mathematical explanations are the most certain, and classifies mechanics under mathematics rather than natural philosophy on the basis of its use of mathematical demonstrations. One can begin to glimpse why Descartes extolled the certainty of mathematics in his *Discourse on the Method* and why he chose geometry as the model for his new method of philosophising. However, Piccolomini's approach, while distinct from the author of the pseudo-Aristotelian text and from Leonico Tomeo, is still very far removed from Cartesian mechanism, as he considers

---

[35] *Ibid.*, A5v.

[36] *Ibid.*, A4v.

[37] Hugh of St Victor 1961, Bk. I, chap. 8, p. 55.

[38] Piccolomini notes that the manner of expression of the *Quaestiones Mechanicae* is consistent with that found in works of Aristotle where he uses mathematical demonstrations. Piccolomini gives Aristotle's discussion of the rainbow in his *Meteorology* as an example; Piccolomini 1565, A6r.

mechanics to be a mixed mathematical science that explains effects that are wondrous or contrary to nature.

Later commentators on the mechanical questions begin to blur the boundary between mechanics and natural philosophy. For example, Girolamo Cardano comments on several of the pseudo-Aristotelian mechanical questions in his *Opus Novum*, first published in 1570.[39] The work consists in an eclectic collection of propositions, dealing with everything from mathematical problems, to mechanical devices, to natural motions, to the parts of the soul. The unifying thread is that all these propositions have to do with proportions. In fact the full title of the work is: *The New Work on the proportions of numbers, and the measuring of motions, weights, sounds, and other things, not only established of things more geometrico, but also by various experiments and observations in the nature of things, illustrated by clever demonstration, suited to many uses, and arranged into five books.*

In his Preface Cardano emphasises the importance of measure or moderation to living a good life and then writes:

> This the ancients called reason, others proportion, not only of course in the trite things, so easy that men struggle against [them], [but also] on the other hand, in other matters granted to be very obscure, and I myself consider it for this reason to be difficult in every respect, and perhaps more so where we do not consider it. Whence we see many fall down with great help and evident hope. What else is the cause but the unknown measures of things, which nevertheless very many deem themselves to have? Thus, since I have determined the greatest good to be situated in this measure (just as is clearly shown by the voices of music which are not able to remain fixed unless in an individual space or place, as I say, in this manner, accordingly in the shapes of paintings and statues, and in decreed days and in civil business) you would value the previous works made by me if all those which were spread out widely were brought back briefly in one...[40]

For Cardano then, the principles of mechanics are just one instance of the proportions underlying all phenomena with which we humans must be concerned in order to live a good life.

Francesco Maurolyco abbot of a monastery near Messina and then lecturer at a Jesuit college in Messina, wrote his *Problemata Mechanica* circa 1569, the same year he began lecturing.[41] These were probably his lecture notes and they were published posthumously in 1613. In his letter of dedication to D. D. M. Anton

---

[39] As well as being notorious for his dispute with Niccolò Tartaglia, over the discovery of the algebraic solution of third-degree equations, Girolamo Cardano (1501–1576), authored over two hundred works on medicine, mathematics, physics, philosophy, religion and music; Schmitt *et al.* 1988, p. 812; Gliozzi 1976.

[40] Cardano 1570, Preface p. [2].

[41] Francesco Maurolyco (1494–1575) was ordained as a priest in 1521 and later became a Benedictine. He spent his whole life in Sicily where he taught mathematics, as well as holding several civil commissions in Messina. He gave public lectures on mathematics at the University of Messina and became a Professor there in 1569. His first work was published in Messina in 1558 and included treatises on the sphere by Theodosius of Bynthinia, Menelaus of Alexandria, and his own treatise on the sphere. His book on Apollonius' *Conics* and his collection of Archimedes' works were published posthumously. Other important extant works by Maurolyco include his *Cosmographia* (1543) and his *Opuscula Mathematica* (1575), which includes treatises on arithmetic, astronomy, optics and music; Masotti 1974.

Amulius, Maurolyco characterises his work as requiring 'Theoretical as well as Mathematical pains'.[42] He indicates that there is a conflict between the philosophers and those engaged in the practical professions, i.e., the architects, craftsmen, sculptors and painters who are accused of being ignorant of geometry. Maurolyco claims to follow neither side and takes the ancient architect Vitruvius as his model, 'Since in his most learned works he offers optimally not only Architecture but *Mathesis* in every way'.[43] Maurolyco appears to take a mixed approach, asserting that neither theories alone nor exercises alone suffice. Moreover, he presents himself as interested in investigating the causes of things and claims to deduce the causes of the magnet and the rainbow. Like Piccolomini, Maurolyco classifies mechanics under the mathematical part of Philosophy; in fact, his account of this is a close paraphrase of Piccolomini's commentary.[44] However, he differs from Piccolomini in holding that, even though Aristotle's mechanics was more ancient than Archimedes' accounts of the lever, the balance and the centre of gravity, it was so obscure that it needed to be clarified by the principles Archimedes had laid down. Thus it seems that Maurolyco intends to take a more rigorous mathematical approach than previous commentators had. In fact, he states 'they seem to me to labour in vain who in several places of this book, try to explain the things demonstrated there by sensible experiments'.[45]

[42] Maurolyco 1613, p. 5.
[43] *Ibid.*, p. 5.
[44] *Ibid.*, p. 7.
[45] *Ibid.*, p. 10.

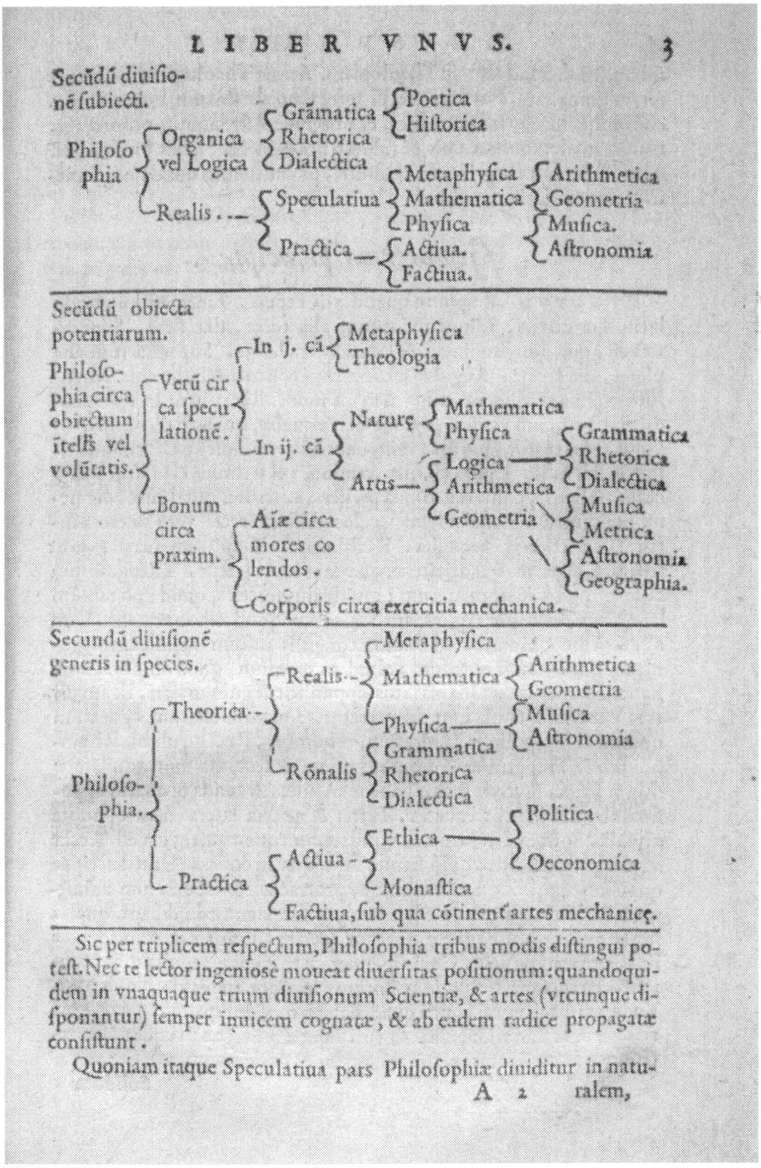

LIBER VNVS.   3

Secūdū diuifio-
né fubiecti.

Philofo- ⎰ Organica ⎱ Gramatica ⎰ Poetica
phia ⎱ vel Logica ⎰ Rhetorica ⎱ Hiftorica
                 ⎱ Dialectica
        ⎱ Realis... ⎰ Speculatiua ⎰ Metaphyfica ⎰ Arithmetica
                              ⎱ Mathematica ⎱ Geometria
                              ⎱ Phyfica ⎰ Mufica.
                ⎱ Practica ⎰ Actiua. ⎱ Aftronomia
                          ⎱ Factiua.

Secūdū obiecta
potentiarum.

Philofo-   ⎰ Verū cir ⎰ In j. cā ⎰ Metaphyfica
phia circa   ⎱ ca fpecu ⎱   ⎱ Theologia
obiectum   ⎱ latione. ⎱ In ij. cā ⎰ Naturę ⎰ Mathematica
itelłs vel   ⎱ Phyfica ⎰ Grammatica
volūtatis.   ⎱ Rhetorica
        ⎱ Bonum   ⎰ Logica ⎱ Dialectica
        circa   ⎰ Aiæ circa   ⎱ Artūs ⎰ Arithmetica ⎰ Mufica
        praxim.   ⎰ mores co   ⎱ Geometria ⎱ Metrica
                  ⎱ lendos .   ⎰ Aftronomia
        ⎱ Corporis circa exercitia mechanica. ⎱ Geographia.

Secundū diuifioné
generis in fpecies.

                              ⎰ Metaphyfica
                  ⎰ Realis... ⎰ Mathematica ⎰ Arithmetica
Theorica-   ⎱ Geometria
                              ⎱ Phyfica ⎰ Mufica
                                        ⎱ Aftronomia
                  ⎱ Rōnalis ⎰ Grammatica
Philofo-   ⎱ Rhetorica
phia.   ⎱ Dialectica
                            ⎰ Ethica ⎰ Politica
        ⎱ Practica ⎰ Actiua- ⎱   ⎱ Oeconomica
                            ⎱ Monaftica
                  ⎱ Factiua, fub qua cōtinenť artes mechanicę.

Sic per triplicem refpectum, Philofophia tribus modis diftingui po-
teft. Nec te lector ingeniosè moueat diuerfitas pofitionum: quandoqui-
dem in vnaquaque trium diuifionum Scientiæ, & artes (vtcunque di-
fponantur) femper inuicem cognatæ, & ab eadem radice propagatæ
confiftunt .

Quoniam itaque Speculatiua pars Philofophiæ diuiditur in natu-
                                        A 2   ralem,

Figure 17. Francesco Maurolico, Opuscula Mathematica De Sphaera Liber Unus,
1575, p. 3.

In the Prologue to his *Opuscula Mathematica* published in 1575 Maurolyco
offers three different classifications of the sciences. [See **Fig. 17**] The first is

secundum divisionem subiecti where music and astronomy are placed under physics
not mathematics. Maurolyco does the same in the third classification, *secundum
divisionem generis in species*. The mechanical arts are classified under the
productive part of practical philosophy. In the second classification, *secundum
obiecta potentiarum* mathematics is classified under philosophy insofar as it
concerns the object of the understanding engaged in speculation of nature. However,
music and astronomy are now classified under the intellectual speculation of art,
music falling under arithmetic and astronomy under geometry. Mechanical exercises
are still classified under practice.[46] While mechanics remains a practical endeavour
for Maurolyco, one can see that mixed mathematical sciences such as music and
astronomy admit of different classifications, and that mathematics itself is, under
one classification, regarded as a study of natural things.

Guidobaldo the Marquis of Monte published his *Mechanicorum Liber* in 1577,
and like Maurolyco, was engaged in the Archimedean revival.[47] He draws on both
Archimedes' works and the works of Pappus in his solutions to the pseudo-
Aristotelian questions on mechanics. In his dedication to the Duke of Urbino,
Francesco Maria II, Guidobaldo characterises mechanics as the noblest of all arts
due to both its subject matter, which belongs to physics, and the logical necessity of
its arguments, which come from geometry. Furthermore mechanics is of great
practical utility as it 'holds control of the realm of nature' and 'operates against
nature or rather in rivalry with the laws of nature'.[48] It is Guidobaldo who expresses
most eloquently the dual nature of mechanics, as both physical and mathematical:

> Thus there are found some keen mathematicians of our time who assert that mechanics
> may be considered either mathematically, removed [from physical considerations], or
> else physically. As if, at any time, mechanics could be considered apart from either
> geometrical demonstrations or actual motion! Surely when that distinction is made, it
> seems to me (to deal gently with them) that all they accomplish by putting themselves
> forth alternately as physicists and mathematicians is simply that they fall between two
> stools, as the saying goes. For mechanics can no longer be called mechanics when it is
> abstracted and separated from machines.[49]

Guidobaldo was a great admirer of the rigour of Archimedean mathematics.
Nevertheless, in his letter of dedication he acknowledges that the science of
mechanics cannot be abstracted from actual motion and praises the practical aims of
the study of mechanics. Thus he seems to combine Archimedean statics with the

---

[46] Maurolyco 1575, p. 3.

[47] Guidobaldo (1545–1607) came from a noble family in the territory of the dukes of Urbino. He studied
mathematics under Federico Commandino (1509–1575), and was a close friend of Bernardino Baldi
(1553–1617), the mathematical historian. In 1588 Guidobaldo oversaw the publication of
Commandino's Latin translation of Pappus. He also secured a position at Padua for Galileo Galilei
and was his patron and friend for twenty years. Guidobaldo's greatest contribution to mechanics is
thought to be his analysis of pulleys in the *Mechanicorum Liber*. He reduces them to the lever, an
analysis Galileo also adopts. Guidobaldo went on to publish a *Paraphrase on Archimedes:
Equilibrium of Planes* (1588), which he sent to Galileo, and a posthumous work *De Cochlea* (1615).
His other works include three manuscript treatises on proportion and Euclid, two astronomical works
and the best Renaissance study of perspective; Rose 1974.

[48] Guidobaldo del Monte *Mechanicorum Liber*, 1577, trans. in Drake and Drabkin 1969, p. 241.

[49] Drake and Drabkin 1969, p. 245.

more dynamical considerations that characterise Aristotelian mechanics. Guidobaldo was one of Galileo Galilei's patrons and corresponded with him; thus there are both historical and conceptual connections between the Aristotelian (as well as the Archimedean) tradition in mechanics, and the new science of motion.[50]

2.2 From mixed science to divine art

By far the most intriguing development is found in the commentary on the *Quaestiones Mechanicae* by Descartes' compatriot, Henri de Monantheuil (which was among those cited by Mersenne).[51] In his letter of dedication to Henri IV of Navarre, Monantheuil effectively erases the division that persisted, however blurred, between nature, the subject matter of physics, and machines, the subject matter of mechanics. He claims that when Plato was asked what God did, he responded, *ἀειγεωμεζεῖν*. Monantheuil translates this with the Latin expression 'always being busy measuring the earth'. He then rejects the view that Geometry is 'ridiculous' 'ridiculous' and 'unworthy of the majesty of God' with the following observation.[52]

> But if from the powers and magnificent things issuing forth from that same art he [God] is estimated ἀειγεωμεζεῖν [to be always busy measuring], that is, to constitute, define and measure out by reason, proportion and similarity the measure of the accessible universe as great as you will, far and wide, of all bodies in it, of surfaces, lines and measurable things, then certainly, to the extent that the whole (finite indeed, but very like the infinite) is more noble and excellent than its tiny and lowest part, certainly, so much nobler than the prior [the whole] will be the action. And he who will have accurately weighed this action with his own weights, will by no means judge it unworthy for God to think about and be occupied with.[53]

Monantheuil then makes a surprising but logical leap to the art that borrows its principles from Geometry, adding,

> if Plato had added ἀειγεωμεζεῖν to καὶ ἀειμηχανᾶσθαι,[54] he would have responded much more brilliantly and more in agreement with the divine majesty and magnificence.

---

[50] Mary Henninger-Voss (2000, p. 237) claims that Guidobaldo is the most influential of the commentators and points out that his texts are the only mechanical texts to regularly show up in Jesuit curricula. Thus Descartes may have encountered his works during the course of his Jesuit education as well as through his friendship with Marin Mersenne. Rose and Drake conclude after connecting Guidobaldo to Galileo, 'It seems evident to us that the closest of links existed directly between the men who studied the *Mechanica* in the sixteenth century and those who gave birth to modern mechanics'; Rose and Drake 1971, p. 102.

[51] Henri de Monantheuil or Henricus Monantholius was born in 1536 and died in 1606. After studying under Petrus Ramus he became a royal professor of medicine in Paris in 1574 and in that same year published an *Oration for the Mathematical Arts*. Apparently he then lost this position and delivered another oration in 1585 asking for reinstatement. From that year on he was a professor of mathematics in Paris. His subsequent works include the *Ludus iatromathematicus* of 1597, and the *Aristotelis Mechanica* of 1599 in addition to some mathematical treatises. See Thorndike 1941, VI, pp. 141–142.

[52] Monantheuil 1599, Dedication, p. 2.

[53] *Ibid.*

[54] *μηχανᾶσθαι* is the infinitive of *μηχανάομαι* which Liddell and Scott (1968) define as follows: 'make by art, construct, build', in a more general sense 'prepare, make ready', and also in a frequently bad sense 'contrive, devise, by art or cunning'. I have translated it as neutrally as possible with 'always making by art'.

> For who would have fashioned this world *ex nihilo*, brought it to completion with all its numbers, balanced it from all sides with its weights, kept it uniform in longitude, latitude, altitude, and constantly retained, stabilised, conserved it in the same state and perfection, in every appearance and respect other than 'the always busy measuring and always making by art' [ἀειγεωμεζεῖν καὶ ἀειμηχανᾶσθαι]? For this world is a machine, and indeed of machines, the greatest, most efficient, most firm, most beautiful. [55]

By describing the world itself as a machine, Monantheuil effectively transforms mechanics from a mixed mathematical science, subordinate to geometry and physics, into the key that will unlock the hidden causes of the world's motions. God, the creator of nature, is now not only 'the most accurate and incessant Geometer' as Plato recognised, but is also 'by the evidence of so many magnificent works, the wisest, best, most powerful mechanic and maker of machines'.[56] Furthermore, since he is made in the divine image, man is bestowed with the capacity to make machines and instruments by virtue of the mind, which Monantheuil characterises as the 'art of arts', and the hand as the 'tool of tools'.[57] Monantheuil even suggests the beginnings of a cosmological argument for the existence of God on this basis:

> Indeed with these great and numerous things which were manifestly in the eyes of all both made and conserved by man ... it is most easy [for] whoever has a mind to believe, know, and grasp that this world, certainly the greatest work of works, was made and conserved, even if when it happened he was absent, not however by any man, but by another 'maker of machines' surpassing man by as much excellence, wisdom and power, indeed infinitely, as the amount by which this machine of the world surpasses and is superior to the machines of all men, even of the Archimedeans.[58]

The remainder of the Letter of Dedication consists in the usual flattery of the powerful patron and an enumeration of the many virtues of mechanics when it comes to the things that matter most to such patrons: the affairs of war and peace. But lest we think that Monantheuil's characterisation of mechanics as the divine art by which God constructs and maintains the *machina mundi* is a mere rhetorical ploy to win over a king who would be more interested in mechanical inventions than philosophical principles, we must turn briefly to the definition and classification of mechanics he gives in the Commentary itself.

After discussing the definitions of 'mechanics' found in ancient authors, like Vitruvius and Pliny, and the traditional division of disciplines into the liberal and mechanical arts, Monantheuil, like Piccolomini, distinguishes the 'mechanical', from what the Greeks called, the 'humble' or 'cheap' arts. Following Aristotle, he explains that nothing that aims at some good ought to be deemed 'cheap' in and of itself, but that some arts are considered more prestigious than others, and so was

---

[55] Monantheuil 1599, Dedication pp. 2–3.
[56] The Greek terms in the text are μηχανικός and μεχανοποιόν; *ibid.*, Dedication, p. 4.
[57] *Ibid.*, Dedication, p. 5.
[58] *Ibid.*, Dedication pp. 6–7. This argument is reminiscent of one of Descartes' replies to Caterus' objection to the first cosmological proof. To show that the objective reality of an idea must have a cause that is at least as perfect, Descartes gives the example of the idea of an intricate machine, which can only be caused by the mind of someone who has seen such a machine, or who at least has the requisite knowledge of machines to imagine it. See *Meditations on First Philosophy* in Descartes 1985, II, pp. 75–76.

born the division between the seven liberal arts, and the vulgar arts of agriculture, hunting, military, arts, craftsmanship, surgery, woolworking and seafaring. As Monantheuil explains, even though the works of the latter are more necessary, useful, certain or excellent (for agriculture produces necessary goods, military victory is useful, craftsmanship is certain and medicine produces the excellence of good health), the liberal arts are the commanding arts that exist in the most successful men. They have the advantage of not requiring the powers of a body and they hold the reasons behind the arts that produce effects. The serving arts, by contrast, 'require youthful powers' and are 'learned and exercised by youth and custom'.[59] Monantheuil explains the nature of those reasons, which only the commanding arts possess, 'The form of all instruments consists in certain shapes, by which some tend to be suited towards a certain use. The reason why such shapes would be the most apt none of the serving [arts] investigates: they have enough if they hold the way of making and using'.[60] Like Piccolomini, he considers the Aristotelian treatise on mechanics to contain principles borrowed from geometry that 'explain the causes of the powers of instruments pertaining to the above mentioned mechanical arts'.[61] Specifically, the treatise explains how their shapes make the mechanical devices more suited to their uses and accomplishments.[62]

But Monantheuil goes further than Piccolomini, associating the mixed science of mechanics with philosophy itself. He connects the 'admirable art' of mechanics with the wonder that inspires philosophy, stating that we begin by resolving doubts about small things and then work our way up to astronomical phenomena and finally the generation of the universe.[63] He claims that the 'Philosopher not only admires rare and huge things, as the masses do, but also frequent and small things if they have hidden causes'.[64] That mechanics is on a par with physics rather than a subordinate science for Monantheuil is clear from his characterisation and classification of mechanics. Taking issue with Leonicus' suggestion that the unnamed general art mentioned in the Aristotelian text in connection with the perplexities generated by mechanical devices is architecture, Monantheuil writes:

> But since the resistance of nature is not only overcome in those things which are subjected to Architecture, [but] truly also in whatever other things [are] subjected to these arts, if there were an art by which [Aristotle] teaches what is in the universe, then it would be far more general than Architecture. And what hinders us from saying that this is Philosophy? Since Philosophy is the cognition of all arts and considers the causes of both divine and human things, properties and effects. And by this division into its parts, one small part among these will be Mechanics, which tends towards the explanation of violent and wondrous motions, on the other hand, Physics [which tends] towards the explanation of natural motions. And under the latter [are] Medicine, Agriculture and others, likewise under the former the art of weaving, Architecture,

[59] Monantheuil 1599, pp. 3–4.
[60] *Ibid.*, p. 4.
[61] *Ibid.*
[62] *Ibid.*
[63] *Ibid.*, p. 5.
[64] *Ibid.*, p. 6.

cobbling and all those which accomplish their work with artificial instruments and activities'.[65]

As we can see Monantheuil firmly maintains the Aristotelian distinction between violent and natural motions, thus retaining the division between mechanics and physics, *techne* and *scientia*. However, by placing mechanics and physics side by side, as the two primary subdivisions of philosophy, Monantheuil's commentary would suggest to those philosophers (like Descartes) who came to reject the violent/natural motion division that the principles of mechanics and physics are one and the same.

3.THE APPLICATION OF MECHANICAL EXPLANATIONS TO NATURE

As mentioned, the *Quaestiones Mechanicae* themselves make use of mechanical principles to answer some questions about natural phenomena. For example, question 15 asks, 'Why is it that the so-called pebbles found on beaches are round, though they are originally formed from stones and shells which are elongated in shape?' and question 35 asks, 'Why is it that an object which is carried around in whirling water is always eventually carried into the middle?'.[66] Interestingly, it is not primarily the balance, lever or wedge so prominent in the *Quaestiones Mechanicae* and other texts in mechanics that become the models for Descartes' mechanistic account of the unobservable parts of the universe. Rather his heavenly vortices are modelled after the eddies found in rivers,[67] and the jagged parts of celestial matter become rounded into globules like the pebbles that have landed on the beach after being tossed around in the sea. Having said that, there is one artificial device that figures prominently as a model in Descartes' physics, namely, the sling.[68] The motion of an object projected by a sling is addressed in question 12 of the *Quaestiones Mechanicae*, which reads, 'Why is it that a missile travels further from a sling than from the hand, although he who casts it has more control over the missile in his hand than when he holds the weight suspended?'.[69]

Before examining the mechanical explanations of these phenomena in some detail, let me first make some more general observations. In the case of questions 12 and 35, different possible causes are proposed and they appear to be of two distinct kinds. On the one hand, there are explanations based on considerations of force and resistance. On the other hand, there are explanations of the same phenomena in terms of shape and figure, which are reducible to the properties of the circle. One can find a similar appeal to powers to persist and resist in Descartes' physics, as well as an appeal to the shape and size of bodies. However, in Descartes' physics these two kinds of causal principles are connected, with the formal geometrical principle

---

[65] *Ibid.*, p. 9.

[66] Aristotle 1984, 2, pp. 1307–1308, 1317–1318.

[67] *Principia*, III, §30, *AT,* VIIIA, p. 96.

[68] See John Schuster's chapter in this volume for a more detailed treatment of the role these analogies play in Descartes' celestial mechanics.

[69] Aristotle 1984, 2, p. 1307.

being foundational and the physical principle being derivative.[70] Nevertheless, scholars have pointed to the apparent gap between Descartes' laws of nature in *Principia* Part II, from which one is supposed to be able to deduce with mathematical certainty all the particular motions in the universe, and the explanations Descartes ends up giving in Parts III and IV of the *Principia,* which appear instead to be based on hypotheses and analogies to everyday things like slings, eddies and screws.[71] The rhetorical advantage of appropriating mechanical explanations of such mundane things is clear because, as we have seen, mechanics had by this time obtained the reputation of combining the clarity and certainty of mathematical demonstration, which only dealt with abstractions, and physics, which explained the motions of existing material substances. I will argue that, in addition, Descartes applies substantive elements of existing mechanical explanations to new areas of inquiry, as seen by the fact that his explanations of the heavens take into account the criticisms later commentators raised in response to the pseudo-Aristotelian solutions of the above-mentioned questions.

3.1Pebbles and heavenly globules

Before delving into the different mechanical explanations of slings and eddies and their relevance to Descartes' physics, let me briefly highlight the more straightforward analogy between Descartes' explanation of how the parts of matter become rounded and the answers commentators give to question 15 of the *Quaestiones Mechanicae* regarding what causes pebbles to become round. Piccolomini offers the following explanation of the pseudo-Aristotelian account in his commentary of the *Quaestiones Mechanicae*:

> Therefore, it will be necessary that the *crocas* (that is, those remains of stones and oysters, which they discover in beaches) at last approach a round shape on account of the perpetual agitation which they suffer from the continual flow, and flowing back of the sea, since, because of said cause the extreme projecting parts are always first pounded and blunted. For the parts are shaken and agitated more quickly and more frequently the more they recede from the centre, [until] some [parts] having been worn down at last, cease to stand out.[72]

In his commentary Baldi objects that the cause has nothing to do with the distance of the parts from the centre, but that the true cause is that since acute angles are weaker than circular shapes, the most prominent parts of these objects are the weakest and will be easily broken down by opposing forces.[73]

Descartes relies on a general analogy between everyday phenomena, like the angles of stones being worn down in water, and the motions leading to the formation of the different elements of matter in the universe.

> Therefore, in order that we may begin to show the efficacy of the laws of nature in the proposed hypothesis, it must be considered that those particles into which we suppose

[70] I argue for this point in Hattab 2004.

[71] See for example Garber 1978 and Clarke 1979.

[72] Piccolomini 1565, q. 15, p. 38v.

[73] Baldi 1621, q. 15, p. 93.

the whole matter of this world was divided in the beginning, could not indeed have been spherical in the beginning, because many globules joined together do not fill a continuous space. But whatever shapes existed then, by the succession of time, they had to become round, since they had various circular motions. For since in the beginning they were moved by a force great enough to separate one from the others, that same force persevering was certainly also great enough to rub away all their angles when afterwards they ran into each other, for this did not require as much [force] as the former. And from this alone, namely, that the angles of a certain body were thus rubbed away, we easily understand that it at last became round, because the name angle in this place must extend to every thing which projects beyond the spherical shape. [74]

Here Descartes seems to apply the explanation Baldi gives of a particular observable phenomenon on a much larger scale to all matter as such, and extends it back in time to the unobservable origins of our world. In other words, the object of explanation has changed from Baldi to Descartes. Furthermore, while Baldi appeals to a principle that was easily observable in architecture, namely that acute angles are weaker than circles, Descartes appeals to the 'efficacy of the laws of nature' and extends the definition of 'angle' to 'every thing which projects beyond the spherical shape'. These are certainly important differences indicating that Descartes is engaged in a very different enterprise. However, it is striking that the difference lies entirely in the scope of the phenomena to be explained and the justification. While the latter have both been expanded and universalised by Descartes, the basic form of the explanation is consistent with Baldi's. Furthermore, the potential for universalising this kind of explanation to the universe at large is already suggested by Monantheuil's claim that the world is a machine constructed and maintained by the divine machine maker and mechanic.

3.2 Slings and vortices

Turning now to the example of the sling, the question posed is: Why does a missile projected by a sling travel further, even though the weight is gripped less firmly than when it is held in the hand, and the slinger has to move two weights as opposed to one, namely, the sling and the object? As Leonico Tomeo comments, two causes are adduced: the first is based on a principle of motion that belongs to physics. The principle is that things already in motion are more easily moved than those at rest, and so the object being rotated in the sling, is more easily thrown from its state of motion than the object that is at rest in the hand, prior to the throwing action. [75] Piccolomini puts this principle in terms of forces, stating that less force is required to add a new motion to something already in motion than to initiate motion in something of the same weight.

Now this is also evident by sense because by a quite moderate force, with respect to a certain weight, while it is in motion, a new motion is also added, which nevertheless would have required a wholly greater force in the beginning of the mutation, since the

[74] *Principia*, II, §48, *AT,* VIIIA, pp. 103–104.
[75] Tomeo 1525, q. 12, p. xxxviiv.

motion of the same weight will be more easily continued by a certain force than the
beginning is initiated.[76]

Maurolyco explains that impetus is acquired from the rotation of the sling in the
same way that someone jumping acquires impetus and is able to jump further by
running before taking the jump.[77] Benedetti also writes, 'And there is no doubt that a
greater impetus of motion can be impressed on the body by the sling, since from
repeated revolutions an ever greater impetus comes to the body in question'.[78]

   While Descartes does not share the medieval notion of impetus that appears to
be at play here, he does base his laws of impact on the fundamental opposition
between motion and rest. According to Descartes, motion is not opposed by another
motion of equal speed. Motion is most strongly opposed by rest, and to a lesser
degree by slower motions insofar as they share in the nature of rest.[79] So while
Descartes does not share the view that impetus must be added for a body to maintain
its motion, he does share the view that motion and rest are in opposition and thus it
is easier to displace a moving body than a body at rest.[80]

   The second cause of the fact that a missile projected from a sling travels
further is based on geometrical principles. The whole action forms a circle, with the
hand as the centre, the sling as the radius, and the missile's circular path becoming
the circumference. One of the basic principles of the circle, introduced at the
beginning of the *Quaestiones Mechanicae*, is that since a point on the radius which
is further from the centre traces a greater circumference in the same time, it must
move at a greater speed than a point on the radius closer to the centre, which
completes a smaller circumference in an equal amount of time.[81] If we apply this
principle to the sling, it clearly extends further from the centre of the circle than the
hand does, and thus its greater distance from the centre explains why the stone will
be carried more quickly in the sling and, as a result, thrown further. Whereas
Leonico, Piccolomini and Maurolyco repeat the pseudo-Aristotelian explanation,
both Benedetti and Baldi question it. Benedetti's alternative explanation is
instructive with respect to Descartes' analysis of the sling. Benedetti points out that
the hand is not the fixed centre but rather also moves in a circle. The circular motion
of the hand causes the projectile to be carried in a circle [see **Fig. 18**].

---

[76] Piccolomini 1565, q. 12, p. 37r.

[77] Maurolyco 1613, q. 12, p. 16.

[78] Benedetti in *Book of Various Mathematical and Physical Ideas* in Drake and Drabkin 1969, p. 189.

[79] *Principia*, II, §44, *AT*, VIIIA, p. 67; Descartes 1983, p. 63.

[80] There is a vast and growing body of literature on Descartes' problematic metaphysical foundation of
    the forces of motion and rest. The classic articles are Gueroult 1980 and Gabbey 1980. For an
    influential alternative view, as well as a discussion of Descartes' theory in relation to impetus theory,
    see Garber 1992, chaps 7–9. More recently, Dennis Des Chene has given a detailed treatment of the
    Cartesian causes of motion and rest in the context of scholastic theories in Des Chene 1996, chap. 8.
    See also Hattab 2004.

[81] Aristotle 1984, 2, q. 1, p. 1300.

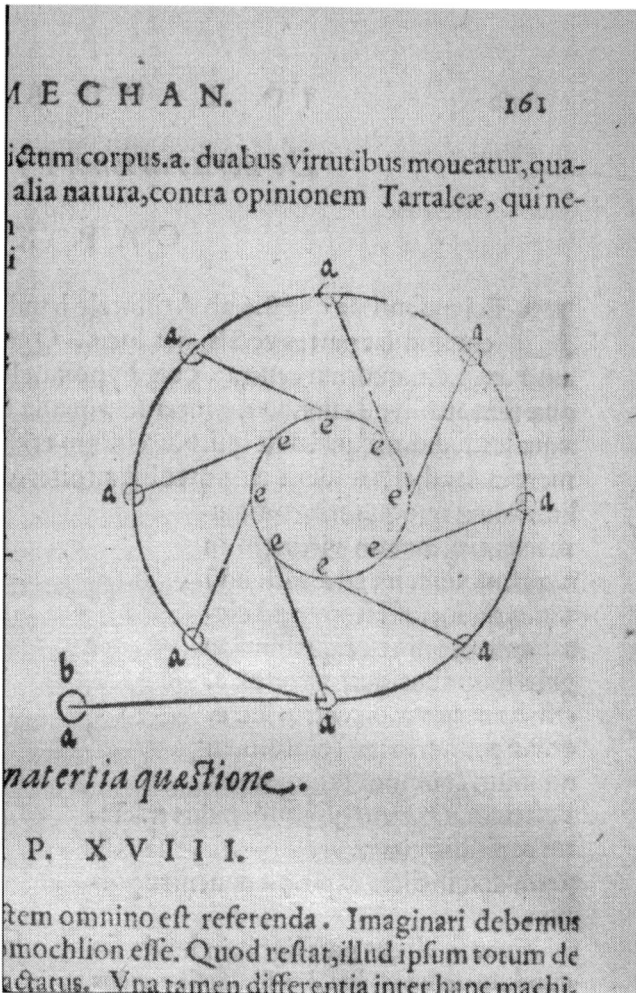

Figure 18. Giovanni Battista Benedetti, Diversarum speculationem mathematicarum & physicarum liber, *1599, p. 161.*

Furthermore, once the body receives a small impetus its natural tendency will be to move in a straight line at a tangent to the circle. Benedetti supposes that the impetus will gradually decrease and thus the downward tendency of the body, caused by its weight, will mix with the impressed force. The combination of the impressed violent motion that is rectilinear at a tangent to the circle, and the natural downward motion causes the motion of the projectile to become curved.[82] Baldi also objects that the hand is not the centre of the circle, but claims the elbow is the centre. He also notes that the line from the earth to the top of the sling is longer than the line from the earth to the hand—presumably this shows that the projectile traces a greater circumference in the sling and thus moves faster.[83]

To illustrate the second law of nature, namely, that the natural tendency of any body is towards rectilinear motion, Descartes uses the example of a stone being rotated in a sling. Even though the stone will move along a circular path while in the sling, it is inclined to move in a straight line along a tangent to the circle, in accordance with the second law of motion. This is confirmed by the fact that if the stone leaves the sling, it will not continue in its circular motion, but will move along a straight line. Descartes concludes from this that 'those things which are moved circularly always tend to recede from the center of the circle which they describe'.[84] Descartes returns to the example of the stone in the sling in Part III of the *Principia*, article 57. He now identifies different tendencies to motion in relation to the different causes of the motion of the stone. He claims that if all the causes are taken together, i.e., the force of the stone's movement as well as the impeding motion of the sling, then the stone tends towards a circular motion. But if we consider only the force of motion of the stone in accordance with the second law of motion, then it tends to move in a straight line at a tangent to the circle. Lastly, if we only consider the part of the stone's force of motion that is hindered by the sling, then the stone tends to recede from the centre of the circle downwards along a straight line.[85] [86]

This tendency of bodies to recede from the centre of a circle plays a central role in Descartes' explanations of celestial phenomena. The globules of the second element of matter strive to recede from the centre of the vortex in which they rotate and are restrained by the globules beyond them as the stone was by the sling.[87] Light itself is defined by Descartes as this tendency of globules to recede from the centre of the vortex.[88]

---

[82] Drake and Drabkin 1969, p. 189.

[83] Baldi 1621, p. 89.

[84] *Principia*, II, §39, *AT,* VIIIA, p. 63; Descartes 1983, p. 60.

[85] *Principia*, III, §57, *AT,* VIIIA, pp. 108–109; Descartes 1983, pp. 112–113.

[86] Note that the tendencies described here are not separate powers inherent in the stone. That would imply that the stone possessed conflicting powers pulling it in one direction and another. Rather the different tendencies are a product of opposing states of motion and are understood in terms of counterfactuals, e.g., this is the way the stone would move absent the opposing motion of the sling. The important point is that a tendency is always understood in terms of an impeding cause. In other words, all tendencies arise from an analysis of the motion dictated by the laws of nature, in relation to opposing motions. Tendencies designate the motions that would occur in the absence of the opposing motions that are always present in a plenum. For an analysis of the collision rules in light of Descartes' more general considerations regarding motion and its determinations see McLaughlin 2000. For an account of the origins of Descartes' view of the tendencies to motion, see Gaukroger and Schuster 2002, especially pp. 568–570.

[87] *Principia*, III, §60, *AT,* VIIIA, p. 112; Descartes 1983, p. 115.

[88] Descartes also refers to this tendency, which constitutes light, as the 'first preparation for motion', *Principia*, III, §63, *AT,* VIIIA, p. 115; Descartes 1983, p. 117.

Descartes does not share Benedetti's conception of a dissipating impetus that is impressed on the body, however, he does break the circular motion of the stone in the sling down into the same counteracting forces that Benedetti identified. What is novel is that Descartes takes this account of the forces at work in the motion of a sling and derives a general principle from it, namely, that any body being moved circularly tends to recede from the centre. He then uses this principle to explain the motion of second element particles in a vortex, which in turn causes light. Once again, the primary difference between Descartes' explanation and the ones developed by later commentators of question 12 in the *Quaestiones Mechanicae* is that Descartes universalises the principle at work and applies it to nature as a whole. Whereas Descartes left the analogy between the matter of the universe and bits of matter being tossed around in the sea implicit, here he makes an explicit connection between his analysis of the different tendencies to move of a stone carried in a sling (an artificial device) and the general tendency of natural bodies to recede from the centre of a celestial vortex.

3.3 Motion in a vortex

Slings and vortices taken on their own have no apparent logical connection to one another but in the *Quaestiones Mechanicae* they are connected by the marvellous properties of the circle. Both phenomena involve objects carried along in a circular motion and in each case the *Quaestiones Mechanicae* give both physical explanations in terms of forces, and explanations based on the mathematical properties of the circle. As in the case of the sling, the analogy between the eddies of rivers and the heavens is made explicit by Descartes:

> And so with every doubt about the motion of the earth removed, we suppose that the whole matter of the heaven, in which the Planets are agitated, gyrates unceasingly in the manner of certain vortices, in the center of which is the Sun, and that those of its parts which are closer to the Sun are moved faster than those further away, and that all the Planets (among their number is Earth) always hover among the very same parts of celestial matter. From this alone, without any machines, all their phenomena are very easily understood. For as in those places of the river in which water whirling around itself makes a vortex, if various straws lie on that water, we will see that they are carried away with it, that some even whirl around their own circle, and that those closer to the center of the vortex complete the whole circle faster.[89]

Descartes does not appear to be merely employing a poetic metaphor here, for once again there are interesting parallels between explanations of what happens to an object caught in a water vortex found in the mechanical texts and Descartes' mechanistic explanations of the motions of celestial bodies being carried in their vortices.

Renaissance commentators identify three answers to question 35 of the *Quaestiones Mechanicae*, regarding what causes objects to end up in the centre of a water vortex. In the pseudo-Aristotelian text and early commentaries the vortex is described as a series of concentric circles which move progressively faster the

---

[89] *Principia*, III, §30, *AT,* VIIIA, p. 92; Descartes 1983, p. 96.

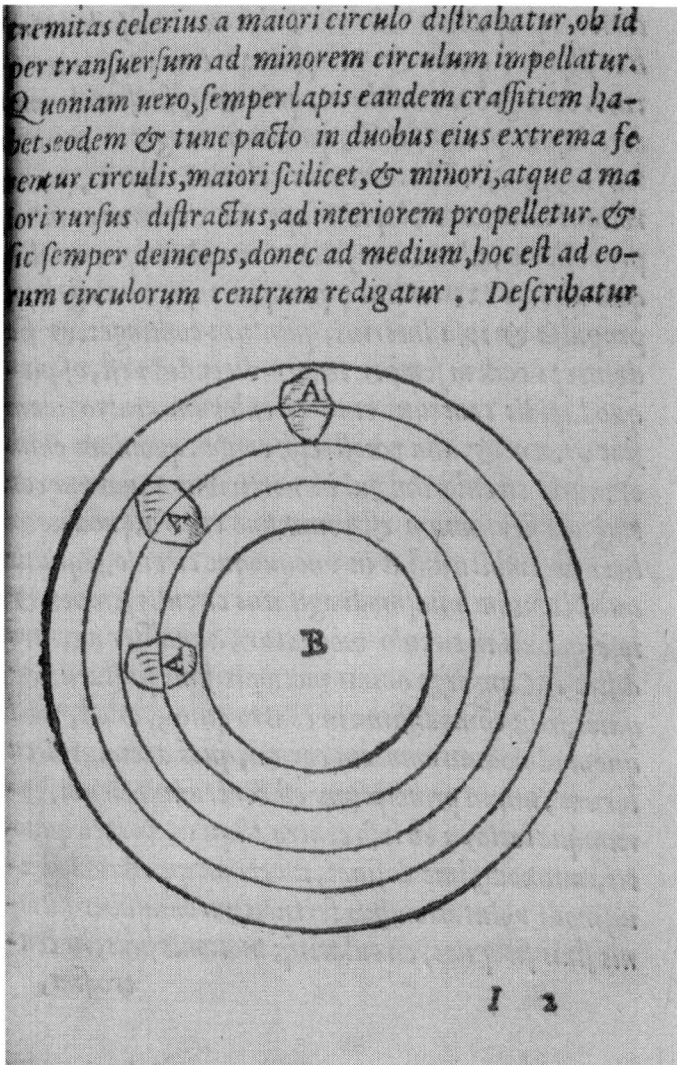

*tremitas celerius a maiori circulo diftrahatur,ob id
per tranfuerfum ad minorem circulum impellatur.
Quoniam uero,femper lapis eandem craffitiem habet,eodem & tunc pacto in duobus eius extrema fe
rentur circulis,maiori fcilicet,& minori,atque a ma
iori rurfus diftractus,ad interiorem propelletur. &
fic femper deinceps,donec ad medium,hoc eft ad eo-
rum circulorum centrum redigatur. Defcribatur*

Figure 19. Alessandro Piccolomini, In mechanicas quaestiones Aristotelis,
Paraphrasis paulo quidem plenior, *1565, p. 66.*

further they are from the common centre. The first cause is based on the properties
of the circle. The object is propelled towards the centre of the vortex because it is
situated between two or more circles, its upper extremity being carried along by the
outermost circle, and its lower extremity by the circle closer to the centre. Since the

outermost circle moves more rapidly, the upper extremity of the object is moved forward more quickly and the whole object is carried down into the next circle. This process repeats itself circle by circle until it reaches the centre of the vortex. [See **Fig. 19**] Piccolomini declares this cause the most certain.[90]

However, Maurolyco, Benedetti and Baldi are not convinced. Benedetti says the true explanation for why objects descend to the centre is because the centres of whirlpools are always more depressed since they are concave and have almost a conical shape.[91] Maurolyco and Baldi add that the motions of vortices are not made up of concentric circles but rather take a spiral form.[92] This, and the fact that vortices are concave, causes objects to move towards the centre of the vortex.

Descartes characterises celestial vortices neither as spirals nor as concave but models them after the older view, depicting them as concentric circles. However, he admits they are not perfect circles.

> Finally, we must not think that all the centers of the Planets are always situated exactly on the same plane, or that the circles they describe are absolutely perfect; let us instead judge that, as we see occurring in all other natural things, they are only approximately so, and also that they are continuously changed by the passing of the ages.[93]

Descartes agrees that objects in the outermost circle move the fastest and that the speed of motion decreases the closer the circular path is to the centre, but he gives a different reason for this, namely, the motion of objects on the outermost circumference most closely approximates rectilinear motion. He then adds the qualification that as one gets closer to the centre, the agitation of the star found there increases the motion of the matter nearest to it, and so the innermost circles of a celestial vortex move faster than the outer ones with the middle ones moving the slowest.[94] Descartes shares the views of Maurolyco, Benedetti and Baldi insofar as he does not think that the swifter motion of outer circles is enough to account for a body ending up in the centre of a vortex. So let us turn to the other two causes, which involve a consideration of forces.

The second cause of a body's descent to the centre of a vortex is that since the centre is equidistant from each point of any given circumference and rest is most complete in the centre, the object will be carried to its natural place of rest in the common centre. According to Piccolomini:

> For since of all the circles that happen to be made in the vortices of water, there is one common centre, which is equally distant from the edge in each given circle, it follows that the stone set in motion in this manner by the circles is itself always moved to whatever place you will in the circle [all of which are] equally distant from the centre.

---

[90] Piccolomini 1565, chap. XL, q. 35, p. 65v.

[91] Drake and Drabkin 1969, p. 195.

[92] Maurolyco 1613, q. 35, p. 26, Baldi, q. 35, pp. 186–187.

[93] *Principia*, III, § 34, translated in Descartes 1983, p. 98.

[94] *Le Monde, AT*, XI, pp. 49–51, p. 57.

> Since, therefore, the end of every motion is a position and rest, and rest is most
> complete in the centre, it is appropriate that in the same way that we said the one
> principle of all these circles is their centre, and their bringing forward ceases most
> completely in that same centre just as in a place of rest, thus something is also carried in
> this manner by rolling motions [*volutationibus*], since the end of whatever source of
> motion is rest, and in circular motions an end occurs in the centre—after all no injury
> would attain to the middle.[95]

While Leonico does not question this explanation, Piccolomini says it is perhaps
more probable than true. Baldi rejects the second cause entirely, claiming that the
fact that all circles are carried equally around the same centre does not suffice to
explain why the object is driven towards the middle, but rather some other outside
force propelling it towards the centre is required.[96] On this point, Descartes'
reasoning is in line with Baldi's as he writes in *Le Monde* that unless something else
prevents them, larger bodies, like planets, will be carried along circularly by the
surrounding matter, just as a boat follows the course of the river.

> For if at first they moved themselves more quickly than this matter, since they could not
> have avoided pushing it when encountering it in their path, they had to transfer a part of
> their agitation to it in a short time; and if, to the contrary, they had in themselves no
> inclination to move, nevertheless, since they were surrounded on all sides by this matter
> of the heavens, they necessarily had to follow its course, just as everyday we see that
> boats and other bodies that float on water (the biggest and the most massive just as
> much as those that are less so) follow the course of the water they are in when there is
> nothing else to prevent them from doing so.[97]

The third cause is presented by Leonico as an argument by dilemma. Depending on
its magnitude and heaviness, the object is either overcome by the speed of the outer
circle in which it is initially thrown, or it overcomes it. If it resists and overcomes
the motion of the water, then it is left behind and carried more slowly than the
outermost circle, causing it to descend into the smaller slower circles that are nearer
to the centre. If it is overcome by the motion of the water, it will be carried around in
the outermost circle for a while. This will cause its resistance to grow and it will
eventually diminish the forces of the outermost circle with the same result that it will
eventually descend via the smaller, slower circles to the centre.[98] Benedetti attributes
the motion towards the centre of the vortex to the weight of the object that causes it
to move downwards.[99] Finally, Baldi is the most critical of this third explanation. He
points out that a body that is able to resist motion to a certain degree will partly
follow the impetus of the water, but it will also be slowed down by its own nature.
Therefore, it will finish the rotation more slowly than the water, however, it will not
on that account be carried towards the centre. For that another cause is required.[100]

[95] Piccolomini 1565, chap. XL, q. 35, p. 66v.
[96] Baldi 1621, q. 35, p. 189.
[97] *Le Monde*, chap. 9, *AT*, XI, pp. 57–58.
[98] Tomeo 1525, q. 35, p. LIIv.
[99] Drake and Drabkin 1969, p. 195.
[100] Baldi 1621, q. 35, p. 189.

Descartes does not subscribe to the Aristotelian view that all things strive towards a natural place of rest and thus he rejects the idea that an object in a vortex naturally descends to the centre due to its heaviness. Instead he holds that the natural tendency of bodies being moved circularly is to recede from the centre. However, despite this difference he appears to accept Baldi's criticisms of the pseudo-Aristotelian account. For example, like Baldi Descartes thinks, absent other causes, the celestial bodies will be carried along circularly in their vortices. He agrees that the larger the body, the more it will resist being moved and the more slowly it will be carried along.

> After this, we must note that, just as we observe that the boats which follow the course of a river never move as fast as the water that carries them along, nor do even the largest among them [move] as fast as the smallest, so too, even though the planets follow the course of the celestial matter without resistance, and move with the same rolling motion as it, this does not mean that they ever move as fast; and even the inequality of their movement must bear some relation to that which is found between the largeness of their bulk [*masse*] and the smallness of the parts of the heaven that surround them.[101]

To explain why planets remain in their orbits Descartes claims that their tendency to recede from the centre is opposed by the equally strong circular motion of the surrounding globules, pushing planets inwards in the way that the sling impedes the rectilinear tendency of the stone. The motions of celestial bodies depend on their solidity, which is a function of their size and surface area. For example, a star that has very little solidity will descend a great deal towards the centre of a vortex because the surrounding globules of matter will overcome its natural tendency to recede from the centre. However, if the star has greater solidity, it will have greater force both to resist the motion of the globules and maintain its own motion away from the centre. If it resists the motion of the surrounding globules to such a degree that it overcomes their force, then it will break out of the circular motion of the vortex in which it finds itself and become a comet.[102] Descartes thus uses similar considerations about forces to those found in the later, more critical discussions of question 35 to account for the motions of various celestial bodies in the heavenly vortices.

While the isomorphism between some of Descartes' mechanistic explanations of the heavens and the explanations developed by later commentators of the *Quaestiones Mechanicae* does not show conclusively that Renaissance mechanics was the sole or even primary inspiration for all the particular explanations Descartes used in his celestial mechanics, it is rather striking that the examples of the sling and river vortices, which were examined in some detail within the Aristotelian mechanical tradition, became the means by which Descartes justified his general principles of motion and his basic approach to explaining celestial phenomena. Moreover, Descartes' appeal to the mechanical principles that account for pebbles

---

[101] *Le Monde*, chap. 10, *AT*, XI, pp. 68–69.

[102] *Principia*, III, §120, *AT*, VIIIA, pp. 169–170; Descartes 1983, p. 151. Again, Schuster has a more detailed treatment of this, as well as the role of the river analogy.

and the motions of objects in slings and vortices does not appear to be a mere rhetorical or illustrative device for he proclaims, 'There really are no reasonings [*rationes*] in mechanics which do not also pertain to physics, of which it is a part or species: nor is it less natural for a clock composed of these or those wheels to tell the time, than it is for a tree originating from this or that seed to produce a certain kind of fruit'.[103] This is confirmed by the above comparison between particular mechanical and mechanistic explanations, which suggests that Descartes' innovation lay in applying the basic forms of explanation already found in the Aristotelian tradition of mechanics, more broadly to the universe at large. In doing so Descartes draws out the implications of Henri de Monantheuil's conception of the world as a machine that God constructed by means of mechanics. In fact, given the reclassification of mechanics and the gradual erasing of the line between physical explanations of natural phenomena and geometrical explanations of machines found in Renaissance commentaries on the *Quaestiones Mechanicae*, Descartes' absorption of mechanics into physics and his extension of mechanical forms of explanation to the unobservable causes of natural phenomena appears less like a rupture from the Aristotelian tradition in mechanics and more like an offshoot that ultimately supplanted the parent tree.[104]

---

[103] *Principia*, IV, §203, *AT*, VIIIA, p. 326.

[104] I was inspired to look at the *Quaestiones Mechanicae* tradition in connection with Descartes during my participation in an NEH Summer Seminar directed by Daniel Garber and Roger Ariew in the summer of 2000. I began my research on Aristotelian mechanics commentaries and textbooks in the summer of 2001 during my participation in an NEH Summer Institute on 'Experience and Experiment in Early Modern Europe' directed by Pamela Long and Pamela Smith. I thank the directors of both the seminar and institute for motivating me to look into this particular topic, and various guest lecturers and participants at the institute for discussions on early mechanics. Much of the research was made possible in the summer of 2001 by my vicinity to the extensive collection in early modern mechanics and physics of the Smithsonian Institution Libraries, The Dibner Library of the History of Science and Technology in Washington DC. Final revisions to this paper were made during a residential fellowship at the Dibner Library in spring 2004. I am most grateful for the generous help and support I received from the director of the library, Ron Brashear, and his staff, in locating and obtaining copies of particular rare texts in their collection. I also thank the Folger Library in Washington DC and its staff for the use of one or two rare texts found in their collection. Parts of this paper were presented in less developed forms at the *Patristic, Medieval and Renaissance Studies* annual meeting at Villanova, PA in September 2001, the International Conference on *The Origins of Modernity: Early Modern Thought 1543–1789* in Sydney, Australia in July of 2002 and the annual meeting of the History of Science Society in November 2003. I thank the organisers and participants of each conference for their interest and feedback. Finally, I thank Steve Walton and the editors of this volume for their comments on this paper.

REFERENCES

Aristotle (1984) *The Complete Works of Aristotle: The Revised Oxford Translation*, ed. J. Barnes, 2 vols, Princeton: Princeton University Press.
Baldi, B. (1621) *Mechanica Aristotelis Problemata Exercitationes*, Moguntiae: Ioannis Albini.
Benedetti, G. B. (1585) *Diversarum speculationum... liber*, Turin.
Boyle, R. (1999–2000) *The Works of Robert Boyle,* eds M. Hunter and E. B. Davis, 14 vols, London: Pickering & Chatto.
Cardano, G. (1570) *Opus Novum*, Basileae.
Clarke, D. (1979) 'Physics and metaphysics in Descartes' *Principles*', *Studies in History and Philosophy of Science,* 10, pp. 89–112.
Dear, P. (1988) *Mersenne and the Learning of the Schools*, Ithaca: Cornell University Press.
de Gandt, F. (1986) 'Les Mécaniques attribuées à Aristote et le renouveau de la science des machines au XVIe Siècle', *Les Etudes Philosophiques*, 3, pp. 391–405.
Descartes, R. (1983) *Principles of Philosophy*, trans. V. R. Miller and R. P. Miller, Dordrecht: D. Reidel.
— (1985) *The Philosophical Writings of Descartes,* trans. J. Cottingham, R. Stoothoff, and D. Murdoch, 2 vols, Cambridge: Cambridge University Press.
— (1996) *Oeuvres de Descartes*, eds C. Adam and P. Tannery, 11 vols, Paris: J. Vrin.
Des Chene, D. (1996) *Physiologia: Natural Philosophy in Late Aristotelian and Cartesian Thought*, Ithaca: Cornell University Press.
Drake, S. and Drabkin, I. E. eds and trans (1969) *Mechanics in Sixteenth-Century Italy: Selections from Tartaglia, Benedetti, Guido Ubaldo, and Galileo*, Madison: University of Wisconsin Press.
Field, J. V. and James, A. J. L. eds (1993) *Renaissance and Revolution: Humanists, Scholars, Craftsmen and Natural Philosophers in Early Modern Europe*, Cambridge: Cambridge University Press.
Gabbey, A. (1980) 'Force and inertia in the seventeenth century: Descartes and Newton' in *Descartes: Philosophy, Mathematics and Physics*, ed. S. W. Gaukroger, Sussex: Harvester Press, pp. 230–320.
— (1993) 'Between ars and philosophia naturalis: reflections on the historiography of early modern mechanics' in *Renaissance and Revolution: Humanists, Scholars, Craftsmen and Natural Philosophers in Early Modern Europe*, eds J. V. Field and A. J. L. James, Cambridge: Cambridge University Press, pp. 133–145.
— (1998) 'New doctrines of motion' in *The Cambridge History of Seventeenth-Century Philosophy*, eds D. Garber and M. Ayers, Cambridge: Cambridge University Press, I, pp. 649–679.
Garber, D. (1978) 'Science and certainty in Descartes' in *Descartes: Critical and Interpretive Essays*, ed. M. Hooker, Baltimore: Johns Hopkins University Press, pp. 115–151.
— (1992) *Descartes' Metaphysical Physics*, Chicago: University of Chicago Press.
— (2002) 'Descartes, mechanics and the mechanical philosophy', *Midwest Studies in Philosophy*, 25, pp. 183–204.
Garber, D. and Ayers, M. eds (1998) *The Cambridge History of Seventeenth-Century Philosophy*, 2 vols, Cambridge: Cambridge University Press.
Garber, D. and Nadler, S. (2004) *Oxford Studies in Early Modern Philosophy*, vol. 1, Oxford: Clarendon Press.
Gaukroger, S. W. ed. (1980) *Descartes: Philosophy, Mathematics and Physics*, Sussex: Harvester Press.
Gaukroger, S. W. and Schuster, J. A. (2002) 'The hydrostatic paradox and the origins of Cartesian dynamics', *Studies in History and Philosophy of Science*, 33, pp. 535–572.
Gaukroger, S. W., Schuster, J. A. and Sutton, J. eds (2000) *Descartes' Natural Philosophy*, London: Routledge.
Gillispie, C. C. ed. (1970–1990) *The Dictionary of Scientific Biography*, 18 vols, New York: Charles Scribner's Sons.
Gliozzi, M. (1976) 'Cardano' in *The Dictionary of Scientific Biography*, ed. C. C. Gillispie, New York: Charles Scribner's Sons, III, pp. 64–67.
Gueroult, M. (1980) 'The metaphysics and physics of force in Descartes' in *Descartes: Philosophy, Mathematics and Physics*, ed. S. W. Gaukroger, Sussex: Harvester Press, pp. 196–229.
Guidobaldo del Monte (1577) *Mechanicorum Liber*, Pisauri: Hieronymum Concordiam.
Hattab, H. (2004) 'Conflicting causalities: The Jesuits, their opponents, and Descartes on the causality of the efficient cause' in *Oxford Studies in Early Modern Philosophy*, vol. 1, eds D. Garber and S. Nadler, Oxford: Clarendon Press, pp. 1–22.

Henninger-Voss, M. (2000) 'Working machines and noble mechanics: Guidobaldo del Monte and the translation of knowledge', *Isis*, 91, pp. 233–259.

Hooker, M. ed. (1978) *Descartes: Critical and Interpretive Essays*, Baltimore: Johns Hopkins University Press.

Hugh of St Victor (1961) *The Didascalicon of Hugh of St Victor: A Medieval Guide to the Arts*, trans. J. Taylor, New York: Columbia University Press.

Laird, W. R. (1986) 'The scope of Renaissance mechanics', *Osiris,* 2, pp. 43–68.

Liddell, H. G. and Scott, R. (1968) *A Greek-English Lexicon*, Oxford: Clarendon Press.

Lüthy, C., Murdoch, J. E. and Newman, W. R. eds (2001) *Late Medieval and Early Modern Corpuscular Theories*, Leiden: Brill.

McLaughlin, P. (2000) 'Force, determination and impact' in *Descartes' Natural Philosophy*, eds S. W. Gaukroger, J. A. Schuster and J. Sutton, London: Routledge, pp. 81–111.

Maccagni, C. ed. (1967) *Atti del Primo Convegni Internazionale di Ricognizione Delle Fonti Per la Storia Della Scienza Italiana: I Sicoli XIV–XVI*, Firenze: G. Barbèra.

Masotti, A. (1974) 'Maurolyco' in *The Dictionary of Scientific Biography*, ed. C. C. Gillispie, New York: Charles Scribner's Sons, IX, pp. 190–194.

Maurolyco, F. (1575) *Opuscula mathematica*, Venetijs.

— (1613) *Problemata Mechanica cum appendice, & ad Magnetem, & ad Pixidem Nauticam pertinentia*, Messina.

Mersenne, M. (1623) *Quaestiones in Genesim*, Lutetiae Parisorum: Sebastiani Cramoisy.

— (1634) *Questions Theologiques*, Paris: Henry Guenon.

Monantheuil, H. (1599) *Aristotelis Mechanica Graeca, emendate, Latina facta, & Commentariis illustrate*, Paris.

Nemore, I. de (1565) *De ratione ponderis*, Venice: Curtium Troianum.

Ovitt, G. Jr. (1983) 'The Status of the mechanical arts in medieval classifications of leaning', *Viator*, 14, pp. 89–105.

Palmerino, C. R. (2001) 'Galileo's and Gassendi's solutions to the *Rota Aristotelis* paradox: a bridge between matter and motion theories' in *Late Medieval and Early Modern Corpuscular Theories*, eds C. Lüthy, J. E. Murdoch and W. R. Newman, Leiden: Brill, pp. 381–422.

Piccolomini, A. (1565) *In Mechanicas quaestiones Aristotelis, Paraphrasis paulo quidem plenior*, Venetijs: 1st edn 1547.

Rose, P. L. (1975) *The Italian Renaissance of Mathematics: studies on humanists and mathematicians from Petrarch to Galileo*, Genève: Librarie Droz.

— (1974) 'Monte, Guidobaldo' in *Dictionary of Scientific Biography*, ed. C. C. Gillispie, New York: Charles Scribner's Sons, IX, pp. 487–489.

Rose, P. L. and Drake, S. (1971) 'The Pseudo-Aristotelian *Questions in Mechanics* in Renaissance culture', *Studies in the Renaissance*, 18, pp. 65–104.

Schmitt, C., Skinner, Q., Kessler, E. and Kraye, J. eds (1988) *The Cambridge History of Renaissance Philosophy*, Cambridge: Cambridge University Press.

Schramm, M. (1967) 'The Mechanical Problems of the "Corpus Aristotelicum", the "Elementa Iordani Super Demonstrationem Ponderum" and the mechanics of the sixteenth century' in *Atti del Primo Convegni Internazionale di Ricognizione Delle Fonti Per la Storia Della Scienza Italiana: I Sicoli XIV–XVI*, ed. Carlo Maccagni, Firenze: G. Barbèra, pp. 151–163.

Thorndike, L. (1941) *A History of Experimental Science*, 8 vols, New York and London: Columbia University Press.

Tomeo, N. L. (1525) *Opuscula Nuper in Lucem Aedita*, Venice: Bernardinus Vitalis.

STEPHEN GAUKROGER

THE AUTONOMY OF NATURAL PHILOSOPHY

From Truth to Impartiality

In the sixteenth and seventeenth centuries we witness a fundamental rethinking of natural philosophy. In place of attempts to reconcile natural philosophy with pre-given truths, we begin to find attempts to establish the autonomy of natural philosophy. My aim in this chapter is to investigate the factors that underlie this shift, factors which, I argue, turn on the basic questions of just what philosophy is and what aims and skills the philosopher brings to bear on the project.

1.THE IDEA OF A CHRISTIANISED ARISTOTELIAN NATURAL PHILOSOPHY

One of the earliest features of the revival of an intellectual culture in Europe was the attempt to establish a systematic theology on a philosophical basis. A continuous tradition of such enquiry was initiated by Anselm in the eleventh century. Anselm believed that a philosophically-based systematic theology would not only be a bulwark against heresy but also a means of convincing Muslims and others of the truth of the basic tenets of Christianity, above all, the distinctive Christian doctrines of the Trinity and the Incarnation. Yet this project was problematic right from the start, with Berengar of Tours questioning the philosophical credentials of the doctrine of transubstantiation in the 1050s and Roscelin questioning those of the Trinity in the 1090s. By the twelfth century it had become clear that what was at issue philosophically was a deep dichotomy between a view of knowledge as something arrived at via abstraction, and a view of knowledge as something arrived at via descent from abstract universals. The latter was the traditional view, embodied in the Christianised metaphysics developed by Augustine, who saw Christianity as the final answer to what earlier philosophers were striving for. Reading the project of ancient philosophy in Neoplatonist terms, and treating Neoplatonism as the culmination of pagan philosophical thought, he argued that the ancient philosophers mistakenly believed they could reach God by purely intellectual means, whereas in fact he can only be reached through the sacraments, which were instituted with the Incarnation.

Augustine's Christianised Neoplatonism synthesised theological and philosophical issues into a harmonious whole, but it proved inadequate to the

P. R. Anstey and J. A. Schuster (eds.), The Science of Nature in the Seventeenth Century, 131-163.

metaphysical complexities of the doctrines of transubstantiation, the Trinity, and the Incarnation, doctrines which philosophically-sophisticated Islamic and Jewish philosophers had balked at. As was clear from Abelard onwards, making philosophical sense of a number of fundamental doctrines of Christian theology required the development of a conception of abstractive knowledge, and thirteenth-century scholastic philosophers realised that Aristotelianism was the only comprehensive philosophical system able to provide the resources for this. But Aristotle's doctrine that all knowledge started with sense perception gave priority to natural philosophy over all other disciplines, and the adoption of Aristotelianism meant that the point of entry into a philosophically-grounded theology was natural philosophy, tying the fate of Christian theology to natural philosophy in an unprecedented way. As a result, natural philosophy became the discipline on which theological and philosophical arguments turned, from the thirteenth century onwards.

The lack of fit between Aristotelian natural philosophy and Christian theology was evident to many theologians in the thirteenth century, and was strongly resisted, not least in the series of condemnations starting in 1210 and culminating in the 1277 Condemnation of 40 theological and 179 philosophical propositions.[1] But the attempts of critics to return to what by this time was a philosophically impoverished Augustinian model, failed to match the advances, made by Aquinas and others, in thinking through philosophical and theological questions driven by Christianised Aristotelianism. With the failure to reconcile the Western and Eastern Churches in the fifteenth century and the establishment of Thomism as the official philosophy of the Western Church, the fate of the Neoplatonism originally favoured by Augustine and represented in the East in a tradition stretching from Michael Psellus to Plethon, was effectively sealed, despite the efforts of della Mirandola, Ficino, and Patrizzi to re-establish it. But the problems of the irreconcilability of Aristotelian natural philosophy and Christian theology came to a head again in the sixteenth and seventeenth centuries, on the questions of the immortality of the soul and on the physical standing of the heliocentric model. I want to focus on the first of these, and to show how it exemplifies a deep and intractable problem at the heart of Christianised Aristotelianism, a problem whose seventeenth-century resolution, I shall argue, involved the prising open of questions of truth and justification.

In his *Theologia Platonica de immortalitate animae* (1469–74),[2] Ficino welded together Christian, Hermetic, and Neoplatonic sources into a syncretic treatise on philosophical theology which offered the first developed alternative to the Aristotelian system. The seemingly marvellous anticipations of Christianity evident in the Hermetic corpus, all the more remarkable in the light of its great antiquity, and the marvellous and natural coherence between Platonism and both the Hermetic doctrines and Christian revelation, seemed to Ficino, as they had seemed to earlier thinkers in the Eastern Church, to suggest the key to the understanding of the link

---

[1] The texts of the various condemnations are given in vol. 1 of Denifle and Châtelain 1889–1897. See Hisette 1977 as well as his 1980.

[2] A new edition of the Latin text of the *Theologia* with full English translation is gradually appearing: Ficino 2001–.

between God and his creation. The aim is not to use Platonism to take pot shots at Aristotelianism, as some scholastic writers had done, nor is it just setting out a Platonic system without regard to the kinds of questions that Aristotelianism had engaged, as Eastern Platonists had done. It is the setting out of a new synthesis which is presented as the answer to problems that Christian Platonists and Aristotelians share, and it forces to the centre a question that the scholastic tradition had certainly taken seriously, but only as one of a number of issues.[3] Ficino makes the doctrine of the personal immortality of the soul the question on which the whole enterprise stands or falls, tying the defence of the personal immortality of the soul to a return to the Augustinian synthesis, one which depends crucially on a commitment to Neoplatonism. But the Church had already moved philosophically in the other direction some one hundred and seventy years earlier at the Council of Vienne, effectively stipulating that any understanding of these questions had to be couched in Aristotelian terms, in which the soul is the form of the body. This is reinforced by the Fifth Lateran Council in 1513, where the doctrine of personal immortality was established as a dogma, albeit one whose philosophical defence is acknowledged to be problematic.[4] The Council's response was to instruct theologians and philosophers to reconcile philosophy—i.e. Aristotelian natural philosophy—with theology on this issue.

Such reconciliation is the core of the Thomist approach. The understanding of metaphysics in Augustine had been premised on the idea that ancient metaphysics lacked something crucial to its success, something which only Christianity could provide. There could be no complete non-Christian metaphysics, for Christianity was integral to metaphysics. Aquinas moves away from this understanding of metaphysics to one in which it is a general science that is able to provide an architectonic for forms of knowledge with different sources. For Aquinas, natural philosophy is not and could not be an intrinsically Christian enterprise: it proceeds from sensation, which is common to Christians and pagans. We have two distinct

---

[3] The doctrine of the personal immortality of the soul derived from the early Church Fathers, rather than from scripture itself and its standard formulation derived from Augustine, in his *De immortalitate animæ* and *De quantitate animæ*: see Heinzmann 1965. In its Augustinian form, it is in effect the Neoplatonic doctrine of immortality stripped of those ingredients incompatible with Christianity, namely the transmigration of souls and pre-existence. Personal immortality was tied in with Aristotelian natural philosophy in an explicit way in 1311 when the Council of Vienne declared the Aristotelian definition of the soul as the form of the body to be an article of faith: see Tanner 1990, 1, p. 361. It is important to note, however, that the Council was not in fact concerned with the question of immortality as such but with Christological questions. Pierre Olivi and other thirteenth-century Franciscans had denied that the soul could be the form of the body because immersion in matter would deprive it of a separate existence, but this had highly heterodox consequences for the doctrine of the Incarnation, and the Council of Vienne rejected it on these Christological grounds. More generally, the doctrine of personal immortality was not an issue for thirteenth-century scholastic philosophers: Scotus was at best pessimistic about the chances of its rational demonstration, and Aquinas, though he defends the soul's incorruptibility in his rejection of the Averroist doctrine of the unity of the intellect—e.g. in *Summa Theologica*, I, q. 57, a. 6. and *On the Unity of the Intellect Against the Averroists*—is silent about personal immortality as such. It is not Thomists, Scotists, Averroists, or any other movement within scholasticism that puts the question of immortality at centre stage, but the Platonist movement.

[4] See Gilson 1961; and Fowler 1999.

sources of knowledge, sensation and revelation, and since knowledge is unitary, there must be some way of bridging these. The only thing that could bridge them is something that covers the natural and the supernatural, and only metaphysics satisfies this description. The project is one of reconciling natural philosophy to Christian belief, rather than vice versa, but the very notion of metaphysics as a medium of reconciliation requires that metaphysics has some degree of independence from either of these enterprises. The idea is not that metaphysics underlies natural philosophy or theology: it cannot do this, since the only sources of knowledge are revelation and sense perception, and these do not underlie metaphysics but the disciplines it seeks to reconcile. Rather, metaphysics underlies the connection between the two, by providing an account (in Aquinas' theory of analogy) of the various kinds of being and the kinds of knowledge appropriate to them. In this sense, metaphysics is not an inherently Christian enterprise, any more than natural philosophy is. It is true that, because revelation is regarded as secure in a way that knowledge derived from sensation could never be, reconciliation will tend to be unidirectional in favour of theology, but the crucial point is that this is not a feature of metaphysics as such; rather, it is a feature of the disciplines that metaphysics seeks to reconcile.

One the greatest challenges for this exercise was the doctrine of personal immortality. Within three years of the challenge being raised by the Lateran Council, the very doctrines that the Council had condemned as the source of problems about the immortality of the soul, and which Thomism was supposed to have answered decisively, were given an articulate and powerful airing which showed that they were far from having been laid to rest. In his *De immortalitate animæ*, published in 1516, Pomponazzi offers an argument that engages Platonist, Thomist, and Averroist positions on the nature of the soul. He argues against the Platonists (he is clearly responding to Ficino) and in agreement with Averroists and Aquinas, that philosophically speaking the soul was the form of the body. There can be no cognition except through the body, and this rules out the kind of knowledge of pure intelligibles that Ficino and other Platonists had postulated. In this connection he also supports the Aristotelian view that there is no such thing as an uninstantiated form. But if this is the case, he argues, then the death and corruption of the body result in the disappearance of the soul. On the other hand, he accepted the Church teaching of the personal immortality of the soul, and he argues that Aquinas decisively refuted Averroës' view that there cannot be individual souls in his doctrine that the human soul is not a form that arises from matter but is the object of a special creation by God. But he also contends that, in terms of Aristotelian metaphysics/natural philosophy, Aquinas' own proposal is not decisive. Aquinas had separated lower functions of the soul, such as growth and sense perception, which he considered do indeed end with the death and corruption of the body, from higher cognitive and intellective functions, which do not. But it is crucial on his Aristotelian account that, for human beings, the activities characteristic of the higher functions, in particular the grasp of universals, must start from sense perception, that is, from something intrinsically corporeal. In particular, all knowledge works from sensory images. In advocating this doctrine, however, Aquinas distinguishes between the kind of intuitive grasp of truth characteristic of the intellect, and the reasoning processes which underlie and accompany sensation. All knowledge starts from sensation, but once the intellect is engaged and has done the work of abstraction, sensory images are no longer needed.[5] This is where Pomponazzi's difficulties with the Thomist account begin, for the idea of a form of cognition that does not involve a representation of the object cognised is just not cognition for Pomponazzi, and the representation can hardly be pure form for no Aristotelian account of cognition could countenance pure forms. Consequently, the mind cannot act in cognition without corporeal representations, that is, without the body. As Pomponazzi realises, this leaves the question of immortality wide open. Philosophically, Pomponazzi advocates the view that the soul is the 'highest form'—and it is interesting how even Pomponazzi has to resort to Neoplatonic notions at this point—but philosophy cannot establish its immortality.

Pomponazzi's dilemma is that two completely different lines of thought, each of which he has every reason to believe to be completely compelling and neither of which he was prepared to renounce, lead to incompatible conclusions. Somehow one must embrace both. Note, however, that Pomponazzi is not claiming that both are true. He clearly holds the truth of the doctrine of personal immortality—it is 'true and most certain in itself' as he puts it in the Preface to *De immortalitate*[6]—and the falsity of its denial. But in pointing out, in the same sentence, that this doctrine 'is in complete disagreement with what Aristotle says', he does not oppose the truth of what Aristotle says: he simply does not discuss Aristotle's doctrine in terms of truth. What he is drawing attention to is the fact that there are two quite different but completely legitimate ways of pursuing the question, and the radical twist is that there is no metaphysics that can reconcile these.

The implications of this failure of Aristotelian natural philosophy to supply appropriate philosophical support for a core doctrine go beyond the issue of personal immortality. At the most fundamental level, what is at stake is the failure of Aristotelian natural philosophy to provide a philosophical basis for a systematic theology. By the early to middle decades of the sixteenth century, such a failure had become deeply problematic, because the need for a philosophical basis was greater than ever, and the need for it to take the form of Aristotelian natural philosophy was greater than ever. The magnitude and ramifications of the failure are evident on the issue of transubstantiation, which became the issue on which everything hinged in the theological disputes of the 1520s onwards, because on it rested the whole question of an ecclesiastical hierarchy, the central issue at stake in the Protestant break with Catholicism, with the Protestants denying any priesthood other than Christ's, and the Catholics insisting on the need for a priesthood and a clerical hierarchy. Catholic theologians in the 1520s were well aware that baptism could be conferred by anyone, that sins could be forgiven in the case of genuine contrition without priestly forgiveness, and that marriage partners conferred the sacrament on one another. Consequently, without transubstantiation, there is no need for a hierarchical priesthood, especially once the newly emerging nation states had

---

[5] See my 1989, pp. 38–47; and Peghaire 1936.

[6] The passage is on p. 281 of the translation in Cassirer, Kristeller and Randall 1948.

assumed responsibility for civic functions.[7] Transubstantiation was a doctrine that had always been formulated and defended in Aristotelian terms, and it would seem that it is impossible to capture the doctrine in a satisfactory way in any other philosophical terms.[8] Never was the defence of Aristotelianism, both in its own right and in terms of its credentials as a foundation for a systematic theology, so necessary, yet it no longer seemed that these roles were reconcilable.

The seventeenth-century non-scholastic response to these issues, I want to argue, involves a complex shift in the relations between truth and justification, relations which shape the context within which both natural philosophy and religious thought are pursued, and bear directly on the questions of legitimation which come to affect both natural philosophy and religion in the seventeenth and eighteenth centuries. Above all they turn on the questions of what natural philosophy does, and what the natural philosopher aims to achieve.

2. SPECULATIVE VERSUS PRODUCTIVE PHILOSOPHERS

The dilemma posed by Pomponazzi's arguments on the immortality of the soul raise the question of what it means to be a philosopher. In particular, the question of the legitimacy of philosophy comes to the fore. In distinguishing the genuine philosopher from the sophist in his early dialogues, Plato offers the image of the sophist as someone who is willing to teach anyone who is prepared to pay to devise arguments to win a case, including making weak arguments appear better than strong ones. The failing of the sophist is ultimately not an intellectual but a moral failing.

The tendency to see philosophical failings along the lines of moral ones will have a long history and it is especially prevalent in the seventeenth century. The attack on Aristotle by Joseph Glanvill, one of the most prominent apologists for the Royal Society, brings out the flavour of the issues:

> Consonant whereunto are the observations of *Patricius* that he carpes at the *Antients* by name in more than 250 places, and without name in more than 1000. [H]e reprehends 46 *Philosophers* of worth, besides *Poets* and *Rhetoricians*, and most of all spent his spleen upon his excellent and venerable Master *Plato*, whom in above 60 places by name he hath contradicted. And as *Plato* opposed all the *Sophisters*, and but two *Philosophers*, viz. *Anaxagoras* and *Heraclitus*; so *Aristotle* that he might be opposite to him in, *this* also, oppos'd all the *Philosophers*, and but two *Sophisters* viz, *Protagoras* and *Gorgias*. Yea, and not only *assaulted* them with his arguments, but *persecuted* them by his *reproaches*, calling the *Philosophy* of *Empedocles*, and all the Antients *Stuttering*; *Xenocrates*, and *Melissus*, *Rusticks*; *Anaxagoras*, *simple* and *inconsiderate;* yea, and all of them in a heap, as *Patricius* testifies, *gross Ignorants*, *Fools* and *Madmen*.[9]

Glanvill's use of Bacon here is pivotal. A crucial part of Bacon's project for the reform of natural philosophy was a reform of its practitioners. One ingredient in this was the elaboration of a new image of the natural philosopher, an image that

---

[7] Levi 2002, p. 353.

[8] See Armogathe 1977.

[9] Joseph Glanvill, *A Letter to a Friend Concerning Aristotle*, appended to Glanvill 1665, pp. 84–85.

conveyed the fact that the natural philosopher is no longer an individual seeker after the arcane mysteries of the natural world, employing an esoteric language and protecting his discoveries from others, but a public figure in the service of the public good, that is, the crown.[10]

The idea that philosophers prefer useless learning to virtue goes back to Petrarch, and indeed is one of the mainstays of Petrarchian humanism.[11] Renaissance humanists raised the question of the responsibilities appropriate to the humanist, in particular whether the life of activity in affairs of state (*negotium*) should be preferred to that of detachment and contemplation (*otium*). The answer almost invariably given—not least by Bacon himself, in the seventh Book of *De Dignitate*[12]—was that *negotium* should be preferred to *otium*. Once this question had been decided, the issue then became not just the appropriate learning but also, given the practical nature of the programme, the appropriate behaviour for such a practical humanist. The choice, in the first instance, is between the active or practical life and the contemplative life, where philosophers had traditionally fallen in the latter category. The explicit shift to the defence of the active or practical life places new requirements on philosophy, for philosophers now had to show that they were able to live up to the aims of the active or practical life. What Bacon effectively does is to transform philosophy into something that comes within the realm of *negotium*. This is completely at odds with the conceptions of philosophy of classical antiquity and the Christian Middle Ages. Promoted through the rhetorical unity of *honestas* and *utilitas*, Bacon presents philosophy as something good and useful, and thus as intrinsic to the active life. Indeed, it starts to become a paradigmatic form of *negotium*, and in this way, it can usurp the claims made for poetry by writers such as Philip Sidney, who argued that poetry can move one to act virtuously, whereas philosophy cannot do this.

In the humanist thought that makes up the source from which Bacon derives much of his inspiration, moral philosophy figures predominantly. There are two respects in which the model of moral philosophy is important here. First, philosophical self-fashioning had always turned on the moral question of the understanding and regulation of the passions, and because of this they have a peculiar centrality, for they have not merely been one object of study among others for philosophers, but something which must be understood if one is to be 'philosophical' in the first place. Mastery of the passions was, in one form or another, not only a theme in philosophy but a distinctive feature of the philosophical *persona* from Socrates onwards, and Renaissance and early modern philosophers pursue the theme of self-control with no less vigour than had the philosophers of antiquity. This is the model around which Bacon wishes to shape his new practitioner of natural philosophy.[13] It is a model inappropriate to the artisan, and it gives the new practitioner a dignity and standing that the collective nature of his

[10] This forms one of the central themes of Gaukroger 2001a: see esp. chaps 2 and 4.

[11] See Gilbert 1971.

[12] Bacon 1859, I, pp. 713–744 [text], V, pp. 3–30 [trans].

[13] This is particularly evident in Bacon's account of his scientific utopia, *New Atlantis*, where self-respect, self-control, and internalised moral authority are central.

work would not otherwise suggest. Second, in a humanist dimension, being virtuous and acting virtuously are the same thing: there is no separate practical dimension to morality. Indeed, this forms the basis for much humanist criticism of traditional moral philosophy: Sidney, for example, in stressing the superiority of the active, practical life over the contemplative one, draws what he takes to be the consequences for moral thought, namely that teaching the nature of virtue is not the same thing as, and indeed is no substitute for, moving people to practice virtue, and that all philosophy has managed is the former.[14] Sidney and Bacon both want to obliterate the distinction between being moral and acting morally. Moreover, it is interesting to note here that Bacon stresses in the *Advancement of Learning* that moral philosophy is a cognitive enterprise, one in which the practical outcome is constitutive of the discipline.[15] If, as I am suggesting, we see natural philosophy as being in some respects modelled on moral philosophy, something which is natural enough in a humanist context, and which is reinforced in the shift from *otium* to *negotium*, then we may be able to delve a little more deeply into why Bacon famously claims that the aim of the natural philosopher is not merely to discover truths, even informative ones, but to produce new works.

For Bacon, the natural philosopher is not simply someone with a particular expertise, but someone with a particular kind of standing, a quasi-moral standing, which results from the replacement of the idea of the sage as a moral philosopher with the idea of the sage as a natural philosopher, whose paradigm is Solomon. And just as the sage as moral philosopher cannot be such unless his grasp of morality is manifest in his behaviour, so the sage as natural philosopher cannot be such unless his grasp of nature is manifest in his behaviour, and the only way in which it can be manifest is in the production of new works.

How does one become such a sage? In the most general terms, at least one ingredient in the answer is a very traditional one: the purging of the emotions. But Bacon puts a distinctive gloss on this. The sage for Bacon must purge not just affective states but cognitive ones as well. This is the core of his doctrine of the 'Idols' of the mind, the need for which he spells out in the Preface to *Novum Organum*,[16] and which provides the platform for setting out, in Book I of *Novum Organum*, an account of the systematic forms of error to which the mind is subject. Here the question is raised of what psychological or cognitive state we must be in to be able to pursue natural philosophy in the first place. Bacon believes an understanding of nature of a kind that had never been achieved since the Fall is possible in his own time. This is because the distinctive obstacles that have held up all previous attempts have been identified, in what is in many respects a novel theory of what might traditionally have been treated under a theory of the passions, one directed specifically at natural-philosophical practice.

---

[14] Sidney 1965, p. 112. Compare Bacon's assessment in *De Augmentis*: 'Moral philosophers have chosen for themselves a certain glittering and lustrous mass of matter, wherein they may principally glorify themselves for the point of their wit, or power of their eloquence; but those which are of the most use for practice, seeing that they cannot be so clothed with rhetorical ornaments, they have for the most part passed over.'; Bacon 1859, I, p. 715 [text], V, pp. 4–5 [trans].

[15] *Advancement of Learning*, Book II: Bacon 1859, III, pp. 432–434.

[16] Bacon 1859, I, pp. 151–152 [text], IV, p. 40 [trans].

Bacon argues that there are identifiable obstacles to cognition arising from innate tendencies of the mind (Idols of the Tribe), from inherited or idiosyncratic features of individual minds (Idols of the Cave), from the nature of the language that we must use to communicate results (Idols of the Market-Place), or from the education and upbringing we receive (Idols of the Theatre). Because of these, we pursue natural philosophy with seriously deficient natural faculties, we operate with a severely inadequate means of communication, and we rely on a hopelessly corrupt philosophical culture. In many respects, these are a result of the Fall and are beyond remedy. The practitioners of natural philosophy certainly need to reform their behaviour, overcome their natural inclinations and passions etc., but not so that, in doing this, they might aspire to a natural, prelapsarian state in which they might know things as they are with an unmediated knowledge. This they will never achieve. Rather, the reform of behaviour is a discipline to which they must subject themselves if they are to be able to follow a procedure which is in many respects quite contrary to their natural inclinations. In short, the reform of one's *persona* is needed because of the Fall: after the Fall it is lacking in crucial ways. Whereas earlier philosophers had assumed that a certain kind of philosophical training would shape the requisite kind of character, Bacon argues that we need to start further back as it were, with a radical purging of our natural characters, in order to shape something wholly new.

One of the great failures of Bacon's project, in his own lifetime, was his inability to find an audience for his work. The transformation of the natural philosopher is necessary for the transformation of natural philosophy, but who was this new natural philosophy written for? After all, if the qualities required by the new natural philosopher were so radically different from those of the old, surely these needs would be paralleled in the readers, but these could no more be the traditional readers than the writers were the traditional writers. If the new natural philosophers were not simply to write for one another, it was crucial that a new kind of audience be constructed for the new kind of natural philosophy.

Bacon, who showed no knowledge of or interest in centres of natural-philosophical research (not even the leading such institution of the day, Gresham College, which had been set up from an endowment from Bacon's own uncle's will), and who engaged in no correspondence on natural-philosophical questions, saw the audience for his natural philosophy as being the monarch, although neither Elizabeth nor James showed any interest in his expensive grandiose schemes.[17] Descartes, by contrast, wrote several letters on natural-philosophical topics each day in his maturity, maintaining extensive contact with the natural-philosophical community through the circle of Mersenne. He even designed his *Principia Philosophiæ* along the model on late scholastic textbooks as something for use in colleges and universities, and in the last weeks of his life he was busy drawing up a plan for an Academy at Stockholm.[18] Descartes certainly had a better sense than did Bacon of the importance of an audience able to respond to the new work in the appropriate way, but his writings were subject to significant censorship in the second half of the

[17] See Gaukroger 2001a, pp. 130–131 and 160–165.
[18] 'Project d'une Académie à Stockholm', 1 Feb. 1650, in Descartes 1996, hereafter *AT*, XI, pp. 663–635.

seventeenth century,[19] whereas Bacon's quite suddenly began to receive an enthusiastic reception in England and in continental Europe, in the years immediately after his death.[20] As a result, it was Bacon, rather than Descartes, who provided the ideology behind the new scientific academies, even in intensely nationalistic France, where Cartesians were excluded from membership of the Académie des Sciences, founded by Colbert, chief minister to Louis XIV, 'in the manner suggested by Verulam [Bacon]'.

But it is the case of Galileo that is the most interesting one in this respect, for here we can discern a process whereby an audience is shaped and the natural-philosophical enterprise legitimated. Galileo was a mathematician, a profession which was associated with the mechanical arts and had a particularly lowly standing in the ranking of university disciplines. His own education in the subject had come not from his university training—following his father's wishes he had trained in medicine at Pisa, although he left before taking his degree—but from the Florentine court instructor Ostilio Ricci, who taught military fortification, mechanics, architecture and perspective, and whom Galileo had invited to his father's house for instruction.[21] When Galileo took up a university teaching post in mathematics at Pisa (1589) and then at Padua (1591), the fact that he was able to do this without having completed a degree indicates that he was teaching not in a philosophical discipline but in a technical one, which was learned through apprenticeship rather than training, and the salary was correspondingly less: about one-sixth to one-eighth that of a philosophy professor.[22] Mathematics was crucial to Galileo's understanding of natural philosophy but there was very little he could do within the university system to further his approach to natural philosophy, or even to build up an audience for it.

The patronage system, by contrast, was structured in an entirely different way, with an entirely different ranking of priorities. Its attraction was not that it had a more sympathetic approach to the practical-mathematical disciplines than the universities, for it didn't, but that there was no inherent fixed ranking of disciplines. The main clients of the Florentine patrons—painters, sculptors, architects and others—had attempted, throughout the sixteenth century, to enhance their social status by developing explicit theories grounded in the liberal arts, attempting to transform their standing as mere artisans into that of artists, thereby setting a model for natural philosophers.[23] The overriding factor in the patron/client relationship was the enhancement of the reputation of the patron, and in the realm of natural philosophy, natural-philosophical discoveries played a key role. Just as in painting,

---

[19] See McClaughlin 1979; Jolley 1992; and Schmaltz 1999. For details of publication of Descartes' works in the seventeenth century, see van Otegem 2002.

[20] For a summary view of his reception see Gaukroger 2001a, pp. 1–5. For more details, see Pérez-Ramos 1988, chap. 2; and Brown 1978. On his posthumous reception in England see Webster 2002. For details of publication of Bacon's works in the seventeenth century, see Gibson 1950.

[21] See Settle 1971.

[22] Biagioli 1989, p. 53.

[23] See Burke 1986, chap. 3. Biagioli notes that Galileo felt similarly obliged to immerse himself in literary and artistic disputes—on Dante's *Inferno*, on the relative priorities of Ariosto and Tasso, on the relative merits of painting and sculpture—to prove his competence with courtly and academic culture: Biagioli 1993, pp. 118–119.

architecture, music, and verse, the client was expected, ideally, to produce something that would dazzle the patron's competitors, so too in natural philosophy what was to be preferred was some dazzling new discovery. Here we have something which in many ways realised Bacon's picture of a successful natural-philosophical practice, in that it was directed towards manifest and concrete results, and had no place for merely contemplative natural philosophy. Moreover, the princes to whom Galileo dedicates his discoveries have for many purposes the same absolutist powers as had the sovereign whom Bacon wishes to oversee his 'great instauration'. Patronage provided a powerful system of legitimation outside the university system, with its own standards of social status and credibility, but one in which the patrons needed the clients as much as the clients needed the patrons, and as a result the natural-philosophical agenda—as long as it produced the goods—could be shaped to a large extent by the client natural philosophers, since the patrons themselves were considered to be above the details, to which their characteristic attitude was one of disinterestedness. In this way, natural philosophy makes a move from the clerical to the civil terrain. The price to be paid, as Biagioli points out, is that 'within court patronage one could gain legitimation as a scientific author only by effacing one's individual authorial voice. To be a legitimate author meant to represent oneself as an "agent" … of the prince'.[24]

The main difference between the patronage model and that advocated by Bacon was that, in Bacon's scheme, gentlemanly behaviour required rejection of adversarial dispute characteristic of scholasticism, since this was considered as ungentlemanly and fruitless, whereas, in the patronage system of the northern Italian states, patrons initiated and managed natural-philosophical disputes to enhance their image. They were part of a social economy of honour and status, much as duels had been, and like duels they had sharply defined rules of etiquette, constraining who should dispute with whom: attacks on Galileo's work on buoyancy derived from someone of lower social standing, for example, and he was advised to have them answered by 'someone young' so that his opponent could be shamed and 'taught a lesson'.[25] What resulted from the patronage model was a radically adversarial mode of dispute, although it functioned in a significantly different way from scholastic dispute. In the first place, the whole adversarial style was different. Galileo criticises those who 'would like to see philosophical doctrines compressed into the most limited space, and would like people always to use that stiff and concise manner, that manner bare of any grace or adornment typical of pure geometricians who would not even use one word that was not absolutely necessary'.[26] Galileo's attacks on the spokesman for Aristotelianism in his dialogues, Simplicio, was, as Biagioli notes, 'not only Galileo's straw man but also a representative of what court culture

---

[24] Biagioli 1993, p. 53.

[25] See *ibid.*, p. 62, and more generally pp. 60–73.

[26] Galileo to Prince Leopold of Tuscany, quoted in Biagioli 1993, pp. 114–115. The gulf between Italian patronage culture and later Royal Society culture could not be greater here. Compare Sprat's instruction to adopt 'a close, naked, natural way of speaking; positive expressions; clear senses; a native easiness, bringing all things as near to mathematical plainness as they can; and preferring the language of Artizans, Countrymen, and merchants, before that, of Wits, or Scholars'; Sprat 1667, p. 113.

perceived itself to be rejecting',[27] and we should remember here that court culture included senior clerics, such as cardinals, who were very different from the scholastic representatives of religious orders.[28] Second, scholastic dispute was above all part of a method of discovery, whereas that is not the case here. Rather, what seems to be at issue in the case of patronage-directed disputes is that they act as a means of defending the dignity, and expanding the standing, of the patron: in the process, they act to legitimate the natural-philosophical programmes pursued under the umbrella of the patronage.

The spectacular development which finally projected Galileo into the public arena, and quickly secured him the patronage of the Grand Duke of Tuscany, came in 1609, with his discovery of four satellites of Jupiter. *Sidereus Nuncius* catapulted Galileo to fame. The spectacular novelty value of the work was not lost on contemporary audiences, which for discoveries of this magnitude were immense. Nor was the concern with novelties restricted to the patronage system. The Jesuits paid extensive attention to novel scientific discoveries in the teaching in their colleges, to the extent that some critics of the Jesuit teaching system have suggested that the Jesuit masters had little genuine scientific interest, but were concerned rather with novelties.[29] In 1611, on the first anniversary of the death of it founder, Henri IV, the *collège* at La Flèche—where Mersenne, Descartes, and Descartes' later collaborator in optics, Claude Mydorge, were all students—engaged in elaborate celebrations.[30] Among the sonnets presented to commemorate the king was one describing how God had made Henri into a celestial body to serve as 'a heavenly torch for mortals'; it is entitled 'On the death of King Henri the Great and on the discovery of some new planets or stars moving around Jupiter, made this year by Galileo, celebrated mathematician of the Grand Duke of Florence'.[31] Galileo's discovery of the moons of Jupiter was indeed widely celebrated, and the Collegio Romano had supported theses defending Galileo in the same year,[32] although they had incorporated the discovery into a Tychonic framework, not a Copernican one. There can be no doubt that the Jesuits encouraged a fascination with novelties in their students, and Descartes was to be no exception. In a manuscript dating from 1621,[33] he describes with evident fascination how to create various optical illusions deriving from della Porta's *Magia naturalis*—a textbook of natural-philosophical illusions, remedies, novelties and much else—which first appeared in 1589, and which he was almost certainly familiar with from his days at La Flèche. In short, the concern with novelty that was so central to the patronage of natural philosophy was not unique to it, but pervaded European culture more widely. And of course it stands to reason that if what the patronage system had prized had attracted no interest and

---

[27] Biagioli 1993, pp. 115–116.

[28] Biagioli notes that Pope Urban VIII, who was behind the 1633 condemnation of Galileo, was not an orthodox Aristotelian at all but held a position closer to Ockhamism; *ibid.*, p. 351.

[29] See for example in Compayré 1879, 1, p. 194.

[30] Theatre and public spectacle were an important ingredient in Jesuit culture; see Bjurstrom 1972.

[31] See de Rochemonteix 1889, 1, p. 147.

[32] Biagioli (1993, pp. 296–297) points out that the Jesuits also supported Galileo's anti-Aristotelian work on buoyancy, and were unhappy with the 1616 condemnation of Copernicanism.

[33] *AT*, X, pp. 215–216.

had no appeal outside that system, then it would hardly be able to further the interests and influence of patrons, or display their grandeur and qualities. Indeed, to a large extent the patronage system was able to present itself both as a source of natural-philosophical novelties, and of the no less remarkable and ingenious taming of these novelties by court natural philosophers, fuelling a kind of interest in natural philosophy which was quite different from the increasingly limited appeal of scholastic textbook natural philosophy.

The same concern to shape a new kind of natural philosopher is evident in Descartes, although the questions are approached differently from both Bacon and Galileo. That Descartes should be concerned with such questions might at first seem somewhat surprising, especially when compared with Bacon. Bacon's purging is targeted very precisely in his doctrine of Idols, and his understanding of what is needed to build on the newly cleared foundations is not abstract and metaphysical but something psychological and practical: in keeping with his conception of the reformed philosophical enterprise. It might seem that Descartes could not countenance a project of this kind, since he has such a rarefied notion of philosophical activity: after all, the *Meditationes* ask us to begin our search for knowledge by imagining that there is no natural world, and that we have no bodies. In his *Disquisitio metaphysica*, Gassendi makes exactly these criticisms of Descartes. Gassendi had been one of those asked by Mersenne to set out a set of objections to Descartes' *Meditationes*, which were published with Descartes' replies, and unsatisfied with the reply, he had elaborated on his own objections, and responded to Descartes' replies at length.[34] Descartes attacks Gassendi for raising objections which are not those that a philosopher would raise,[35] thereby opening up the question of what it is to be a philosopher. Amongst other things, he charges Gassendi with using debating skills rather than philosophical argument; with being concerned with matters of the flesh rather than those of the mind; and with failing to recognise the importance of clearing the mind of pre-conceived ideas. The dispute pits Descartes the advocate of a complete purging of the mind against Gassendi the defender of legitimate learning. But in fact matters are not quite so simple and, in the broad outlines of what he seeks to achieve, Descartes' aims are similar to those of both Bacon and Galileo.

To understand how, it is crucial that we distinguish between two kinds of enterprise. The first, which is largely legitimatory, is set out in the *Principia Philosophiæ*,[36] and the route it follows is that of a radical purging of the mind of anything that can conceivably be doubted, establishing clarity and distinctness (manifested paradigmatically in the *cogito*) as the only criterion by which to establish the veridicality of our ideas, and then, having established that our understanding of the natural world must begin with quantitatively and mechanistically formulated ideas, building up a novel cosmology. This is the way to establish the truth of Cartesian natural philosophy, but Descartes does not claim that

---

[34] Gassendi 1658, 3, pp. 269–410. Descartes himself wrote a counter-reply, which appeared in the French translation of the *Meditationes*; *AT*, IXA, pp. 198–217.

[35] *AT*, VII, pp. 348–349.

[36] For a detailed discussion see Gaukroger 2002, chaps 1 and 3. We shall return to these questions below.

it is the way to pursue this natural philosophy. It is the route to be followed by someone who wishes to be convinced of the truth of Cartesian natural philosophy, but it is not the path of discovery to be followed by the natural philosopher. This path, and the requisite state of mind and character of the natural philosopher who wishes to pursue it, are formulated in quite different terms, ones that involve psychological and moral considerations as much as epistemological ones.[37]

Descartes' discussion of this path occurs in *La Recherche de la verité par la lumière naturelle*, which contrasts the fitness for natural philosophy of three characters: Epistemon, someone well versed in scholasticism; Eudoxe, a man of moderate intelligence who has not been corrupted by false beliefs; and Poliandre, who has never studied but is a man of action, a courtier, and a soldier (as Descartes himself had been). Epistemon and Poliandre are taken over the territory of sceptical doubt and foundational questions by Eudoxe, but in a way that shows Poliandre's preparedness for, or capacity for, natural philosophy, and Epistemon's lack of preparedness. Preparedness here is in effect preparedness for receiving instruction in Cartesian natural philosophy. The *honnête homme*, Descartes tells us,

> came ignorant into the world, and since the knowledge of his early years rested solely on the weak foundation of the senses and the authority of his teachers, it was close to inevitable that his imagination should have been filled with innumerable false thoughts before his reason could guide his conduct. So later on, he needs to have either very great natural talent or the instruction of a very wise teacher, to lay the foundations for a solid science.[38]

The thrust of Descartes' discussion is that Poliandre has not had his mind significantly corrupted, because, in his role as an *honnête homme*, he has not spent too much time on book-learning, which 'would be a kind of defect in his education'. The implication is that Epistemon has been corrupted in this way, and so is not trainable as the kind of natural philosopher Descartes seeks. It is only the *honnête homme* who can be trained, and it is Poliandre whom Eudoxe sets out to coax into the fold of Cartesian natural philosophy, not Epistemon. It is true that we might think of the procedure of radical doubt and the purging that results as a way of transforming everyone into an *honnête homme*, and to some extent it is, although in his account of the passions Descartes makes it clear that, once we leave the programmatic level, ridding ourselves of prejudices and pre-conceived ideas is not so simple, and it requires the cultivation of a particular mentality, which is really what we witness in *La Recherche*.

In the *Recherche*, the *honnête homme* alone is identified as the kind of person who uses his natural faculty of forming clear and distinct ideas to the highest degree: or, at least, it is he who, when called upon, uses it to the highest degree. This does not mean that the *honnête homme* alone is able to put himself through the rigours of hyperbolic doubt and discover the true foundations of knowledge: in theory everyone is able to do that, scholastics included. After all, hyperbolic doubt erases our beliefs—everyone's beliefs—to such an extent that everyone becomes a natural-philosophical *tabula rasa*:

---

[37] See *ibid.*, pp. 239–246.

[38] *AT*, X, p. 496. On *La Recherche* see Ranea 2000.

> An examination of the nature of many different minds has led me to observe that there are almost none at all so dull and slow as to be incapable of forming sound opinions or indeed of grasping all the most advanced sciences, provided they receive proper guidance. And this may be proved by reason. For since the principles in question [namely, those of the *Principia*] are clear, and nothing is permitted to be deduced from them except by very evident reasoning, everyone has enough intelligence to understand the things that depend upon them.[39]

But if the aim is to develop and refine natural-philosophical skills as one progresses, then we require something different:

> As for the individual, it is not only beneficial to live with those who apply themselves to [the study of philosophy]; it is incomparably better to undertake it oneself. For by the same token it is undoubtedly much better to use one's eyes to get about, but also to enjoy the colours of beauty and light, than to close one's eyes and be led around by someone else. Yet even the latter is much better than keeping one's eyes closed and having no guide but oneself.[40]

'Using one's eyes to get about' is not something that everyone finds equally easy, however. What Descartes is seeking are those who can develop his system to completion:

> the majority of truths remaining to be discovered depend on various particular observations/experiments which we can never happen upon by chance but which must be sought out with care and expense by very intelligent people. It will not easily come about that the same people who have the capacity to make good use of these observations will have the means to make them. What is more, the majority of the best minds have formed such a bad opinion of the whole of philosophy that has been current up until now, that they certainly will not apply themselves to look for a better one.[41]

We must recognise that some are more fitted than others to follow the path of instruction/enlightenment in natural philosophy. And in the *Recherche*, Descartes realises, practically, that people come to natural philosophy not with a *tabula rasa* but with different sets of highly developed beliefs which are motivated in different ways and developed to different degrees. These rest upon various things, and this is what leads him, in the *Recherche*, to construct an image of the *honnête homme* as a model in which the moral sage and the natural philosopher meet, for, as he puts it in the Prefatory Letter to the French translation of the *Principia*, 'the study of philosophy is more necessary for the regulation of our morals and our conduct in this life than is the use of our eyes to guide our steps'.[42]

In the *Principia*, Descartes set out to reform philosophy in its entirety, but he does not see the project as establishing the kind of stagnant system that scholasticism had become, where what has caused the decline of the system was clearly in large part due, in his view, to the slavish adherence of its proponents to Aristotle. In this respect, Descartes is not in the slightest bit interested in winning over scholastic philosophers to his system: they are simply not the kind of people who can develop it, and would only lead it to the kind of stagnation to which they

---

[39] *AT*, IXB, p. 12.
[40] *AT*, IXB, p. 3.
[41] *AT*, IXB, p. 20.
[42] *AT*, IXB, pp. 3–4.

have led Aristotelianism. *A fortiori*, they cannot act as paradigm philosophers, as sages whose wisdom can guide the rest. This role falls instead to those who, reflecting upon the current state of philosophy, have formed a low opinion of it, and have avoided taking it up. This low opinion, wholly merited, is what makes them *honnêtes hommes*, and it is precisely these whom Descartes sees as being potentially the new paradigm philosophers, marked by an intellectual honesty which rescues philosophy from the intellectual disgrace into which it has fallen.

This concern overlaps very significantly with those of Bacon and Galileo, and what underlies it is above all is the rejection of the idea of coming to natural philosophy with pre-conceived ideas. Bacon's doctrine of Idols is dedicated to removing such pre-conceived ideas, and this informs the whole outlook of the Royal Society. Robert Hooke, in his Preface to Robert Knox's history of Ceylon, describes to the reader that the ideal reporter: 'I conceive him to be no ways prejudiced or byassed by Interest, affection, hatred, fear or hopes, or the vain-glory of telling strange Things, so as to make him swerve from the truth of Matter of Fact'.[43] In his history of the Royal Society, Sprat stresses that the 'histories' collected by the Royal Society 'have fetch'd their Intelligence from the constant and unerring use of *experienc'd Men* of the most unaffected, and most unartificial kinds of life'[44] and that:

> If we cannot have sufficient choice of those that are skill'd in all *Divine* and *human* things (which was the antient definition of a Philosopher) it suffices, if many of them be plain, diligent, and laborious observers: such, who, though they bring not much knowledg, yet bring their hands, and their eyes uncorrupted: such as have not their Brains infected by false Images; and can honestly assist in the *examining*, and *Registring* what the others represent to their view.[45]

Galileo uses the charge that his opponents have pre-conceived ideas as a rhetorical ploy, and he links this with their failure to control their passions. This is clear in his attacks on Grassi in *Il Saggiatore*,[46] where Grassi's failure to appreciate the novel hypotheses on the nature of comets that Galileo presents to him is taken as 'a sign of a soul altered by some passion.'[47] Pre-conceived ideas are construed there as a form of vested interests, and Grassi, as a supporter of Aristotelianism, is presented as someone with an axe to grind, someone who is unable to argue a case on its merits and so has to rely on a philosophical system, which is construed as a form of intellectual dishonesty and a lack of objectivity. In fact, Galileo is far from being entirely fair to Grassi, and, twelve years earlier, Galileo had done exactly what he is now accusing Grassi of doing. In a dispute with delle Colombe over buoyancy

---

[43] Knox 1681, Preface, p. xlvii.

[44] Sprat 1667, p. 257.

[45] *Ibid.*, pp. 72–73.

[46] Galileo Galilei, *Il Saggiatore* (Rome, 1623), trans. in Drake 1960, pp. 151–336. There is an excellent discussion of the controversy in Biagioli 1993, chap. 5. As Biagioli notes, the situation is complicated, for Galileo's argument in *Il Saggiatore* is not anti-system per se, but rather a response to the 1616 condemnation of the Copernicanism system. Worried that the Tychonic system might replace the condemned Copernican one (as indeed it was doing among Jesuit astronomers), Galileo responds by trying to put the whole question of astronomical reality on hold, denying validity to any system.

[47] Cited in Biagioli 1993, p. 308.

that began in 1611, it is Galileo who, when faced with recalcitrant evidence, tries (ultimately with success) to turn the dispute away from particular observations to systems of natural philosophy.[48] In this case, Galileo had maintained that whether a body floats on the surface of water or sinks depends on the specific weight of the body and not its shape. Delle Colombe was able to show, however, that whereas a sphere of ebony sank to the bottom of a container of water, a shaving of ebony floated on the surface. Galileo envelops the questions in basic hydrostatics, trying to turn the focus away from delle Colombe's experiment, and arguing for its irrelevance when seen in the context of the larger theory.

Whatever the rights and wrongs of the dispute, however, the crucial point is that, whereas earlier disputes in natural philosophy automatically involved competing systems (for that was what was ultimately at stake), there is now a new ingredient in the brew, as charges of intellectual dishonesty are brought against those who argue from the standpoint of a purported systematic understanding. This anti-system view will take a variety of forms. One will be the kind of radical stand against system-building that we find in Voltaire and Hume. Another is eclecticism. Lipsius, who was one of the first to use the term 'eclecticism' in the modern era, takes Seneca as his model, and advises that we should 'not strictly adhere to one man, nor indeed one sect' and that the only sect we should follow 'is the Eclectic (let me translate it "Elective") which was founded by one Potamo of Alexandria'.[49] The English natural philosopher Walter Charleton spells out his debt to this 'school' in no uncertain terms, telling us that eclectics

> adore no Authority, pay a reverend esteem, but no implicite Adherence to Antiquity, nor erect any Fabrick of Natural Science upon Foundations of their own laying: but, reading all with the same constant Indifference, and æquanimity, select out of each of the other sects, whatever of Method, Principles, Positions, Maxims, Examples, &c. seems in their impartial judgements, most consentaneous to *Verity*; and on the contrary, refute, and, as occasion requires, elenchically refute what will not endure the Test of either right *Reason*, or faithful *Experiment*.[50]

Boyle set out his preference for a form of syncretism in a no less explicit way, telling us approvingly that eclectics do 'not confine themselves to the notions and dictates of any one sect, but in a manner include them all, by selecting and picking out of each that, which seemed most consonant to truth and reason, and leaving the rest to their particular authors and abettors.'[51] The connection between the character of eclecticism and the character of the philosopher is if anything reinforced in the eighteenth century, for example in d'Alembert's entry on eclecticism in the *Encyclopédie*, where the claim that 'the eclectic is a philosopher who, riding roughshod over prejudice, tradition, antiquity, universal consent, authority, in a

---

[48] On this dispute see Drake 1970, chap. 8 and Biagioli 1993, chap. 3.

[49] Justus Lipsius 1604, p. 10; cited in Blackwell, p. 53. Potamo[n] was an Alexandrian living at the end of the first century BC, who attempted to reconcile the doctrines of Plato, Aristotle and the Stoics. On the history of eclecticism in the early modern era, see Albrecht 1994.

[50] Charleton 1654, p. 4.

[51] *Appendix to the First Part of The Christian Virtuoso* in Boyle 1999–2000, 12, p. 405.

word, everything that subjugates the mass of minds, and dares to think for himself',[52] stresses the dignity of the philosopher.

In sum, the figures we have focused on—Bacon, Galileo, and Descartes—each saw philosophy as being in desperate need of radical reform, and each of them saw this reform as being carried out by a wholly new kind of philosopher. This wholly new kind of philosopher was not simply someone who carried out investigations in a different way from his predecessors: he had, and needed to have, a wholly different *persona*. The techniques of self-examination and self-investigation encouraged both by the wholesale attempt to transfer monastic religious values to the population at large during the sixteenth century, and by the sense that one was responsible for the minute details of one's daily life in the form of new norms of appropriate behaviour, [53] opened up the possibility of a new understanding of one's psychology, motivation, sense of responsibility, and shaped one's personal, moral, and intellectual bearing. Bacon, Galileo, and Descartes used this—in rather different ways, but with the same broad aims—to transform our understanding of what qualities, including personal qualities, one needs to be a philosopher.

3.TRANSCENDENCE VERSUS IMMANENCE

In looking, as we have just done, at the widespread early-modern concern with the *persona* of the natural philosopher, we can identify a number of strands which, once thought of in the light of the problem of justification and truth, can be seen not only to offer distinctive ways of dealing with aspects of the problem, but as helping us identify those various aspects in the first place. Bacon's construal of truth as something essentially productive, for example, is a way of engaging the question of truth as something revealed. The difference is that, whereas on the traditional Platonist understanding, what is revealed is another realm—that of the reality underlying the appearances—Bacon shifts the whole question of truth from a contemplative to a practical exercise, so that the required outcome is dominion over nature. The aim of the exercise is no longer the discovery of truth conceived as the outcome of contemplation, and which leads nowhere, but the discovery of relevant, informative truth, where the criteria of relevance and informativeness derive from the ability of that truth to take us beyond our present state of engagement with natural processes to one in which our degree of control over those processes is increased. Here the goals of natural-philosophical enquiry and the justificatory procedures which it engages are referred to something outside it, no longer something in the realm of theology or metaphysics however, but in our practical relation with the natural realm. A different aspect of the question is revealed most clearly in Descartes' sense that what lies beyond the justificatory procedures of natural-philosophical enquiry is an expectation about the intellectual morality of the

---

[52] Diderot 1751–1765, 5, p. 270 col. 1–col. 2. The article on eclecticism runs from p. 270 col. 1 to p. 293 col. 2, and the treatment is comprehensive.

[53] For details see Jean Delumeau's tetralogy, 1978, 1983, 1989 and 1992. See also Delumeau 1971; R. Po-Chia Hsia 1989; and Oestreich 1982, chap. 11.

natural philosopher. This, as I have indicated, takes us back to what I have identified as the original Platonic need for truth as something over and above the procedures of argument distinctive of the philosophical project: Plato seeks a way of preventing what he considers the misuse of these forms of argumentation being included in the genuinely philosophical repertoire, and seeking truth satisfies what is really a requirement of intellectual morality. Descartes' notion of the *persona* of the natural philosopher is most easily identifiable as a form of intellectual morality. But Bacon's idea that the natural philosopher must subject himself to the dictates of an externally-imposed method, and the notion we find in looking at Galileo's career that it is the intellectual disinterestedness of the patron that validates the work of the courtier natural philosopher, both also offer something over and above the procedures of justification which vindicate the natural-philosophical enterprise.

Note that the kinds of questions raised in Bacon's defence of the productive nature of truth, and the attempts to shape the *persona* of the natural philosopher and by extension to establish the standing of the natural-philosophical enterprise, are not competing accounts but complementary ones. They reflect different aspects of a complex of factors blended together in the general question of the relation between truth and justification as it comes to a head in the Pomponazzi affair, and later in the condemnations of Copernicanism. A third kind of factor that plays a part in this complex is epistemological, and this is what I now want to focus on.

As a first—schematic but necessary—approximation, we can identify the range of orthodox positions on the relation between natural-philosophical justification and pre-given religious truth as lying between two poles. At one extreme, we have the view that God wholly transcends our knowledge, so cannot be reached by any form of natural reasoning. The strongest version of this view has the consequence that divine truths and natural truths are different kinds of things, the former having a rationale which is incomprehensible to us. If natural-philosophical reasoning is to have any vindication, it must somehow be guaranteed by God, otherwise it lacks any legitimacy. This, I shall argue, is Descartes' view, and the requisite vindication comes through God's guarantee of the criterion of clear and distinct ideas. The Cartesian position is modified in various ways in the course of the seventeenth and eighteenth centuries, most importantly in the hands of Malebranche, where it begins to form the basis for a phenomenalist view of the natural realm. At the other pole, we have a view that God can be known by means of natural reasoning: here, natural philosophy becomes in effect a form of natural theology.[54] This is a far more widespread view than the first, and we can find it clearly set out in Gassendi and apologists for the Royal Society, for example, although it has earlier precedents. It becomes increasingly important in the course of the seventeenth century, and with the shift of focus in natural philosophy in the eighteenth century from cosmology to natural history, it takes on a very central role. Note, however, that God is still transcendent in very significant respects for both Gassendi and Boyle and the Royal Society apologists, and none of them identify him with his creation. This is particularly important for Gassendi, since he wants to revive and Christianise Epicurean atomism, and it is crucial for this that he be able to purge it of its

---

[54] See Peter Harrison's chapter in this collection.

naturalistic elements.[55] The construal of God as immanent in nature remains the preserve of naturalism, and only Spinoza will defend such a view in the seventeenth century.

Turning first to the transcendence view, in works such as *L'Impiete des Deistes* (1624), Mersenne argued that the pantheism, mortalism, and other dangers he saw inherent in Renaissance naturalism had their root cause in a blurring of the separation between the natural and the supernatural. His response to this was to make matter completely inert, and locate all activity in the realm of the supernatural, an approach which had the added advantage that it made possible a quantitative account of the natural realm, which could now be measured in terms of sizes and speeds of constituent corpuscles, and was not called upon to deal with apparently unquantifiable forces and powers. It was Descartes who took up this approach in its fullest form, however, and it was Descartes who offered the first mechanist cosmological system. Here our concern is with the way in which Descartes deals with the Pomponazzi problem of justification and truth.

There are two basic ingredients in Descartes' account. The first is the doctrine that what we are able to imagine or conceive of does not impose any limit on the powers of an omnipotent and omniscient God. In a letter to Mersenne of 6 May 1630, Descartes tells him that 'eternal truths' are 'true or possible only because God knows them as true or possible, and they are not known as true by God in any way that would imply that they are true independently of Him'.[56] The doctrine of God's creation of eternal truths advances the claim that God not only made things so that certain propositions were true of them, he also created the true propositions and, because he created their content, he could have made 'eternal truths', such as the truths of mathematics for example, different from what they are.[57] Note that Descartes is not claiming that we can make sense of these other possible mathematical truths: on the contrary, he makes it perfectly clear that we cannot make sense of such a thing. Rather, the point is that the fact that we cannot make sense of something is not a constraint on what God can do.

Clearly, on such a view, natural philosophy could, by means of its own resources, generate nothing that would match divinely instituted truths. Descartes' very strong reading of divine transcendence seems to exacerbate rather than resolve the Pomponazzi problem, in effect allowing a gap between what is true for us and what is true for God.[58] But in forcing open this gulf, what Descartes emphasises is the fact that it cannot be bridged, but must be closed in some other way. Metaphysics is not sufficient, or even appropriate: only something divinely guided could play this role. He defends this view by employing a doctrine which, in its early development, looks decidedly unpromising: the doctrine of clear and distinct ideas.[59] The doctrine has its roots in the rhetorical-psychological theories of the Roman rhetorical writers, especially Quintilian. He was concerned with the qualities of the

---

[55] See Osler 1991a.

[56] *AT*, I, p. 149.

[57] Descartes to Mersenne, 27 May 1630: *AT*, I, p. 152.

[58] See Gaukroger 1989, pp. 60–71.

[59] For details of the development of this doctrine in Descartes see Gaukroger 1992.

'image', with the search for and presentation of images that are distinctive in their vividness and particularity, above all with the question of what features or qualities they must have if they are to be employed effectively in convincing an audience. Whether one is an orator at court or an actor on stage, Quintilian tells us, our aim is to engage the emotions of the audience, and perhaps to get it to behave in a particular way as a result, and what one needs in order to do this to employ images that have the quality of *evidentia*—vivid illustration. The core of Quintilian's account is that unless we are already convinced by our own images, we will not be in a position to use them to convince others. So self conviction is a prerequisite for the conviction of others. And self conviction, like the conviction of one's audience, depends on the qualities of the image, amongst which must figure clarity and vividness.

In taking up this model of self conviction, Descartes transforms it from a rhetorical doctrine, in which we amplify some emotion or belief by presenting that belief clearly and distinctly to ourselves, into a cognitive doctrine, in which we assess the truth or falsity of an idea by presenting it to ourselves clearly and distinctly. The first version of this new cognitive doctrine is evident in the *Regulae*, in Rule 3 for example, where we are told that what we must seek is something we can clearly and evidently intuit, and that the mind that is 'clear and attentive' will be able to achieve this.[60] The early *Regulae* draws its model of knowledge almost exclusively from mathematics, and the doctrine of clear and distinct ideas is applied to the case of mathematics in what becomes a paradigmatic way. The thrust of the doctrine is that there is a way of representing mathematical operations such that the truth or falsity of the operation so represented is evident. The truth of the operation $2 + 2 = 4$ is not immediately evident in this form of representation, but it is evident if we represent the operation of addition as the joining of one pair of points, :, with another, :, for then we see that the result must be ::. It turned out that this way of representing mathematical operations (Descartes actually uses line lengths rather than points but the principle is the same) broke down when it came to the complex algebraic operations that Descartes was working on in the late 1620s,[61] but it remained his model of cognitive grasp throughout his career.

The doctrine is transformed yet again in the early 1630s in response to the Church's condemnation of Galileo's *Dialogo* for its advocacy of Copernicanism. The *Dialogo* was withdrawn shortly after its publication in Florence in March 1632, and it was officially condemned by the Roman Inquisition on 23 July 1633. The condemnation focused on the question of the physical reality of the Copernican hypothesis, and a core issue was whether the heliocentric theory was 'a matter of faith and morals' which the second decree of the Council of Trent had given the Church the sole power to decide.[62] Galileo and his defenders denied that it was, maintaining that the motion of the Earth and the stability of the Sun were covered by the first criterion in Melchior Cano's handbook of post-Tridentine orthodoxy, *Locorum Theologicorum Libri Duodecim*, namely that when the authority of the

---

[60] *AT*, X, p. 368.

[61] See Gaukroger 1995, pp. 178–181.

[62] For details see Blackwell 1991.

Church Fathers 'pertains to the faculties contained within the natural light of reason, it does not provide certain arguments but only arguments as strong as reason itself when in agreement with nature'. Opponents of Galileo argued that the case was covered by different criteria, such as the sixth, which states that the Church Fathers, if they agree on something, 'cannot err on dogmas of the faith'. In the 1633 condemnation, the latter interpretation was effectively established, and this meant that the physical motion of the earth could not be established by natural-philosophical means. Galileo had advocated heliocentrism on natural-philosophical grounds, and had generated a conclusion at odds with biblical teaching. It is not too difficult to see this aspect of the condemnation as in some respects a re-run of the Pomponazzi problem, and the Inquisition's decision was that astronomical and natural-philosophical arguments, no matter how compelling in their own right, could not decide the matter.

Descartes' response is to transform the doctrine of clear and distinct ideas so that natural-philosophical arguments which can be formulated clearly and distinctly receive a divine sanction. In the first instance, he gradually shifts to a sceptically-driven metaphysics in which clarity and distinctness in themselves do not even guarantee the truth of simple arithmetical operations because we have no way of correlating what is true for God with what is true for us. The hyperbolic doubt of the mature metaphysical writings is simply an epistemologised version of the doctrine of the divine creation of truths.[63] Having opened up the gap even wider than the Inquisition could have imagined, however, Descartes now proceeds to show how a divine sanction must work. In understanding how this strategy operates, it is important to note that this is a sanction, not a bridge, and it depends on the nature of God in a way that a bridge, of the kind provided in Thomist metaphysics, does not. Descartes' mature metaphysical writings—the *Discours de la méthode*, *Meditationes*, and *Principia*—subject sense perception, the point of entry into Aristotelian natural philosophy, to intense sceptical doubt, removing it as a source of knowledge. If sensation is ruled out, we need to ask whether any part of our cognitive life is reliable as a guide to how the world is. The answer for Descartes lies in a form of reflective judgement about our perceptions and beliefs, which is no less natural than sensation, but which, unlike sensation, is genuinely concerned with veridicality. God does not guarantee the veridicality of sensation because on Descartes' radically anti-Aristotelian view the function of sensation is not to inform us of the nature of the world, but God does guarantee the veridicality of clear and distinct ideas because the function of these *is* to inform us about the world (as well as about purely intellectual matters, like mathematics and metaphysics). We might think of clear and distinct ideas as the epistemological analogue of conscience, as something that God has given us in order to guide us in cognitive matters, just as he has given us conscience to guide us in moral matters. The fact that he has given us these faculties for these purposes settles any question of whether they might deceive us. Providing we exercise these faculties with the requisite care—and for Descartes, this means that we only make judgements about a question once we have formulated it in clear and distinct terms—then we know that the good God of Christianity, who

---

[63] See Gaukroger 1995, pp. 316–318.

has provided us with these faculties in the first place, would not deceive us, for there would be no point in providing us with these faculties (which, unlike sense organs, are genuinely cognitive on Descartes' account) unless their exercise yielded truths which corresponded to those which God had created. This does not necessarily put natural philosophy (pursued in the appropriate way) on a superior footing to theology, but it does at least put them on an equal footing: much as the dictates of conscience were, for many thinkers, on at least an equal footing with decrees of the Church on moral matters.

In one respect, this conception makes our cognitive processes dependent upon God, but in other respects it frees them from divine regulation. Our cognitive processes in no way mirror divine ones, and divine ones can in no sense act as a model for human cognition. In deducing truths on a clear and distinct basis, we deduce something that corresponds to divinely ordained truths, but we have no grasp of the rationale of these divinely ordained truths. Note also that natural philosophy emerges from this conception in one respect very dependent upon God, who is called upon to sanction certain kinds of reasoning in natural philosophy, but in another respect natural philosophy acquires a significant degree of autonomy. Clear and distinct natural-philosophical reasoning is no longer under threat from theology: quite the contrary, once the doctrine of clear and distinct reasoning is accepted as providing a divinely-sanctioned criterion by which to identify 'genuine' truths, the way is open, for those so inclined, to its being used as a criterion more generally, and indeed it was used in some quite radical ways, the most radical being in theology, a move that Spinoza will make, opening up the question of the legitimacy of theology in a way parallel to that in which the legitimacy of natural philosophy had been opened up.

The doctrine of clear and distinct ideas was, however, not widely accepted. Gassendi, one of the earliest critics of the doctrine, admonishes Descartes for failing, in Meditation III, to follow the 'royal path' of philosophising:

> First, as I said before, you have strayed from the royal path, which is open and level, and which leads to the knowledge of God's existence, power, wisdom, goodness, and other qualities: namely, the excellent work of this universe, which exalts its author through its immensity, its divisions, its variety, its order, its beauty, its constancy, and its other attributes.[64]

The path Descartes has abandoned is, in short, that which leads to the discovery of design and purpose in nature. In his *Syntagma*, Gassendi collapses metaphysics into natural philosophy. He rejects the idea of metaphysics (which he calls 'theology') as a discipline separate from natural philosophy, a separation he traces back to Plato, and he follows the Hellenistic division of philosophy into logic, natural philosophy, and ethics:

> The Stoics, Epicureans, and others combined theology with physics. Since the task of theology is to contemplate the natures of things, these philosophers considered that the contemplation of the divine nature and of the other immortal beings was included,

---

[64] Gassendi 1658, 3, p. 337 col. 2. Note that this passage does not appear in the set of objections that was appended to Descartes' *Meditationes,* but in the much expanded and more elaborate version that appeared in 1644 as *Disquisitio Metaphysica.*

especially since the divine nature reveals itself in the creation and government of the universe.[65]

Gassendi's idea was that, in pursuing natural philosophy, one automatically pursues questions about the nature of God because one finds abundant evidence of divine purpose, and hence of the nature of divine causation.

Gassendi was a Catholic priest, and clearly saw his project as operating under the constraints of Catholic theology, but the idea of a natural-philosophical road to natural theology had an obvious radical edge, for natural philosophy was relatively uncontrolled compared to theology, and home-grown natural philosophies might lead to home-grown natural theologies. The phenomenon of 'enthusiasm', zealous sectarianism, in its early sixteenth-century manifestations was primarily associated either with poetic inspiration or claims to the gift of prophecy and direct divine inspiration,[66] but the term was extended, particularly in seventeenth-century England, to those who claimed access to divine inspiration through the study of nature (such as Paracelsians and alchemists), as well as those who practised popular medicine.[67] The problem was far more evident in Protestant countries than Catholic ones, since mystical and miraculous occurrences could be incorporated into Catholic culture in a way that they could not in the case of Protestant culture.[68]

This was a phenomenon evident in England in the rise of a radical Puritanism in the 1580s, and it was targeted by Bacon in *An Advertisement touching the Controversies of the Church of England* of 1589, where he attacks the substitution of enthusiasm or zealotry for learning.[69] Most zealots, he tells us in his *Advertisement*, are 'men of young years and superficial understanding, carried away with partial respect of persons', and their contentions 'either violate truth, sobriety, or peace'.[70] They 'leap from ignorance to a prejudicate opinion, and never take a sound judgement in their way'.[71] They are, in short, incapable of assessing and making sound judgements on the cases they consider, and yet they not only come to conclusions on such cases, but do not consider their lack of learning a handicap. It is for these reasons that Bacon insists that 'the people is no meet judge nor arbitrator, but rather the quiet, moderate, and private assemblies of the learned'.[72]

Bacon was not alone among natural philosophers in perceiving the danger. In the preface to *De Magnete*, Gilbert complains of the 'Ocean of Books' published in his time,

---

[65] Gassendi 1658, 1, p. 27 col. 1.

[66] The first explicit target was the Anabaptist movement; see Bullinger 1560.

[67] For a general coverage of enthusiasm, see Heyd 1995.

[68] Nevertheless, as the phenomenon of Jansenism shows, even within Catholic cultures things could get out of hand; see Kreiser 1978.

[69] See the excellent discussion in Martin 1992, pp. 42ff.

[70] *Advertisement* in Bacon 1859, VIII, p. 82.

[71] *Ibid.*, pp. 82–83.

[72] *Ibid.*, p. 94.

through which very foolish productions the world and unreasoning men are intoxicated,
and puffed up, rave and create literary broils, and while professing to be philosophers,
physicians, mathematicians, and astrologers, neglect and despise men of learning.[73]

It was not just that there was a movement afoot which had eschewed learning in
favour of some special form of insight to which Puritans had claimed access. There
was also an extensive undergrowth of literature, in the form of self-help and self-
improvement books—some of it based on the *problemata* model of frequently asked
questions with answers, following prototypes traditionally ascribed to Aristotle,
Alexander of Aphrodisias, and Plutarch, and some of it apparently *sui generis*—
which was beginning to replace traditional learning. This literature was strongly
associated with the Puritan movement in England, and the problem was that it
contained much of practical value, mixed in with much that was unremarkable, and
much rubbish. This posed a challenge for natural philosophers of Bacon's
generation. Unlike the Aristotelian tradition, it actually produced material of some
worth and use, especially in the area of natural philosophy, but in a completely
unsystematic way, and it was the lack of system that permeated its production that
was responsible for the indiscriminate mixture of wheat and chaff that resulted, a
mixture that its proponents were unable to sort out for themselves, and from which
they were as a consequence unable to learn any lessons. The real challenge was not
so much how to rebut the Puritan threat, but rather how to extricate what was
valuable in it, and to create the environment in which it could be nurtured so as to
maximise its yield.

This was achieved in part through a harnessing of the energies and novel
practical skills produced within this enthusiast movement,[74] by incorporating
elements of the project into a tightly-controlled natural-philosophical programme.
Glanvill, in his defence of the Royal Society in 1668, gives us an idea of how this
project was conceived, distinguishing the kind of chemistry pursued there from that
of Paracelsians and others, for example, telling us that:

> its late *Cultivators*, and particularly the Royal Society, have refined it from its *dross*,
> and made it *honest, sober,* and *intelligible,* an excellent *Interpreter* to *Philosophy,* and
> *help* to *common Life.* For *they* have laid aside the *Chrysopoietick,* the *delusory Designs,*
> and *vain Transmutations,* and *Rosicrucian vapours, Magical Charms* and *superstitious
> Suggestions,* and formed it into an *Instrument* to know the *depths* and *efficacies* of
> Nature.[75]

Such sentiments played a crucial role in the natural-philosophical programme that
emerged in England in the seventeenth century, which was somewhat different from
that we find in France, Italy, or the Netherlands. For one thing, it retains a very
practical view of what it is to do natural philosophy, compared for example with
early seventeenth-century France. But it also involved the transformation of a
disorganised,   highly   individualistic,   practically-oriented   form   of   natural-

---

[73] Gilbert 1958, Preface, sig. *ijr. Pumfrey (1987, pp. 14–73) argues that one of the main aims of
publishing *De Magnete* was to reclaim magnetism for natural philosophy from unlicensed magical
writers. More generally on the threat of printing in England at this time, see Johns 1998, esp. chap. 2.

[74] By the middle of the century, some members the Hartlib circle can be included in this group: see
Webster 1994.

[75] Glanvill 1668.

philosophical practice into something in which enthusiast excesses can be reshaped or curbed.

Tudor and Elizabethan England had raised practical above theoretical learning,[76] and practical knowledge was very much part of the attack on scholasticism. By the 1590s it has taken a distinctly astringent form in writers like Thomas Blundeville in his *Exercises* (London, 1594), in William Barlow's *The Navigator's Supply* (1597),[77] but most of all in *The newe Attractiue* (1581) of Robert Norman, seaman turned instrument-maker, who attacks those who seek knowledge from Latin and Greek texts—they are referred to as pedants who promise much and perform little—and offers an empirically-based, as opposed to a textually-based, procedure:

> I meane not to vse barely tedious coniectures or imaginations, but briefly as I maie to passe it ouer, foundyng my arguements only vpon experience, reason, and demonstration, whiche are the groundes of Artes.[78]

The first attempt to harness this kind of approach and incorporate it into a broad natural-philosophical programme was that of Bacon, who, as we have seen, contrasted contemplation of natural processes with the invention of artificial means of establishing dominion over nature and making it more productive. The project is given a new direction in Boyle's emphasis on uncovering facts and providing nothing but the lowest level theories—he was, as he put it himself, 'no Admirer of the Theorical Part of [the chemists'] Art'[79]—and his insistence on the collective witnessing of experiments,[80] helping to undermine the possibility of drawing contentious natural-theological consequences from natural philosophy by making that natural philosophy as uncontentious as possible. This does not mean that it removes as much as possible from the natural realm, deferring to the supernatural. Quite the contrary, such a move is taken as characteristic of the enthusiast, who confuses the private and the public, the natural and the supernatural. In the latter case, the threat comes not from the collapsing of the supernatural into the natural, as with the naturalists, but with the mistaking of the natural for the supernatural, with the result, as Sprat points out, that the enthusiast 'goes neer to bring down the price of the True and Primitive Miracles, by such a vast, and such a negligent augmenting of their number'.[81]

Sprat makes it an issue of intellectual morality and intellectual honesty. Natural philosophy, as practised by the Royal Society, far from harming Christian values, he tells us, reinforces them,

> seeing many duties of which it is compos'd, do bear some resemblance to the qualifications that are requisite in *Experimental Philosophers*. The spiritual *Repentance* is a careful survay of our former Errors, and a resolution of amendment. The spiritual *Humility* is an observation of our Defects, and a lowly sense of our own weaknesses.

---

[76] See Gaukroger 2001a, pp. 14–18.

[77] See the discussion in Bennett 1991.

[78] I have used the 1614 edition; 'To the Reader', sig. A1r.

[79] Boyle 1999–2000, 2, p. 213.

[80] These aspects of Boyle's programme are discussed in detail in Shapin and Schaffer 1985 and Shapin 1994. But see Shapiro 2000, for a corrective to many aspects of these accounts, especially the latter.

[81] Sprat 1667, p. 362.

> And the *Experimenter* for his part must have some Qualities that answer to these: he
> must judge aright of himself; he must misdoubt the best of his own thoughts; he must be
> sensible of his own ignorance, if ever he will attempt to purge and renew his Reason …
> it may well be concluded, that the doubtful, the scrupulous, the diligent *Observer* of
> *Nature*, is neerer to make a modest, a severe, a meek, an humble *Christian*, than the
> man of *Speculative Science*, who has better thoughts of himself and his own
> *Knowledge*.[82]

The point is echoed in Glanvill, who tells us that 'the *Philosophy* of the *Virtuosi*'
deals 'with the *plain Objects* of *Sense*, in which, if any where, there is *Certainty*; and
teacheth *suspension* of *Assent* till what is *proposed*, is well proved; and so is equally
an Adversary to *Scepticism* and *Credulity*'.[83]

The Royal Society to a large extent institutionalised this approach in its early
years, as least as far as presentation of results was concerned, and Sprat, presenting
the case that it provides a firm foundation for the social order,[84] makes no secret of
its power to curb enthusiasm:

> So that it is now the fittest season for *Experiments* to arise, to teach us a Wisdome,
> which springs from the depths of *Knowledge*, to shake off the shadows, and to scatter
> the mists, which fill the minds of men with a vain consternation. This is a *work* well-
> becoming the most *Christian Profession*. For the most apparent effect, which attended
> the passion of *Christ*, was the putting of an eternal silence, on all the false oracles, and
> dissembled inspirations of *Antient Times*.[85]

Indeed, not only is it able to curb enthusiasm, it is able to harness it as well, in
the form of natural philosophy, irrespective of the religion of the participants. If
enthusiasm was manifested in private interpretations of scripture, then the antidote
lay in a public, co-operative and universally valid enterprise.[86] Since it operates via a
procedure that avoids disputes, it offers a means of 'abolishing or restraining the
fury of *Enthusiasme*'.[87] The Royal Society 'freely admitted men of different
Religions, Countries, and Professions of Life … For they openly profess, not to lay
the Foundation of an *English, Scotch, Irish, Popish*, or *Protestant* Philosophy; but a
Philosophy of *Mankind*'.[88] This was a radical claim, and some critics of the Royal
Society, such as Meric Casaubon and Henry Stubbe believed that it had not only
misunderstood and underestimated the threat of enthusiasm, but had become a centre
for it.[89] In particular, Sprat had spelled out the consequences of a properly pursued
natural philosophy for a properly constructed natural theology, maintaining that the
experimental natural philosopher:

> will be led to admire the wonderful contrivance of the *Creation*, and so to apply, and
> direct his praises aright: which no doubt, when they are offer'd up to *Heven*, from the

---

[82] Sprat 1667, pp. 366–367.
[83] Glanvill 1671, pp. 143–144.
[84] See Wood 1980.
[85] Sprat 1667, pp. 362–363.
[86] See Heyd 1995, p. 152.
[87] Sprat 1667, p. 428.
[88] *Ibid.*, p. 63.
[89] See Heyd 1995, chap. 5; and Spiller 1980.

mouth of one, who has well studied what he commends, will be more sutable to the
Divine Nature, than the blind applauses of the Ignorant.[90]

Stubbe is incensed that it could be suggested that there could be a route to salvation
that did not rely on the mediation of Christ. Commenting directly on this passage in
Sprat, he writes:

> The former part of the passage is contrary to the *Analogy of Faith* and *Scripture*, in that
> it makes the acceptableness of mens prayers to depend more or less on the study of
> natural Philosophy. Whereas the *Apostle* suspends the *acceptableness of all Prayers
> unto God,* in being made unto him *in the name,* and *for the mediation of Christ Jesus,*
> applied by *faith.*[91]

Here we have is the nub of the issue. The Royal Society apologists were
concerned to find a middle ground that enabled them to pursue natural philosophy in
such a way that natural-theological consequences could be drawn from it. In this
way, not only would natural philosophy and natural theology no longer stand in need
of reconciliation, but radically conflicting religious beliefs would not be able to enter
the picture and destroy any theological consensus before the process had even got
off the ground. This way of proceeding—which critics such as Casaubon and Stubbe
rejected on the grounds that theology can only be grounded in revelation—required
subjection to a mode of pursuing natural philosophy (or at least a mode of
representing how natural philosophy was to be pursued) which placed constraints on
both natural philosophy and natural theology. Indeed, to a large extent it shaped
natural-philosophical practice in mid seventeenth-century England, and it provided
fertile ground for a form of natural theology whose connection to Christianity
became looser as time progressed.

In sum, there are two kinds of response to the dilemma brought to a head in
Pomponazzi. One is to consolidate the gulf between theological and natural-
philosophical truths, making these completely different kinds of truths, but
introducing a divine guarantee so that the pursuit of natural philosophy in the
appropriate way could be allowed to produce natural truths which in fact
corresponded to divine or absolute truths. The other is to abolish the gulf by guiding
natural philosophy in the direction of natural theology, making it the 'royal road' to
an understanding of God's purposes in creating the world. Whatever their original
rationale, both of these developments point in the direction of significantly increased
autonomy for natural philosophy, where what 'autonomy' means in this context is
that there is no discourse external to natural philosophy by which it must judge its
results. In the Cartesian version of the absolute transcendence project, this is
because, while there is a discourse which lays claim to absolute truth, we have no
access in principle to such discourse so no form of bridging is possible. All we can
rely on is our faculty of clear and distinct ideas, something which is purely natural
and which is constitutive of our reasoning processes, providing internal criteria for
cognitive judgement. On this view, however, once we appreciate the nature of God,
we realise that this faculty is given to us by God and that he guarantees that its use

---

[90] Sprat 1667, p. 349.
[91] Stubbe 1670, p. 36.

generates absolute truths, even though we have no understanding of the rationale behind these absolute truths. In contrast, on the approach whereby natural philosophy is a form of natural theology, autonomy is not quite so strong in the first instance, since the natural theology generated is subject to correction by revelation, although with the development of deism towards the end of the seventeenth century this requirement is heavily qualified, as fuller autonomy is achieved.

4.THE AIMS OF ENQUIRY

We have been concerned with a number of factors that shape natural-philosophical projects, and what has emerged is a complex set of considerations within which questions of the relation between justification and truth are embedded. This complexity is in part constitutive of the problem, not an added extra, and to cope with it, we need to approach the question in terms of an expanded set of resources. In particular, as well as truth and justification, we also need to consider the issues of objectivity and legitimation. If it was no longer a constraint on the newly-emerging natural philosophies that they judge their results against a criterion of truth which derives from something external, then there is some sense in which it is unhelpful to view such programmes solely in terms of truth. In particular, it is appropriate to ask whether consideration of the issues of objectivity and legitimation might ease some of the pressure on the idea of truth.

An enquiry is objective to the extent that it does not depend upon any features of the particular subject who studies it. An objective account is, in this sense, impartial, one which could ideally be accepted by any subject, because it does not draw on any assumptions, prejudices, or values of particular subjects.[92] Objectivity has two features that make it especially attractive as something that might regulate natural-philosophical enquiry. First, unlike truth, objectivity comes in degrees—some procedures can be more objective than others—and it is something that can be improved upon through practice. Second, the idea that natural-philosophical enquiry aims at truth might hold for what Aristotle called the theoretical sciences, but this does seem particularly inappropriate as a characterisation of the aims of a discipline such as medicine, which was increasingly incorporated into natural-philosophical enquiry in the course of the sixteenth and seventeenth centuries. By contrast, the idea that objectivity, to the extent to which it is part of a natural-philosophical enterprise, should regulate an area such as medical enquiry is as unproblematically appropriate as the demand that it should regulate cosmological enquiry.

The ways in which early modern philosophers tried to reformulate the project of natural philosophy bears this out. When Bacon advocated the purging of 'idols' from the mind, when Galileo presented his arguments in the context of a patronage system which was disinterested, when Descartes argued that scholastics should be replaced by men of the world as natural philosophers, when Boyle and the members of the

---

[92] This is not to deny that the extent to which full objectivity is possible in any particular case, and just how objectivity should be secured, are not going to be uncontentious in every case, although many of the supposed problems here turn out not to be such on closer examination. See Gaukroger 2001b.

Royal Society attempted to present their findings in the closest way to bare 'facts',[93] what they were all seeking, in their different fashions, was a way of securing objectivity, not a means of securing truth. Purging the mind of idols does not produce truth as such: it rids the mind of those features that would impair its objectivity. Disinterestedness does not produce truth, but manifests a form of freedom from pre-conceived ideas and prejudices (both of which could in fact be true). Men of the world are no more seekers after the truth than scholastic clergymen are, but they bring an intellectual honesty to the task because they are free from prejudice and from an education which prevents them from thinking for themselves. Presenting bare facts rather than grand theories does not produce truth but rather favours one kind of truth over another because one kind manifests objectivity whereas this is more difficult to display in the other. The presentation of results in terms of bare facts, for example, which often exasperated Boyle's continental contemporaries,[94] was not for Boyle a provisional record of research which was at a stage too early to merit systematisation, so much as a way of manifesting the legitimacy of his whole natural-philosophical project.

English natural philosophy, at least from the middle of the seventeenth century, is dominated, in the areas of natural history and matter theory, by the notion of objectivity, and this is pursued not externally, in terms of truth, but internally, in terms of impartiality. Truth is not contextual, and so cannot change with contextual changes, but objectivity is contextual and can and does change with contextual changes. This is why tracking fundamental epistemological changes in seventeenth-century natural philosophy is impossible in terms of truth,[95] but why objectivity is promising. Indeed, the attempt to generate internal criteria of objectivity acts not as a second best option in the absence of a means of establishing truth, but as a way of legitimating any cognitive claims a discipline might make.

The concern with objectivity, in the form of impartiality and lack of bias, is reflected in the fact that there is at least as much concern with the character of the natural philosopher as there is with questions of method. It is not just the procedure one follows, it is the qualities that the natural philosopher brings to bear on the enterprise, just as—especially in the wake of the Reformation—it is the personal qualities of the cleric that manifest the authenticity of his religion as much as do his

---

[93] Shapiro 2000, p. 31, notes that, in the English legal tradition from which the natural-philosophical usage derives, a 'fact' originally meant an alleged act whose occurrence was in contention. As she notes, 'One of the great changes that occurred in the course of two centuries in some cultural areas was the transformation of "fact" from something that had to be sufficiently proved by appropriate evidence to be considered worthy of belief to something for which appropriate verification had already taken place'.

[94] Leibniz, for example, wrote to Oldenburg asking him impatiently to urge Boyle to produce a systematic exposition of his views on chemistry; Leibniz to Oldenburg, 5 July 1674 and 10 May 1675; Oldenburg 1965–1975, 11, p. 46 and p. 306 respectively. Spinoza also expressed reservations to Oldenburg on this score in a letter of April 1662; *ibid.*, 1, p. 462.

[95] *Contra* Shapin 1994. Although some of the details of Shapin's account are questionable—especially his account of gentlemanly modes of witnessing (see for example Shapiro 2000, pp. 25–33)—he draws attention to crucial epistemological changes in English natural philosophy. But he construes these in terms of shifts in understanding what constitutes truth, whereas they are more helpfully seen as moves to establish what constitutes objectivity and how it is manifested.

theological beliefs. In this way, objectivity becomes a distinctive way of pursuing key disciplines such as natural philosophy and history, in which a new set of cognitive values are shaped around the notion of impartiality. The procedures originate in the Renaissance models of history and law, gradually become transferred to natural philosophy via natural history, and then become extended to all discourses making cognitive claims, most notably theology, where not only does natural theology take on a new significance, but biblical chronology and biblical narrative come to subjected to these same general criteria.

REFERENCES

Albrecht, M. (1994) *Eklektik. Eine Begriffsgeschichte mit Hinweisen auf die Philosophie- und Wissenschaftsgeschichte*, Stuttgart/Bad Cannstatt: Fromann-Holzboog.

Aquinas, T. (1968) *On the Unity of the Intellect Against the Averroists*, trans. B. H. Zedler, Milwaukee: Marquette University Press.

Armogathe, J.-R. (1977) *Theologia cartesiana: l'explication physique de l'Eucharistie chez Descartes et Dom Desgabets*, The Hague: M. Nijhoff.

Bacon, F. (1859) *The Works of Francis Bacon*, 7 vols, eds J. Spedding, R. L. Ellis and D. D. Heath, London.

Barlow, W. (1597) *The Navigator's Supply*, London.

Bennett, J. A. (1991) 'The challenge of practical mathematics' in *Science, Culture and Popular Belief in Renaissance Europe*, eds S. Pumfrey, P. L. Rossi and M. Slawinski, Manchester: Manchester University Press, pp.186–9.

Biagioli, M. (1989) 'The social status of Italian Mathematicians', *History of Science*, 27, pp. 41–95.

— (1993) *Galileo Courtier*, Chicago: University of Chicago Press.

Bjurstrom, P. (1972) 'Baroque Theater and the Jesuits' in *Baroque Art: The Jesuit Contribution*, eds R. Wittkower and O. B. Jaffe, New York: Fordham University Press, pp. 99–110.

Blackwell, C. W. T. (1995) 'The case of Honoré Fabri and the historiography of sixteenth and seventeenth century Jesuit Aristotelianism in Protestant history of philosophy: Sturm, Morhof and Brucker', *Nouvelles de la Republique des Lettres*, pp. 49–77.

Blackwell, R. J. (1991) *Galileo, Bellarmine, and the Bible*, Notre Dame: University of Notre Dame Press.

Blundeville, T. (1594) *M. Blundevile his exercises containing sixe treatises*, London.

Boyle, R. (1999–2000) *The Works of Robert Boyle*, 14 vols, eds M. Hunter and E. B. Davis, London: Pickering and Chatto.

Brown, T. M. (1978) 'The rise of Baconianism in seventeenth-century England: a perspective on science and society during the Scientific Revolution' in *Science and History: Studies in Honor of Edward Rosen*, Studia Copernica, 16, Wroclaw, pp. 501–522.

Bullinger, H. (1560) *Der Widertoeufferen ursprung*, Zurich.

Burke, P. (1986) *The Italian Renaissance: Culture and Society in Italy*, Princeton: Princeton University Press.

Cassirer, E., Kristeller, P. O. and Randall, J. H. Jnr, eds (1948) *The Renaissance Philosophy of Man: Selections in Translation*, Chicago: University of Chicago Press.

Charleton, W. (1654) *Physiologia Epicuro-Gassendo-Charltoniana: or A Fabrick of Science Natural, upon the Hypothesis of Atoms*, London.

Compayré, G. (1879) *Histoire critique des doctrines de l'éducation en France*, 2 vols, Paris: Hachette.

Cottingham, J. ed. (1992) *The Cambridge Companion to Descartes*, Cambridge : Cambridge University Press.

Delumeau, J. (1971) *Le Catholicisme entre Luther et Voltaire*, Paris: Presses universitaires de France.

— (1978) *La Peur en occident (XIVe–XVIIIe siècles): Une cité assiégée*, Paris: Fayard.

— (1983) *Le Péché et la peur: La culpabilisation en occident, XIIIe–XVIIIe siècles*, Paris: Fayard.

— (1989) *Rassurer et protéger: Le sentiment de sécurité dans l'occident d'autrefois*, Paris: Fayard.

— (1992) *L'Aveu et le pardon*, Paris: Fayard.

Denifle, H. and Châtelain, E. eds (1889–1897) *Chartularium Universitatis Parisiensis*, 4 vols, Paris: Fratrum Delalain.

Descartes, R. (1996) *Oeuvres de Descartes*, eds C. Adam and P. Tannery, 11 vols, Paris: J. Vrin.

Diderot, D. (1751–1765) *Encyclopèdie ou Dictionnaire Raisonné des Sciences, des Arts et des Métiers*, 17 vols, Paris.

Drake, S. (1960) *The Controversy on the Comets of 1618*, Philadelphia: Pennsylvania University Press.

— (1970) *Galileo Studies*, Ann Arbor: University of Michigan Press.

Ficino, M. (2001–) *Platonic Theology*, eds T. Hankins and W. Bowen, trans. M. J. B. Allen and J. Warden, 6 vols, Cambridge, MA: Harvard University Press.

Fowler, C. (1999) *Descartes on the Human Soul: Philosophy and the Demands of Christian Doctrine*, Dordrecht: Kluwer.

Gassendi, P. (1658) *Opera Omnia*, 6 vols, Lyon.

Gaukroger, S. (1989) *Cartesian Logic: An Essay on Descartes's Conception of Inference*, Oxford: Oxford University Press.

— (1992) 'Descartes' early doctrine of clear and distinct ideas', *Journal of the History of Ideas*, 53, pp. 585–602.

— (1995) *Descartes, An Intellectual Biography*, Oxford: Oxford University Press.

— (2001a) *Francis Bacon and the Transformation of Early Modern Philosophy*, Cambridge: Cambridge University Press.

— (2001b) 'Objectivity, History of' in *International Encyclopedia of the Social and Behavioural Sciences*, 26 vols, eds N. J. Smelser and P. B. Baltes, Amsterdam and New York: Elsevier, 16, pp. 10785–10789.

— (2002) *Descartes' System of Natural Philosophy*, Cambridge: Cambridge University Press.

Gaukroger, S., Schuster, J. A. and Sutton, J. eds (2000) *Descartes' Natural Philosophy*, London: Routledge.

Gibson, R. W. (1950) *Francis Bacon: A Bibliography of his Works and of Baconiana, to the Year 1750*, Oxford: Scrivener Press.

Gilbert, N. W. (1971) 'The early Italian humanists and disputation' in *Renaissance Studies in Honor of Hans Baron*, eds A. Molho and J. Tedeschi, Dekalb: Northern Illinois University Press, pp. 203–236.

Gilbert, W. (1958) *On the Magnet*, Preface, trans. S. P. Thompson, New York: Basic Books.

Gilson, E. (1961) 'Autour de Pomponazzi: problématique de l'immortalité de l'âme en Italie au début du XVIe siècle', *Archives d'histoire doctrinale et littéraire du moyen âge*, 18, pp. 163–279.

Glanvill, J. (1665) *Scire/i tuum nihil est; or, The Authors Defence of the Vanity of Dogmatizing*, London.

— (1668) *Plus Ultra*, London.

— (1671) *A Præfatory Answer to Mr. Henry Stubbe*, London.

Greengrass, M., Leslie, M. and Raylor, T. eds (1994) *Samuel Hartlib and Universal Reformation*, Cambridge: Cambridge University Press.

Heyd, M. (1995) *'Be Sober and Reasonable': The Critique of Enthusiasm in the Seventeenth and Early Eighteenth Centuries*, Leiden: Brill.

Hisette, R. (1977) *Enquête sur les 219 articles condamnés à Paris le 12 Mars 1277*, Louvain/Paris: Publications universitaires.

— (1980) 'Etienne Tempier et ses condemnations', *Recherches de théologie ancienne et médiévale*, 47, pp. 231–270.

Heinzmann, R. (1965) *Die Unsterbliche der Seele und die Auferstehung des Leibes*, Münster: Aschendorff.

Hsia, R. Po-Chia (1989) *Social Discipline in the Reformation*, London: Routledge.

Johns, A. (1998) *The Nature of the Book*, Chicago: University of Chicago Press.

Jolley, N. (1992) 'The reception of Descartes' philosophy' in *The Cambridge Companion to Descartes*, ed. J. Cottingham, Cambridge: Cambridge University Press, pp. 393–423.

Knox, R. (1681) *An Historical Relation of the Island Ceylon, in the East-Indies*, London.

Kreiser, B. R. (1978) *Miracles, Convulsions, and Ecclestiastical Politics in Early Eighteenth-Century Paris*, Princeton: Princeton University Press.

Levi, A. (2002) *Renaissance and Reformation: The Intellectual Genesis*, New Haven: Yale University Press.

Lipsius, J. (1604) *Manducationis ad Stoicam philosophiam libri tres*, Antwerp.

McClaughlin, T. (1979) 'Censorship and defenders of the Cartesian faith in France (1640–1720)', *Journal of the History of Ideas*, 40, pp. 563–581.

Martin, J. (1992) *Francis Bacon, the State, and the Reform of Natural Philosophy*, Cambridge: Cambridge University Press.

Mersenne, M. (1624) *L'Impiete des Deistes*, Paris.

Molho, A. and Tedeschi, J. eds (1971) *Renaissance Studies in Honor of Hans Baron*, Dekalb: Northern Illinois University Press.

Norman, R. (1614) *The nevve, attractive shewing the nature, propertie, and manifold vertues of the loadston*, London.

Oestreich, G. (1982) *Neostoicism and the Early Modern State*, Cambridge: Cambridge University Press.

Oldenburg, H. (1965–1986) *The Correspondence of Henry Oldenburg*, 13 vols, eds A. R. Hall and M. B. Hall, Madison, Milwaukee and London.

Osler, M. J. (1991a) 'Fortune, fate, and divination: Gassendi's voluntarist theology and the baptism of Epicureanism' in *Atoms, Pneuma, and Tranquillity: Epicurean and Stoic Themes in European Thought*, ed. M. J. Osler, Cambridge: Cambridge University Press, pp. 155–174.

— ed. (1991b) *Atoms, Pneuma, and Tranquillity: Epicurean and Stoic Themes in European Thought*, Cambridge: Cambridge University Press.

Peghaire, J. (1936) *Intellectus et ratio selon S. Thomas d'Aquin*, Paris: Vrin.

Pérez-Ramos, A. (1988) *Francis Bacon's Idea of Science and the Maker's Knowledge Tradition*, Oxford: Clarendon Press.

Pumfrey, S. (1987) 'William Gilbert's Magnetic Philosophy, 1580–1684: The Creation and Dissolution of a Discipline,' Ph.D. thesis, The Warburg Institute, London.

Pumfrey, S., Rossi, P. L. and Slawinski, M. eds (1991) *Science, Culture and Popular Belief in Renaissance Europe*, Manchester: Manchester University Press.

Ranea, A. G. (2000) 'A "science for *honnêtes hommes*": *La Recherche de la Vérité* and the deconstruction of experimental knowledge' in *Descartes' Natural Philosophy*, eds S. Gaukroger, J. A. Schuster, and J. Sutton, London: Routledge, pp. 313–329.

Rochemonteix, C. de (1889) *Un Collège des jesuits au XVIIe et au XVIIIe siècles*, 4 vols, Le Mans: Leguicheux.

Schmaltz, T. M. (1999) 'What has Cartesianism to do with Jansenism?', *Journal of the History of Ideas*, 60, pp. 37–56.

Shapiro, B. (2000) *A Culture of Fact: England, 1550–1720*, Ithaca NY: Cornell University Press.

Sidney, P. (1965) *An Apology for Poetry*, ed. G. Shepherd, London: T. Nelson.

Settle, T. B. (1971) 'Ostilio Ricci, a bridge between Alberti and Galileo', *Actes du XIIe Congrès International d'Histoire des Sciences*, Paris: Blanchard, pp. 121–126.

Shapin, S. (1994) *A Social History of Truth: Civility and Science in Seventeenth-Century England*, Chicago: University of Chicago Press.

Shapin, S. and Schaffer, S. (1985) *Leviathan and the Air Pump: Hobbes, Boyle, and the Experimental Life*, Princeton: Princeton University Press.

Spiller, M. R. S. (1980) *'Concerning Natural Philosophy': Meric Casaubon and the Royal Society*, The Hague: M. Nijhoff.

Sprat, T. (1667) *The History of the Royal-Society of London for the Improving of Natural Knowledge*, London.

Stubbe, H. (1670) *A Censure upon Certain Passages contained in the History of the Royal Society*, Oxford.

Tanner, N. P. ed. (1990) *Decrees of the Ecumenical Councils*, 2 vols, London: Sheed & Ward.

Van Otegem, M. (2002) *A Bibliography of the Works of Descartes (1637–1704)*, 2 vols, Department of Philosophy, University of Utrecht.

Webster, C. (1994) 'Benjamin Worsley: engineering for universal reform from the Invisible College to the Navigation Act' in *Samuel Hartlib and Universal Reformation*, eds M. Greengrass, M. Leslie, and T. Raylor, Cambridge: Cambridge University Press, pp. 213–235.

— (2002) *The Great Instauration: Science, Medicine and Reform (1626–1660)*, 2nd edn, Berne: Peter Lang.

Wood, P. B. (1980) 'Methodology and apologetics: Thomas Sprat's *History of the Royal Society*', *British Journal for the History of Science*, 13, pp. 1–26.

PETER HARRISON

PHYSICO-THEOLOGY AND THE MIXED SCIENCES

The Role of Theology in Early Modern Natural Philosophy

The last decade has witnessed an energetic discussion amongst historians of the early modern period concerning the identity of natural philosophy. Two concerns have dominated the agenda: first, the manner in which mathematical explanations were imported into what was traditionally a qualitative enterprise;[1] second, the extent to which natural philosophy admitted theology and theological modes of explanation.[2] This chapter is primarily concerned with the second of these—to do with the relation of 'science' and 'religion' during this period—but will also suggest that the two issues are related in important ways. Three specific claims will be made: (1) that the emergence of the disciplinary category 'physico-theology' was an explicit attempt to address the issue of the place of theology in early modern natural philosophy; (2) that this category is analogous in certain respects to the 'physico-mathematics' inasmuch as both represent attempts to renegotiate traditional disciplinary boundaries; (3) that physico-theology resolved vocational tensions specific to this period concerning the extent to which it was legitimate for naturalists to be engaged in theology, and conversely, for clerics to be engaged in the study of nature.

1. GOD AND EARLY MODERN NATURAL PHILOSOPHY: GALILEO, BACON, AND DESCARTES

In an influential paper that appeared in 1988 Andrew Cunningham suggested that natural philosophy was fundamentally about 'God's achievements, God's intentions, God's purposes, God's messages to man'.[3] This claim has much to commend it, not least because it poses in an acute way the question of the nature of natural philosophy and of how it differs from modern science. The thesis also has wide-ranging implications. Thus, if Cunningham is correct in asserting that natural philosophy is essentially about God and his creation, then it would seem to follow that much of the discussion about 'science and religion' during this period is misplaced, since in natural philosophy we have a discipline that comprehended both

---

[1] Cunningham 1991; Dear 1995, chap. 6 and 1998; Gaukroger and Schuster 2002.

[2] Cunningham 1988; Osler 1997; Grant 2000; Dear 2001; Cunningham 2001.

[3] Cunningham 1988, p. 384.

P. R. Anstey and J. A. Schuster (eds.), The Science of Nature in the Seventeenth Century, 165-183.

topics within its scope.[4] One might also wonder why the discrete new disciplinary category of physico-theology arose, if natural philosophy already included a significant theological dimension. Some analysis of the relationship between physico-theology and natural philosophy is clearly called for. At the outset, then, we need to consider the 'Cunningham thesis' in order to determine whether, and in what sense, natural philosophy can be regarded as being uniformly about God's achievements, intentions, and so on. In this section I shall suggest that in fact the role of God and of theological explanations in early modern natural philosophy are not taken for granted by the major protagonists, and that there was considerable scope for a range of positions on this question. The positions I shall briefly consider here are those of Galileo, Descartes, and Bacon.

Early modern discussions of the place of God and of theological explanation in natural philosophy take place against the background of two aspects of Aristotle's influential conception of *scientia*. First, Aristotle had proclaimed that each of the sciences has its own class of objects, and that the methods appropriate for one science should not be transposed to another.[5] Aristotle had identified three speculative sciences: mathematics, which dealt with unchanging, immaterial objects that were dependent on the human mind; natural philosophy which was concerned with changeable material objects that were independent of the human mind, and the 'divine science' or metaphysics, which dealt with unchangeable, immaterial objects that were independent of the human mind.[6] The relation of the divine science to the other sciences, as we might expect, exercised the imaginations of Aristotle's medieval heirs to a considerable degree.[7] Thomas Aquinas, for example, considered the question of whether the divine science dealt with all objects. He concluded: 'Sacred doctrine does not treat of God and creatures equally, but of God primarily, and of creatures only so far as they are referable to God as their beginning or end. Hence the unity of this science is not impaired'. Aquinas thus argued that inasmuch as the creatures have God as their final cause, their study is part of the divine science. Aquinas also made reference to the notion of 'subordinate sciences': 'Nothing prevents inferior faculties or habits from being differentiated by something which falls under a higher faculty or habit as well ... Similarly, objects which are the subject-matter of different philosophical sciences can yet be treated of by this one single sacred science.'.[8] For Aquinas, then, it was final causes that provided the key to understanding how the sciences of created objects were subordinate to theology.

---

[4] See, e.g., Dear 2001, p. 378.

[5] Aristotle, *Posterior Analytics*, 75a–b; *Metaphysics*, 989b–990a; *On the Heavens*, 299a–b. See also Funkenstein 1986, pp. 35–37, 303–307.

[6] Aristotle, *Metaphysics*, 1025b–1026a.

[7] Of course, medieval thinkers also operated with a modified conception of 'divine science', which for them was informed by revelation as well as reason.

[8] *Summa theologiae*, 1a. 1, 3. *Scientiae mediae* (middle sciences) is the expression that Aquinas used for other mixed sciences such as optics, mechanics, and music. Aquinas spoke of 'habits' in this context because for him, as indeed for Aristotle, *scientia* was in its primary sense not a body of knowledge but a mental habit. See, e.g., *Summa theologiae*, 1a2ae. 49, 1; 1a2ae. 50, 3; 1a2ae. 52, 2; 1a2ae. 53, 1.

The other significant element of the Aristotelian legacy concerned the ideal of scientific knowledge as demonstrative and certain.[9] Clearly the truths of revelation, some of which were contingent historical facts, did not admit of the demonstrative certainty of logically deduced propositions. How, then, could theology be a true science, far less the queen of the sciences? Aquinas addressed this problem by suggesting that truths of revelation derive their certainty from their source—God or the Church—thus providing alternative grounds for holding propositions to be certain.[10] In due course, Protestant theologians were to add that scriptural propositions also counted as 'scientific' knowledge. Mathematician and astronomer Georg Joachim Rheticus (1514–1576), for example, insisted that all passages of scripture, without distinction, bore the force of demonstration.[11] The allocation of equal weight to the words of scripture and the findings of properly constituted science raised the prospect of irreducible conflict between two demonstrative truths. Such a possibility had been anticipated by Augustine, who established the hermeneutical principle that scriptural assertions about physical reality could not be in conflict with demonstrated truths of natural philosophy.[12] The multi-layered system of biblical interpretation favoured by Augustine and his medieval successors allowed for non-literal readings of scriptural passages in such instances.[13] An alternative solution was that of the 'double truth', a notion commonly attributed to Averroës and some of the schoolmen, according to which what is true in philosophy may be false in theology, and vice versa.[14] These considerations, admittedly presented here in a somewhat simplified form, represent part of the medieval background to early modern discussions of the relation between theology and natural philosophy. They relate in particular to the legitimacy of explanation in terms of final causes, to how the Aristotelian model of 'subordinate sciences' might work in the case of sciences being subordinated to theology; and to how 'demonstrative knowledge' remained a significant scientific ideal.

When we consider the stated positions of Galileo, Bacon, and Descartes, the relevance of these concerns becomes immediately apparent. In his celebrated 'Letter to the Grand Duchess Christina' (1615), Galileo responded to those of his critics who had exploited theological or exegetical arguments against heliocentrism. While Galileo was mostly concerned to elaborate his view of how scripture was to be interpreted on matters relating to natural philosophy, he also sought to clarify the

---

[9] Aristotle, *Posterior Analytics*, 71b–72b.

[10] Aquinas, *Summa contra gentiles*, 1.6. On Aquinas' view of *scientia* see MacDonald 1993. Cf. Descartes, who suggested that 'what has been revealed by God is more certain than any knowledge, since faith in these matters, as in anything obscure, is an act of the will rather than an act of the understanding'; *Rules for the Direction of the Mind*, §370; Descartes 1985, I, 15.

[11] 'For it is written that one shall not diverge from the words of the Lord, either to the right or to the left, and that the Word itself has the force of demonstration, since it has been given to us by God'; Rheticus 1984, pp. 65f. See also Lohr 1988, p. 633. The circularity of this piece of reasoning seems to have escaped Rheticus, but his view was not uncommon, particularly amongst Protestant thinkers for whom the authority of scripture was paramount.

[12] Augustine 1844–1905, vol. 34, p. 270.

[13] For Augustine's hermeneutics see Harrison 1998, pp. 25–33.

[14] It is doubtful, however, that any medieval theologian advocated such a position. See, e.g., Dales 1984 and MacClintock 1972.

relation of theology, as the queen of the sciences, to the 'subordinate' or 'inferior' sciences. The relation of superior to subordinate sciences can be understood in one of two ways, he suggests. First, the 'pure' science is superior to the 'applied', as in the case of arithmetic and geometry in relation to accounting and surveying, for the rules 'are more excellently contained' in the former than in the latter. The theoretical sciences are thus more comprehensive than their applied versions. Second, a science may be superior on account of the special dignity of its object and the manner in which its truths are communicated. Theology, Galileo insisted, is a superior science in this second sense. This account of the relation means that geometry, astronomy, music and medicine, though inferior to theology on account of their objects, are not more excellently contained in the pages of scripture than they are in the writings of the philosophers.[15] To crown his argument Galileo invoked Augustine's assertion that there can be no conflict between the truths of scripture and the demonstrated truths of the sciences. Galileo thus contended that 'truly demonstrated physical conclusions need not be subordinated to biblical passages, but the latter must rather be shown not to interfere with the former'.[16] Galileo seems here to adopt the position that philosophy should be independent of theology, and indeed if there is a dependent relationship, it is biblical interpretation that should rely on the demonstrated conclusions of philosophy, rather than the reverse.[17]

Francis Bacon also appears to insist on a number of occasions that natural philosophy remain pure from admixture with heteronymous disciplines, including theology. In *Novum Organum* he complained that: 'We have as yet no natural philosophy that is pure; all is tainted and corrupted; in Aristotle's school by logic; in Plato's by natural theology; in the second school of Platonists, such as Proclus and others, by mathematics, which ought only to give definiteness to natural philosophy, not to generate or give it birth'.[18] Natural theology, then, corrupts natural philosophy, or at least it did in the case of Plato. *The Advancement of Learning* contains Bacon's best-known admonition against 'unwisely mingling' divinity and philosophy:

> ... let no man upon a weak conceit of sobriety or an ill-applied moderation think or maintain, that a man can search too far, or be too well studied in the book of God's word, or in the book of God's works, divinity or philosophy, but rather let men endeavour an endless progress or proficiency in both; only let them beware that they

---

[15] Galileo, 'Letter to the Grand Duchess Christina', in Drake 1957, pp. 191–193. It might be thought that Galileo is mostly concerned with the relation of astronomy to theology, but in claiming that astronomy dealt with physical truths (rather than with mathematical hypotheses) Galileo was seeking to bring astronomy within the ambit of natural philosophy. He thus distinguishes 'mathematical astronomers' who seek to save the appearances, and 'philosophical astronomers who, going beyond the demand that they somehow save the appearances, seek to investigate the true constitution of the universe...', *Letters on Sunspots* in *ibid.*, p. 97.

[16] *Ibid.*, pp. 194f.

[17] '... having arrived at certainties in physics, we ought to utilize these as the most appropriate aids in the true exposition of the Bible and in the investigation of those meanings which are necessarily contained within'; *ibid.*, p. 183. The issue of which philosophical truths might have demonstrative weight is skirted by Galileo.

[18] *Novum Organum*, I, xcvi; Bacon 1859, IV, p. 93.

apply both to charity, and not to swelling; to use, and not to ostentation; and again, that they do not unwisely mingle, or confound these learnings together.[19]

Knowledge, Bacon goes on to say, 'is first of all divided into divinity and philosophy', implying that this is a boundary that ought to be respected.[20]

In the same work Bacon revisited these issues in his treatment of final causes, arguing that these causes should not be admitted into physics (i.e. natural philosophy): 'For the handling of final causes, mixed with the rest in physical enquiries, hath intercepted the severe and diligent inquiry of all real and physical causes, and given men the occasion to stay upon these unsatisfactory and specious causes, to the great prejudice of further discovery'. The appropriate venue for a discussion of final causes, he insisted, is metaphysics and not physics.[21] Finally, if philosophical theology is to be excluded from physics, so too, is biblical exegesis. Bacon thus censured 'the school of Paracelsus, and some others, that have pretended to find the truth of all natural philosophy in the scriptures'.[22] One cannot, he insisted in *Novum Organum*, found a system of natural philosophy on the first chapter of Genesis.[23] On the face of it, all of this is suggestive of a position that would largely exclude theological considerations from the scope of natural philosophy.

Descartes was another who sought to set out the explicit boundaries of natural philosophy and theology in such a way as to avoid confusion. In his *Comments on a Certain Broadsheet* (1648) he distinguished three kinds of knowledge with a view to clarifying the proper relation of philosophy to religion:

> First, some things are believed through faith alone—such as the mystery of the Incarnation, the Trinity, and the like. Secondly, other questions, while having to do with faith, can also be investigated by natural reason: among the latter, orthodox theologians usually count the questions of the existence of God, and the distinction between the human soul and the body. Thirdly, there are questions which have nothing whatever to do with faith, and which are the concern solely of human reasoning, such as the problem of squaring the circle, or of making gold by the techniques of alchemy, and the like.

Problems arise, Descartes observed, when the methods of philosophy are erroneously applied to revealed truths, and when putatively revealed truths are applied to things properly the subject matter of philosophy alone: 'Just as it is an abuse of Scripture to presume to solve problems of the third sort on the basis of some mistaken interpretation of the Bible, so it diminishes the authority of Scripture to undertake to demonstrate questions of the first kind by means of arguments derived solely from philosophy'.[24] The existence of problems of the third sort suggests a realm of knowledge that ought to be quarantined from theological interests. The second category of knowledge, however, concerning what can be

---

[19] *Of the Advancement of Learning*, I. i. 3; Bacon 1974, pp. 9f.

[20] *Ibid.*, II. v. 1, p. 83.

[21] *Ibid.*, II. vii. 7, p. 94. Bacon explains the difference in *Novum Organum*. The investigation of forms belongs to metaphysics and of efficient causes to physics. Each has a subordinate science, respectively magic and mechanics: *Novum Organum*, II, ix; Bacon 1859, IV, p. 126.

[22] *Of the Advancement of Learning*, II, xxv. 16; Bacon 1974, p. 207.

[23] *Novum Organum*, I, lxv; Bacon 1859, IV, p. 66.

[24] *Comments on a Certain Broadsheet*; Descartes 1985, I, p. 300.

known of God and the soul through reason alone, Descartes allowed to be the proper province of philosophers.[25]

As Galileo before him, Descartes did nonetheless subscribe to a particular understanding of the subordination of natural philosophy to theology. In the *Replies to Objections* he notes that in philosophy 'we must begin with knowledge of God, and our knowledge of all other things must then be subordinated to this single initial piece of knowledge'. This, he maintains, he had explained in the *Meditations*.[26] By this Descartes meant simply that knowledge of God provides the epistemological foundation upon which natural philosophy is constructed. This, of course, is a notion of subordination quite different from the traditional understanding.

Descartes also shared Bacon's view that the consideration of final causes was not appropriate for natural philosophers. The weakness of human reason compared with the immensity of the divine mind meant that the purposes of the Deity were 'impenetrable'. 'And for this reason alone', Descartes concluded, 'I consider the customary search for final causes to be totally useless in physics'.[27] When Gassendi raised objections to this uncompromising position, Descartes pointed out that while it might be appropriate to praise God on account of the functioning of various creatures, we admire him as their efficient cause (i.e., as their creator), not their final cause.[28] Guessing at the purposes of the Deity was relegated to the realm of 'ethics' where, Descartes seems to imply, less restrained conjectures were permissible.[29]

This brief discussion of passages from Galileo, Bacon and Descartes, is not intended to provide a comprehensive account of their final position regarding the relation of theology and physics. Nonetheless it is revealing when we reconsider Cunningham's claim that natural philosophy at this time was about 'God's achievements, God's intentions, God's purposes, God's messages to man'. That natural philosophy should legitimately be about God's intentions and purposes was strongly denied by Descartes and to a lesser extent Bacon. Moreover, in none of these instances does there seem to be a single straightforward understanding of the role that God or theological claims might play in the sphere of natural philosophy. This suggests that the relation between the two disciplines was not a settled one and was under negotiation, as was also the case, for example, with the relation of mathematics to natural philosophy. Each of these individuals found it necessary to grapple with the place of theological argument in the sphere of natural philosophy, and each took a negative view of particular kinds of theological incursions into the realm of physics.

---

[25] 'As to questions of the second sort, not only do they [theologians] not regard them as being resistant to the natural light, but they even encourage philosophers to demonstrate them to the best of their ability by arguments which are grounded in human reason', *ibid.*

[26] *Objections and Replies*; Descartes 1985, II, p. 290.

[27] *Meditations* §55; Descartes 1985, II, p. 38.

[28] *Objections and Replies*; Descartes 1985, II, p. 258; cf. *Principles of Philosophy*, Pt I, §15; Descartes 1985, I, p. 202.

[29] 'In ethics, then, where we may often legitimately employ conjectures, it may admittedly be pious on occasion to try to guess what purpose God may have had in mind in his direction of the universe; but in physics, where everything must be backed up by the strongest arguments, such conjectures are futile', *ibid.*

By the same token, each also found some place for theological claims. Galileo thus invoked the commonplace metaphor of 'the book of nature' to assert that 'the glory and greatness of Almighty God are marvellously discerned in all his works and divinely read in the open book of heaven'. The glory of God becomes more apparent through 'the ingenuity of learned men', and thus astronomy and natural philosophy have a religious role.[30] Moreover, his assertion that the subject matter of the bible is salvation and not natural philosophy did not prevent him from interpreting passages of scripture in such a way that they would support his claims about the movement of the earth.[31] As for Bacon, the whole point of engaging in natural philosophical investigations was to participate in a redemptive process aimed at restoring the mastery of nature which the human race had lost at the Fall.[32] So that while Bacon strived for a natural philosophy that was free from the corrupting influences of 'religious zeal', nonetheless the whole goal of natural philosophy was presented as a religious one, and part of God's providential plan.[33] If the content of natural philosophy was to be innocent of theological considerations, for all that, its whole justification as a useful enterprise was intimately connected with divine intentions. We need also to treat with some caution Bacon's contention that philosophy is corrupted and tainted by admixture with 'natural theology', for Bacon is most probably using the term 'natural theology' in a quite restricted sense. There is a pejorative use of 'natural theology' in both Augustine and Aquinas—probably derived from Varro—and it is likely that Bacon is using the expression in this traditional sense to mean something like 'pagan theology'.[34] For his part, Descartes made theology the beginning rather than the end of natural philosophy, grounding the reliability of clear and distinct ideas, the existence of the material world, and the constancy of the laws of nature in the existence of a perfect and immutable being. Neither did Descartes take issue with the notion of God as an efficient cause, a view that was subsequently adopted with considerable enthusiasm by Malebranche, who famously claimed that God was in fact the only genuine efficient cause. As for the role of scripture, while Descartes clearly did not derive his cosmogony from the book of Genesis, he seems on occasion to have accepted that the biblical narratives of creation could delimit what could be true in the sphere of physics.[35]

---

[30] Galileo, 'Letter to the Grand Duchess Christina'; Drake 1957, pp. 196f.

[31] *Ibid.*, pp. 214f.

[32] *Novum Organum*, II, lii; Bacon 1859, IV, p. 247; Harrison 1998, pp. 211–249. The power over nature which man derives from philosophy, moreover, is to be governed by religion: 'Only let the human race recover that right over nature which belongs to it by divine bequest, and let power be given it; the exercise thereof will be governed by sound reason and true religion', *Novum Organum*, I, cxxix; Bacon 1859, IV, p. 115.

[33] *Novum Organum*, I, xciii; Bacon 1859, IV, p. 287. On Bacon's attempts to establish natural philosophy as an independent enterprise see Gaukroger 2001, pp. 91–95 and Gaukroger's chapter in this volume.

[34] See e.g., Augustine, *City of God*, VI. 5, VI. 8. In Aquinas, *physicam theologiam* (usually rendered 'natural theology') refers to the erroneous theology of the philosophers. 'Natural theology', 'mythical theology' (essentially euhemerism, the worship of dead heroes) and 'civil theology' (state-sponsored worship of images) were all forms of 'superstitious idolatry'; Aquinas, *Summa theologiae*, 2a2ae, 94, 1.

[35] Harrison 2000, esp. pp. 181f.

All of this is suggestive of the fact that various solutions were offered to the question of how theology was to relate to natural philosophy or 'physics'. While God undoubtedly had a place in early modern natural philosophy, the fact that no single role was uncontroversially allotted him suggests that natural philosophy is perhaps best understood not as a monolithic entity in which God has a central and specific place. It is also apparent that the framing of this discussion still bore the imprint of the Peripatetic understanding of how the discrete sciences are to be related to each other—hence the notions of subordination that determine the appropriate conditions under which distinct sciences might be 'mixed'. And, of course, Aristotle's four-fold division of causes remained at the forefront of these discussions. The virtue of the Cunningham thesis is that it encourages consideration of the ways in which the natural philosophy of the period is distinctive, but does not quite tell the whole story. Part of this untold story concerns the emergence of the category 'physico-theology'. In what follows I shall propose that the appearance of the terms 'physico-theological' and 'physico-theology' in the second half of the seventeenth century represent one attempt to stabilise the relationship between theology and natural philosophy by establishing a particular mode of explanation and a specific field of enquiry that represented, to its advocates at least, a legitimate admixture of two distinct enterprises. In this respect, the category 'physico-theology' is akin to 'physico-mathematics', in that both were attempts to define new hybrid disciplines that redefined the boundaries of the traditional Aristotelian sciences.[36] That this is the case is most evident in the writings of Robert Boyle.

2.BOYLE ON PHYSICO-THEOLOGICAL EXPLANATION

The earliest use in English of the term 'physico-theological' occurs in Walter Charleton's *The Darknes of Atheism Dispelled by the Light of Nature. A Physico-Theologicall Treatise* (1652).[37] But if the philosopher-physician Charleton coined the compound term, he took no credit for inventing a new disciplinary category in the pages of his book. The *Physico-Theologicall Treatise* was one of the first in the genre of English natural theology, and indeed, Charleton seems to use the adjective to describe natural theology as we would understand it—and not in the Baconian sense—his self-described task being 'the *Demonstration of the Existence of God*, by beams universally deradiated from that Catholick *Criterion*, the *Light of Nature*'.[38] The arguments of the book include logical and metaphysical cases for existence of God—that is, the 'ontological' and 'cosmological' arguments—as well as the

---

[36] Admittedly, there were different uses of 'physico-mathematics', and the term did not necessarily imply a single understanding of the relationship between physics and mathematics. Some who used the expression intended it to connote the subordination of one science to another. Descartes and Beeckman, however, seem to use to term to describe a mixed discipline with no connotations of the subordination. See Gaukroger and Schuster 2002, p. 536.

[37] The *Oxford English Dictionary* postpones the appearance of the term until Boyle's *Physico-Theological Considerations* in 1675.

[38] Charleton 1652, To the Reader. The prefix 'physico' for Charleton thus refers not to the discipline 'physics' but to nature (Gk. φύσις).

argument from design, which would later become known as the 'physico-theological' argument.

Robert Boyle subsequently used the expression in his *Physico-Theological Considerations about the Possibility of the Resurrection* (1675), and in the *Disquisition about the Final Causes of Natural Things* (1688). However, the general question of the relation between theology and natural philosophy was broached in *The Excellency of Theology, Compar'd with Natural Philosophy*, a work first published in 1674, but written in 1665.[39] The thesis of the book is clear enough from the title. For Boyle, the superiority of theology lay in 'the Excellence and Sublimity of the Object we are invited to contemplate'—the argument of Galileo—and in the unrivalled utility of divine knowledge.[40] More important for our present purpose, however, Boyle also considers some of the ways in which theology and natural philosophy might be combined to their mutual advantage. Thus, we learn in divinity many things relevant to natural philosophy—that the world had a beginning, the approximate age of the earth, that the earth will come to an end, and other 'discoveries' about angels, the universe, and our souls.[41] The first three chapters of Genesis alone, Boyle suggests, relate 'divers particulars, in reference to the Origine of things, which though not *unwarily* or *alone* to be urg'd in Physics, may yet afford very considerable *Hints* to an attentive and inquisitive Peruser'.[42] Conversely, Boyle also cherished the belief that there were many more mysteries of divinity to be disclosed to one possessed of 'a philosophical eye'. Philosophy, in other words, could help unlock hitherto unknown secrets of theology. A combination of both disciplines could provide the foundations of a new pansophia:

> The Gospel comprises indeed, and unfolds the whole Mystery of Man's Redemption, as far forth as 'tis necessary to be known for our Salvation: And the *Corpuscularian* or Mechanical Philosophy, strives to deduce all the *Phænomena* of Nature from Adiaphorous Matter, and Local Motion. But neither the Fundamental Doctrine of Christianity, nor that of the Powers and Effects of Matter and Motion, seems to be more than an Epicycle (if I may so call it) of the Great and Universal System of God's Contrivances, and makes but a part of the more general Theory of things, knowable by the Light of Nature, improv'd by the Information of the Scriptures: So that both these Doctrines, though very general, in respect of the subordinate parts of Theology and Philosophy, seem to be but members of the Universal Hypothesis, whose Objects, I conceive, to be *the Nature, Counsels, and Works of God, as far as they are discoverable by us* (for I say not *to us*) *in this Life.*[43]

In a sense, the *Physico-Theological Considerations*, which appeared in the following year, represent Boyle's first essay in this ambitious task.

The title term 'physico-theological', is here used quite unreflectively, but seems to refer to an application of the methods of physics to a single theological doctrine. Thus Boyle was concerned to investigate possible natural mechanisms—in particular that of a 'plastick power'—that might account for the post-mortem reconstitution of

---

[39] I am grateful to Peter Anstey for pointing out the relevance of this work.

[40] Boyle 1674, p. 2. Boyle admits another argument, that theology is duty in a way that natural philosophy is not; see *ibid.*, p. 66.

[41] *Ibid.*, pp. 13–25.

[42] *Ibid.*, p. 22.

[43] *Ibid.*, pp. 51f. 'Adiaphorous' in this context means 'neutral or theologically indifferent'.

bodies. By the same token, he puzzlingly insists that resurrection 'shall be effected, not by or according to the ordinary course of Nature, but by his [God's] own Power'.[44] This would imply that resurrection was miraculous—that is, above or against the powers of nature—in which case deliberations at the level of natural philosophy would seem to be largely irrelevant. Elsewhere he was to state that, considered from the perspective of reason alone, 'the Resurrection of the Dead' seemed an 'Absolute Impossibility'. Thus he also placed the doctrine beyond the bounds of natural theology.[45] Boyle's apparent equivocation on this point is probably owing to a genuine uncertainty as to what kinds of events, if any, were in principle beyond the explanatory framework of natural philosophy. Boyle's approach implies that this is not something that can be known in advance. To put it another way, there can be no *a priori* division of subject matter between the natural philosopher and the theologian. Some events may successfully resist naturalistic explanation, and thus eventually be delivered over to the theologian. Equally, other events previously regarded as inherently miraculous might well succumb to naturalistic explanation, provided that one prosecutes the investigation with sufficient diligence. Any robust approach to explanation, on this analysis, must in principle allow for both physical and theological accounts, and thus the conscientious investigator must adopt a physico-theological frame of mind.

Boyle's ruminations on this topic were not unrelated to a controversy in the late seventeenth and early eighteenth centuries over the appropriateness of naturalistic accounts of biblical miracles and eschatological events—the creation of the world, the Deluge, the miracles of Moses, and the final destruction and restoration of the earth. The debate was sparked by the appearance in 1681 of Thomas Burnet's *Telluris Theoria Sacra*, a work that attempted to provide a naturalistic account of Noah's flood and of the final conflagration of the world in terms of Descartes' cosmogony. Burnet's work inspired a number of imitators, most notably William Whiston who, in his *New Theory of the Earth* (1696), substituted the more potent natural philosophy of Newton for that of Descartes. John Ray's *Three Physico-Theological Discourses* (1693) covered similar territory. Common to these works was the attempt to use the resources of natural philosophy to provide explanations for apparently miraculous events in biblical narratives and prophecies. At issue was whether such events lent themselves to explanation in terms of laws discovered in natural philosophy, or whether they were to be thought of as wholly miraculous, and hence beyond the bounds of natural philosophical speculation.[46] This was the issue that Boyle had broached with his essay on resurrection.[47]

These discussions are relevant for an understanding of the subject matter not only of natural philosophy, but of natural theology. The latter has typically been regarded as being concerned with theological doctrines that can be known through

---

[44] Boyle 1675, p. 3. Cf. '... the Christian doctrine doth not ascribe the *Resurrection* to *Nature*, or any created Agent, but to the peculiar and immediate operation of *God*....'; *ibid.*, p. 29; see Boyle 1674, pp. 23f.

[45] Royal Society Boyle Papers, vol. VII, fol. 23.

[46] For an outline of the controversy see Harrison 2000.

[47] Subsequent writers were also to attempt naturalistic explanations of resurrection; see, e.g., Bonnet 1769.

reason alone: God's existence, immortality of the soul, moral values, and so on. Generally, resurrection is not regarded as a topic of natural theology, neither are such doctrines as the Trinity and the Incarnation, for these are said to be known only through revelation typically in scripture, and cannot be ascertained through the exercise of reason alone.[48] Boyle's treatment of resurrection as amenable in certain respects to a physical treatment, however, suggests that it might well be dealt with within the scope of natural theology. Around the time of the publication of the *Physico-theological Considerations* some of the Cambridge Platonists were also suggesting that such doctrines as the Trinity could be known through reason. This was because versions of Trinitarian theology had supposedly been espoused by Platonic philosophers ignorant of the Christian revelation.[49]

One implication of the contentions of the Cambridge Platonists was that the traditional division of subject matter between natural and revealed theology was really a matter of historical contingency. The mark of whether a particular doctrine belonged to natural or revealed theology was whether it had been embraced by any culture beyond the pale of Christendom. Thus, neither the Greeks nor anyone else had subscribed to a doctrine of resurrection, and it was inferred on this basis that such notions could not be arrived at through the exercise of reason alone. However, with the early modern challenge to ancient systems of natural philosophy, it became possible to suggest that the Pagan philosophers' ignorance of, say, resurrection—traditionally ascribed to a lack of access to the revealed truths of scripture—might as easily be attributed to the deficiencies of their natural philosophy. This, in turn, meant that no subject matter could in principle be ruled outside the boundaries of natural philosophy. In short, the most robust methodological approach to phenomena allowed for philosophical explanation *and*, in the event of its failure, theological explanation. The kinds of subject matters that particularly lent themselves to such an approach included the beginning and end of the world, animal and human generation (that is, reproduction and embryology), along with death, immortality, and resurrection.[50] All of this meant that revealed doctrines could be considered as legitimate topics for a physico-theological treatment in a way that the traditional topics of natural theology could not. As late as 1749 the Comte de Buffon could still describe the cosmological and eschatological speculations of Burnet, Whiston, and others, as 'physical theology'.[51] For this period, then, physico-theology was not a sub-set of natural theology.

Boyle's more mature reflections on physico-theology came to fruition in his *Disquisition about the Final Causes of Natural Things* (1688). Here, for the first time, we find a formal account of what physico-theological explanation might entail. Boyle's point of departure was the issue that had occupied both Bacon and

---

[48] But cf. Sibiuda [i.e. Raymond Sebonde] 1966, who in his fifteenth-century work *Theologia naturalis seu liber creaturarum* [*Natural Theology, or the Book of the Creatures*], asserted that all of the central tenets of Christian doctrine were evident in the natural world, and could be known from a reading of the book of nature.

[49] See especially Cudworth 1845, II, pp. 312f.

[50] For examples of early modern treatments of some of these topics, see Harrison 2001, pp. 199–224.

[51] Georges Louis Leclerc, Comte de Buffon 1812, I, p. 131.

Descartes—that of the place of final causes in natural philosophy.[52] The crux of Boyle's case, which follows a number of helpful distinctions between different senses of the term 'final cause', is that 'arguments in Physicks should be grounded in solid reasons, but those reasons need not themselves be physical'. As an example Boyle cites the Cartesian principle that the quantity of motion in the universe remains constant. This axiom of physics, he points out, was grounded by Descartes in the immutability of God. Descartes thus admitted a metaphysical principle as a component of physical explanation.[53] Peripatetic strictures on the exportation of specific methods across disciplinary boundaries Boyle addresses in this fashion: 'And to me 'tis not very material, whether or no, in Physicks or any other Discipline, a thing be prov'd by the peculiar Principles of that Science or Discipline; provided it be firmly proved by the common grounds of Reason'.[54] Boyle thus suggests that physics (along with other disciplines) is distinguished by its subject matter, but not by specific methods. This bears a direct analogy to the Renaissance understanding of the division of labour within a 'middle science' (*scientia media*) such as mixed mathematics, which was understood as a subject in which the *res considerata* belongs to natural philosophy, but whose *modus considerandi* belongs to mathematics.[55]

Boyle subsequently introduced terminology that reflects this distinction. There are, he noted, two kinds of arguments from ends—'*Physical* Ones' and '*Physico-Theological* Ones'. Purely physical explanations from final ends refer to the means by which 'the End design'd by Nature may be best and most conveniently attain'd'. Physico-theological arguments, by way of contrast, 'relate to the Author of Nature, and the General Ends he is suppos'd to have intended in things Corporeal: As, when from the manifest usefulness of the Eyes, and all its parts, to the Function of Seeing, Men infer, that at the Beginning of Things the Eye was fram'd by a very Intelligent Being…'.[56] This, then, was not simply the familiar 'argument from design', but rather reflected Boyle's ambitious vision, originally articulated in *The Excellency of Theology*, for a study of 'the Great and Universal System of God's Contrivances' that would constitute a component of 'the more general Theory of things, knowable by the Light of Nature, improv'd by the Information of the Scriptures'.[57]

For all this, Boyle did not go on to propose a distinct discipline 'physico-theology'. As far as I know, he never used the noun. On the face of it, then, he seems to equivocate on the extent to which physico-theological arguments are allowable in

---

[52] Boyle refers to the issue as 'the grand Difficulty that has, ever since *Aristotles* time, and even before that, Perplex'd those that allow in Natural Philosophy, the Considerations of *Final Causes*'; Boyle 1688, p. 87.

[53] Boyle 1688, pp. 24f. Cf. Descartes, *Principles of Philosophy*; Descartes 1985, I, p. 240.

[54] Boyle 1688, pp. 23–24.

[55] Jardine 1991, p. 102. Thus physico-mathematics 'The Mixed [Mathematics] consist of physical subjects investigated and explained by mathematical reasoning'. *Oxford English Dictionary*, *sv* 'physico-mathematics'.

[56] Boyle 1688, pp. 104f. Boyle, following Bacon, also suggests that we can call such arguments 'metaphysical'. But unlike Bacon, he does not draw the implication that they should be excluded from physics.

[57] Boyle 1674, p. 51.

physics. He notes for example, that physico-theological arguments amount to what Bacon had called 'metaphysical arguments', but makes no reference to Bacon's objection to the deployment of such arguments in the sphere of natural philosophy.[58] Did he mean to imply here that physico-theological arguments, despite being essentially metaphysical, do in fact have a legitimate place in physics? And if so, why did he not directly address Bacon's concerns?

Two aspects of Boyle's approach clarify this position. First, he argued that previous abuses of teleological explanation do not provide sufficient grounds for their exclusion from natural philosophy. If the admission of final causes in philosophy had occasioned lazy thinking and absurdities in the past, the appropriate solution lay in the establishment of adequate safeguards against these abuses rather than in a wholesale abandonment of final causes as a mode of explanation. Boyle thus cautioned against allowing the quest for final causes to displace the more immediate task of the philosopher—the discovery of efficient causes.[59] By the same token, he provided clear illustrations of cases in which the quest for final causes has resulted in important advances in knowledge, as for example in Harvey's discovery of the circulation of the blood.[60] Equally importantly, however, Boyle's concern was not simply with the legitimacy of particular modes of explanation and their relative importance, but also with who it is that is providing the explanation. At issue was not only the identity of the discipline, but also the identity of the investigator.

3.PHYSICO-THEOLOGY AND THE CHRISTIAN NATURAL PHILOSOPHER

At least part of Boyle's concern in the *Disquisition* is with what he refers to as the duties and responsibilities of 'the *Christian* philosopher'. The work is explicitly addressed to 'Christian philosophers', and to benefits that relate 'as well to Philosophy as Piety'.[61] Boyle suggests that Christian philosophers, having access to revealed truths concerning God's ends, ought not to ignore this information. ''tis plain', he writes, 'that I suppose the Naturalist to discourse meerly upon Physical Grounds. But if the Revelations contain'd in the *Holy Scriptures*, be admitted, we may rationally believe More, and speak less Hæsitantly, of the Ends of God, than bare Philosophy will warrant us to do'. There is a clear admission, then, of what is permissible in 'bare philosophy', but Boyle suggests that Christian naturalists need not confine their conclusions to what is sanctioned by the methodological requirements of 'bare philosophy'. This amounts to an argument against the Cartesian position, or at least against those Cartesians who identify themselves as Christian. Boyle writes: 'those *Cartesians,* that being Divines, Admit the Authority of Holy Scripture; should not reject the Consideration of such Final

---

[58] *Ibid.*, pp. 104f.
[59] *Ibid.*, pp. 229–237.
[60] *Ibid.*, p. 157.
[61] *Ibid.*, Preface, p. 29.

Causes, as *Revelation* discovers to us'.[62] He went on to express the hope that his arguments 'may justly serve to Recommend the Doctrine about Final Causes that we embrace, to Philosophers *that are truly pious…*'. These pious Christian philosophers are thus in a different position to 'the Ancient Aristotelians, who look'd upon the World as Eternal and Self-existent in a Condition like its present System; [and who] did not use to Thank *God* for the Benefits they receiv'd from things Corporeal'.[63] The absence of robust doctrines of creation and providence in the Aristotelian scheme accounted for both the absence of the theistic teleology and for the strict division of labour that kept natural philosophy and the sacred science distinct. On Boyle's analysis, the traditional division of labour, in which theistic explanations were banished from the study of nature, was an unwelcome vestige of Aristotle's ignorance of the true origin and destiny of the world.

Boyle thus held that the *Christian* philosopher has a 'duty' to progress beyond mere physics to the more sublime reaches of the physico-theological. To adopt any other procedure would be needlessly to mimic the approach of the ancients who, bereft of a Christian view of the world, were blind to the fact that the operations of nature were in fact the operations of the providential Deity. On this analysis, Cartesian philosophers were unconsciously re-enacting a groundless Peripatetic prejudice when they proposed to eschew explanations that invoked divine purposes. Theistic teleology was legitimate, Boyle argued, for anyone who acknowledged 'a most wise author of things'.[64] At any rate, for Boyle the status of physico-theological arguments was as much to do with the religious commitments of the natural philosopher as with the relevant disciplinary boundaries. In these religious commitments, moreover, lay one of the chief justifications for the pursuit of natural philosophy in the first place—namely, the extension of its utility beyond the sphere of mere material gain to the more exalted realm of spiritual benefits.[65]

Boyle's emphasis on the duties of the Christian natural philosopher relates to the larger issue of the vocation of the early modern naturalist. The lack of a specific profession 'scientist' during this period generated considerable difficulties for those who considered themselves called to the study of nature.[66] There were few socially-sanctioned roles for the profession of natural philosophy, and neither was it obvious that scientific proclivities could be comfortably incorporated into one of the three official professions of medicine, law, or theology.[67] In this context Boyle's argument amounts to an assertion that the study of nature, on account of its theological implications, was closest to the clerical vocation. There is, of course, something distinctively Protestant about Boyle's conception of the role of the Christian natural philosopher, for it is related to the Reformation principle of 'the priesthood of all

[62] *Ibid.*, p. 80.
[63] *Ibid.*, pp. 100f.
[64] *Ibid.*, p. 16. The Cartesians, for their part, might reasonably argue that they differed from the Aristotelians in that they identified the efficient cause as God.
[65] This argument parallels to some degree Stephen Gaukroger's contentions about Bacon's concern with 'the office of the natural philosopher' as one whose knowledge contributes to the public good. See Gaukroger 2001, pp. 44–57.
[66] Ross 1962.
[67] Feingold 2002.

believers'. According to this doctrine, which replaced the medieval division of society into vertical 'estates', any individual, in principle, is capable of fulfilling the priestly role.[68]

It is in this context that we are to understand Boyle's identification of natural philosophers as 'priests of nature'—a designation that the Protestant astronomer Kepler had also adopted.[69] For figures such as Kepler and Boyle, the realm of theological speculation was one that was no longer restricted to the priestly classes. Their explorations into theological territory represent not only the breakdown of traditional disciplinary boundaries—as evidenced by the new category of physico-theological explanation—but also signal the disintegration of the traditional vocational demarcations of the middle ages, a process which, in respect of the ontological status of the clergy, took place far more quickly in Protestant than in Catholic countries. It is significant, then, that neither Galileo nor Descartes ventured much into what we might call physico-theological territory, subscribing to a strong view of the integrity of theology and of the unique status of clerical theologians. At the same time, they clearly did not wish to invite reciprocal incursions by theologians into the realm of natural philosophy, and this policy could thus serve to maintain the independence of philosophy. Taking a high view of the unique status of clergy was a means of demarcating theology and philosophy, and for keeping the respective roles of theologian and philosopher distinct.

If linking theology with the pursuit of natural philosophy served to elevate the status of the latter, equally importantly, it sanctioned the pursuit of natural philosophy for those committed to the clerical vocation. Boyle argued, against those who wished to restrict divines to the study of theology, that 'nothing hinders, but that a man who values and inquires into the Mysteries of Religion, may attain to an Eminent degree in the knowledge of those of Nature'.[70] Copernicus, he pointed out, was a 'Churchman', Gassendi a Doctor of Divinity, a number of Jesuits, 'have as prosperously addicted themselves to Mathematicks as Divinity'. Moreover, a number of English clergymen who were 'not onely solid Divines, but Excellent Preachers, have yet been so happily conversant with Nature'.[71] The resolution of these apparently conflicting vocational commitments was to be resolved through a realisation that the ends of natural philosophy served the more noble ends of theology. Thus:

> Those Religious Naturalists, who are invited to Attention and Industry, not onely by the pleasantness of the Knowledge it self, but by a higher and more ingaging Consideration; namely, that by the Discoveries they make in the Book of Nature, both themselves and others may be excited and qualifi'd the better to admire and praise the Authour, whose Goodness does so well match the Wisdom they celebrate.[72]

[68] For the reformers' rejection of hierarchical estates see Luther, *To the Christian Nobility of the German Nation* (1520); Luther 1970, p. 12; Calvin 1960, I, p. 502, II, p. 1473.

[69] Boyle 1688, p. 34; Kepler 1938–, VII, p. 25; Letter to Herwath von Hohenburg, 26 March 1598, *ibid.*, XIII, p. 193; see also Fisch 1953.

[70] Boyle 1674, p. 217.

[71] *Ibid.*, p. 217.

[72] *Ibid.*, p. 220.

Considerations such as these were to inform the vocational choices of a number of prominent clergymen-naturalists. John Ray, for example, had lamented in 1658 that he 'must of necessity enter into orders or else live at great uncertainties'. He thus resolved 'to make it my business to execute the priest's office'. That office, as he then understood it, required the relinquishing of his natural philosophical pursuits: 'I shall bid farewell to my beloved pleasant studies and employments, and give myself up to the priesthood'.[73] As his publishing history testifies, however, Ray subsequently managed to reconcile his priestly calling with the pursuit of natural philosophy, producing in the last decades of the seventeenth century pioneering works of natural history and plant taxonomy. Significantly, though, his most celebrated work was the physico-theological classic, *The Wisdom of God Manifested in the Works of Creation* (1691). This was followed two years later by his *Three Physico-Theological Discourses* (1693). It was the new notion of a legitimate physico-theological approach that enabled Ray to combine his scientific interests with a priestly vocation in a way that those with a more traditional understanding of the relation of theology to natural philosophy would have found difficult.

Thus it was that the overlapping offices of 'the Christian natural philosopher' as epitomised by Boyle, and 'the clergyman-naturalist' as exemplified by Ray, produced the relevant enterprise, 'physico-theology'. The title of the mixed discipline first appears, appropriately enough, in the title of the Boyle Lectures given by the clergyman William Derham in 1711/12: *Physico-Theology: Or A Demonstration of the Being and Attributes of God from the Works of his Creation* (1713). This work assured a place for the term in the English lexicon, and paved the way for hybrid disciplines that represented the increasing specializations of natural philosophy—Derham's own *Astro-theology* (1715), Friedrich Lesser's *Insecto-Theologia* (1738), Peter Ahlwart's *Bronto-Theologie* (1745), and John Balfour's *Phyto-Theology* (1851).[74]

These physico-theological works should not be regarded simply as examples of 'natural theology' or as mere rehearsals of the 'argument from design'. From the outset, physico-theology was intended to represent a combination of natural philosophy and theology—a form of theologising, yes, but one that could only be conducted by those with expertise in natural philosophy and, increasingly, natural history. The familiar form that it took over the course of the eighteenth century consisted in endless rehearsals of instances of organic 'contrivances', listed seriatum, all of which were designed to lead to a final conclusion—the existence of a divine designer. These compilations aimed at providing a watertight cumulative argument and, crucially, one based on induction. Increasingly, then, physico-theology was represented by its exponents as an inductive science, indeed perhaps the one form of theological argument that could lay claim to be such a science. As one of its nineteenth-century exponents was later to express it, natural theology in this form was 'open to no objection', 'in strict conformity with the rules of the inductive philosophy', and 'consistently denied by those only who reject the "Principia" of Newton'. This writer went on to claim that:

---

[73] Ray 1928, p. 16; quoted in Feingold 2002, p. 95.

[74] Lesser 1738; Ahlwardt 1745; Balfour 1851.

Our knowledge of the existence of God, as far as that knowledge is traceable by the light of nature, is acquired by an intellectual process strictly analogous, and exactly similar, to the intellectual process by which we acquire our knowledge of the laws of the physical world.... Newton discovered the true system of the heavens; and it is only by this reasoning that the theist can ascertain, from the light of Nature, the existence and attributes of Him who made the heavens. The proof of a divine intelligence ruling over the universe is as full and, as perfect as the proof that gravitation extends throughout the planetary system.

The great physico-theological works eighteenth and nineteenth century represented, for this author, nothing less than 'inductive philosophy ... applied to theology'.[75]

In sum, physico-theology became a unique enterprise, bearing the dignity of its ultimate object God, and bolstered by the 'scientific' authority of induction. Thus understood, the discipline was true to Boyle's original conception of a theological enterprise that relied on the methods of natural philosophy.

4.CONCLUSION

The emergence of the mixed science 'physico-theology' is symptomatic of the disciplinary flux that was characteristic of the early modern period, and the existence of this term signals an attempt to arrive at a solution to the question of how the new forms of natural philosophy related to theology. If in the seventeenth-century term 'physico-theological' had referred largely to a mode of investigation, the eighteenth century was to deploy this method within a discrete discipline—physico-theology. More specifically, however, the existence of this new hybrid discipline suggests that for at least some early modern figures natural philosophy per se was not considered to be essentially theological in orientation (*pace* Cunningham), for if this were the case there would be no requirement for the introduction of a distinct kind of explanation that went beyond the methodological limits of 'bare philosophy', to use Boyle's expression.

The subsequent history of the term 'physico-theology' has conspired against a proper understanding of its historical significance. Since Kant, it has been customary to associate physico-theology with the argument from design, for in the *Critique of Pure Reason* (1781) Kant appropriated the term 'physicotheological' to label that specific proof for God's existence.[76] In modern analytical philosophy of religion, 'the physico-theological argument' is simply an inelegant synonym for 'the teleological argument' and is treated as one of the three classical arguments for God's existence. Typically, this argument is traced back to the ancient Greeks.[77] Clearly this standard usage overlooks the historical origins of the expression and masks its significance as a marker for an important phase in the evolving and overlapping boundaries of natural philosophy and theology.

[75] Anon. 1834, pp. 216, 217.

[76] *Critique of Pure Reason*, Transcendental Logic, Second Division: Transcendental Dialectic, II. iii. 6; Kant 1965, p. 518.

[77] Thus, e.g., 'Physicotheology is the aspect of natural theology that seeks to prove the existence and attributes of God from the evidence of purpose and design in the physical universe. The argument is very ancient...'; Carré 1967, p. 300.

REFERENCES

Ahlwardt, P. (1745) *Bronto-Theologie, oder vernünftige und theologische Betrachtungen über den Blitz und Donner*, Greifswalde and Leipzig.

Anon., (1834) 'Crombie's Natural Theology', *Quarterly Review*, 51, March and June, pp. 216–218.

Aristotle (1984) *The Complete Works of Aristotle: The Revised Oxford Translation*, ed. J. Barnes, 2 vols, Princeton: Princeton University Press.

Augustine (1844–1905) in *Patrologia Latina*, vols 32–47, ed. J.-P. Migne, Paris.

Bacon, F. (1859) *The Works of Francis Bacon*, 14 vols, eds J. Spedding, R. Ellis and D. Heath, London.

— (1974) *Of the Advancement of Learning and New Atlantis*, ed. A. Johnston, Oxford: Clarendon Press.

Balfour, J. (1851) *Phyto-Theology; or, Botanical Sketches intended to illustrate the works of God in the structure, functions, and general distribution of plants*, London and Edinburgh.

Bonnet, C. (1769) *Palingenesie philosophique*, Geneve.

Boyle, R. (1674) *The Excellency of Theology*, London.

— (1675) *Physico-Theological Considerations about the Possibility of the Resurrection*, London.

— (1688) *Disquisition about the Final Causes of Natural Things*, London.

Buffon, Comte de (1812) *Natural History, General and Particular*, trans. W. Smellie, 20 vols, London.

Burnet, T. (1681) *Telluris Theoria Sacra*, London.

Calvin, J. (1960) *Institutes of the Christian Religion*, 2 vols, ed. J. T. McNeill, trans. F. Battles, Philadelphia: Westminster Press.

Carré, M. (1967) 'Physicotheology' in *The Encyclopedia of Philosophy*, 8 vols, ed. P. Edwards, New York: Macmillan, 6, pp. 300–305.

Charleton, W. (1652) *The Darknes of Atheism Dispelled by the Light of Nature. A Physico-Theologicall Treatise*, London.

Cudworth, R. (1845) *The True Intellectual System of the Universe*, 3 vols, ed. J. Harrison, London: 1st edn 1678.

Cunningham, A. (1988) 'Getting the game right: some plain words on the identity and invention of science', *Studies in History and Philosophy of Science*, 19, pp. 365–389.

(1991) 'How the Principia got its name: or, taking natural philosophy seriously', *History of Science*, 28, pp. 377–392.

— (2001) 'A response to Peter Dear's "Religion, science, and philosophy"', *Studies in History and Philosophy of Science*, 32A pp. 387–391.

Dales, R. C. (1984) 'The origin of the doctrine of the double truth', *Viator*, 15, pp. 169–179.

Dear, P. (1995) *Discipline and Experience: The Mathematical Way in the Scientific Revolution*, Chicago: University of Chicago Press.

— (1998) 'The Mathematical Principles of Natural Philosophy: toward a heuristic narrative for the Scientific Revolution', *Configurations*, 6, pp. 173–193.

— (2001) 'Religion, science, and natural philosophy: thoughts on Cunningham's thesis', *Studies in History and Philosophy of Science*, 32, pp. 377–386.

Derham W. (1713) *Physico-Theology: Or A Demonstration of the Being and Attributes of God from the Works of his Creation*, London.

Descartes, R. (1985) *The Philosophical Writings of Descartes*, trans. J. Cottingham, R. Stoothoff, and D. Murdoch, 2 vols, Cambridge: Cambridge University Press.

Edwards, P. ed. (1967) *The Encyclopedia of Philosophy*, 8 vols, New York: Macmillan.

Feingold, M. (2002) 'Science as a calling: the early modern dilemma', *Science in Context*, 15, pp. 79–119.

Fisch, H. (1953) 'The scientist as priest: a note on Robert Boyle's natural theology', *Isis*, 44, pp. 252–265.

Funkenstein, A. (1986) *Theology and the Scientific Imagination*, Princeton: Princeton University Press.

Galileo, G. (1957) *Discoveries and Opinions of Galileo*, ed. and trans. S. Drake, New York: Doubleday.

Gaukroger, S. W. (2001) *Francis Bacon and the Transformation of Early-Modern Philosophy*, Cambridge: Cambridge University Press.

Gaukroger, S. W. and Schuster, J. A. (2002) 'The hydrostatic paradox and the origins of Cartesian dynamics', *Studies in History and Philosophy of Science*, 33, pp. 535–572.

Gaukroger, S. W., Schuster, J. A. and Sutton, J. eds (2000) *Descartes' Natural Philosophy*, London: Routledge.

Grant, E. (2000) 'God and natural philosophy: the late Middle Ages and Sir Isaac Newton', *Early Science and Medicine*, 6, pp. 279–298.

Harrison, P. (1988) *The Bible, Protestantism and the Rise of Natural Science*, Cambridge: Cambridge University Press.

— (2000) 'The influence of Cartesian cosmology in England', in *Descartes' Natural Philosophy*, eds S. W. Gaukroger, J. A. Schuster and J. Sutton, London: Routledge, pp. 168–92.

— (2001) 'Scaling the ladder of being: theology and early theories of evolution' in *Religion, Reason, and Nature in Early Modern Europe*, ed. R. Crocker, Dordrecht: Kluwer, pp. 199–224.

Jardine, N. (1991) 'Demonstration, dialectic, and rhetoric in Galileo's Dialogue' in *The Shapes of Knowledge from the Renaissance to the Enlightenment*, eds D. R. Kelley and R. H. Popkin, Dordrecht: Kluwer, pp. 101–122.

Kant, I. (1965) *Critique of Pure Reason*, trans. N. Kemp Smith, London: Macmillan: 1st edn 1781.

Kelley, D. R. and Popkin, R. H. eds (1991) *The Shapes of Knowledge from the Renaissance to the Enlightenment*, Dordrecht: Kluwer.

Kepler, J. (1938–) *Johannes Kepler Gesammelte Werke*, 20 vols, eds W. von Dyck and M. Caspar, Munich: C. H. Beck.

Kretzmann, N. and Stump, E. eds (1993) *The Cambridge Companion to Aquinas*, Cambridge: Cambridge University Press.

Lesser, F. C. (1738) *Insecto-Theologia*, Franckfurt and Leipzig.

Lohr, C. (1988) 'Metaphysics' in *The Cambridge History of Renaissance Philosophy*, eds C. Schmitt, Q. Skinner, E. Kessler, and J. Kraye, Cambridge: Cambridge University Press, pp. 537–638.

Luther, M. (1970) *Three Treatises*, Philadelphia: Fortress Press.

MacClintock, S. (1972) 'Averroës' and 'Averroism' in *The Encyclopedia of Philosophy*, 8 vols, ed. P. Edwards, New York: Macmillan, 1, pp. 220–226.

MacDonald, S. (1993) 'Theory of knowledge', in *The Cambridge Companion to Aquinas*, eds N. Kretzmann and E. Stump, Cambridge: Cambridge University Press, pp. 160–195.

Osler, M. J. (1997) 'Mixing metaphors: science and religion or natural philosophy and theology in early modern Europe', *History of Science*, 35, pp. 91–113.

Ray, J. (1693) *Three Physico-Theological Discourses*, London.

— (1928) *Further Correspondence of John Ray*, ed. R. Gunther, London.

Rheticus, G. J. (1984) *G. J. Rheticus' Treatise on the Holy Scripture and the Motion of the Earth*, ed. and trans. R. Hooykaas, Amsterdam: North-Holland Publishing.

Ross, S. (1962) '"Scientist": the story of a word', *Annals of Science*, 18, pp. 65–86.

Sibiuda, R. [Raymond Sebonde] (1966) *Theologia naturalis seu liber creaturarum [Natural Theology or, the Book of the Creatures]*, ed. F. Stegmüller, Stuttgart-Bad Canstatt.

Thomas Aquinas, St. (1924) *Summa contra gentiles*, trans. The English Dominican Fathers, New York: Benziger Brothers.

— (1964–1976) *Summa theologiae*, 60 vols, ed. Thomas Gilby, trans. by Fathers of the English Dominican Province, London: Eyre & Spottiswood.

Whiston, W. (1696) *New Theory of the Earth*, London.

LUCIANO BOSCHIERO

THE SATURN PROBLEM

Natural Philosophical Reputations and Commitments on the Line in 1660 Tuscany

1.INTRODUCTION

During the 1970s, Albert van Helden published a series of important papers about the observations made of Saturn in the seventeenth century. In two of these publications, van Helden recounted the involvement of the Tuscan Accademia del Cimento (1657–1667) in a debate regarding the possibility of a ring surrounding Saturn.[1] To confirm or deny the existence of a ring, the academicians constructed models of Saturn in its proposed forms, and observed these models from a distance and through their telescopes, supposedly re-creating the experience of observing the actual planet. According to van Helden, this demonstrated the deep commitment of the Cimento to what he, and many other historians, described as 'the experimental method'.[2]

As enlightening as van Helden's research was, the results of his work must now be re-interpreted with regard to some pertinent historiographical issues currently dominating the study of the history of seventeenth-century science. The reason that the Cimento's astronomical work is so important from an historiographical perspective is that the manuscript evidence reveals the political, religious, and intellectual complexities that existed behind the experimentalist façade of early modern scientific institutions. Contrary to what has been written by some authors, there is little evidence in this particular historical episode which suggests that the Cimento academicians were simply concerned with establishing a modern experimental method. Instead, the Cimento's involvement in the Saturn dispute of 1660 can be used to analyse the dominance that contentious natural philosophical

---

[1] Van Helden 1970; van Helden 1973.

[2] Van Helden 1970, p. 247. The experimental workings of these institutions, and the supposed birth of an experimental method, have been discussed by several historians during the past twenty years, including Steven Shapin, Simon Schaffer, Peter Dear and John Henry. Publications by these authors include: Shapin and Schaffer 1985; Shapin 1994; Dear 1995; Henry 2002.

P. R. Anstey and J. A. Schuster (eds.), The Science of Nature in the Seventeenth Century, 185-213.

debates, in existence since the beginning of the seventeenth century, continued to hold in the activities of Europe's early scientific institutions.[3]

In this case study regarding Saturn's appearance, Leopoldo de'Medici (1617–1675), the princely patron of the Accademia del Cimento, censored the academicians' from publicly discussing the controversial nature of their work in their only publication, *Saggi di naturali esperienze* (1667). The Prince, and his elder brother, the Grand Duke of Tuscany, Ferdinando II (1610–1670), relied upon a policy of censorship to create a rhetoric suggestive of the use of an unbiased, uncontroversial, experimental programme. Such a rhetorical façade assisted in avoiding controversy with the Catholic Church and promoted the professional status of the academy under the control of the Medici Court.[4] So the Cimento's handling of this controversy reveals the complexities of knowledge-making during this period and the need for historians to look well beyond mere stories about the origins of modern science.

2. THE BIRTH OF MODERN SCIENCE?

Traditional accounts of the birth of experimental science have often begun with, or have at least featured, Galileo Galilei (1564–1642). The reason for this seems obvious: Galileo is well remembered for his recordings of countless astronomical telescopic observations relating to the problem of planetary motion and his experiments regarding terrestrial mechanics. The enduring images of Galileo include his enticing of others to peer at the stars through his telescope, his dropping of heavy objects from the leaning tower of Pisa, and his observations of the pendulum-like movements of the lamp inside Pisa's cathedral.[5]

According to some authors, Galileo's achievements also served as an example for his students and followers who took up the task of preserving and enhancing Galileo's experimentalist image. For example, one may get this impression from reading Giovanni Targioni Tozzetti's preface to his eighteenth-century publication regarding the academicians' activities.[6] In this work, Targioni Tozzetti was interested in exploring the progress of Tuscan celestial and terrestrial science. This included the productive discoveries and diligent experiments carried out firstly by Galileo, then his disciples, and finally the Accademia del Cimento.[7] Meanwhile, Giovanni Batista Clemente Nelli, also writing during the eighteenth century, regarded Galileo's followers, including the members of the Cimento, virtually as the

---

[3] As an example, the presence of natural philosophical concerns and contentions in the early Royal Society of London is discussed in Schuster and Taylor 1997.

[4] See Boschiero 2002; Galluzzi 1981.

[5] For an analysis of the accuracy of these stories supposedly depicting Galileo's experimental prowess, see Segre 1989, p. 230.

[6] Targioni Tozzetti 1780.

[7] *Ibid.*, I, p. 5

ambassadors of Galileo's experimental philosophy during the mid to late seventeenth century.[8]

This image of the birth of experimental science in Tuscany established even firmer roots in twentieth-century writings. In 1903, Stefano Fermi argued that Galileo's students freed themselves from traditional natural philosophical theorising and adopted an inductive experimental method. According to Fermi, the members of the 'Galilean School' used an 'inductive method' to avoid 'metaphysical deductions', 'subtle discriminations', and 'a priori demonstrations of the stale philosophical and scientific school'.[9] Similarly, Gustavo Barbensi also claimed that Galileo's followers in the Accademia del Cimento, perfected the 'experimental method of the Galilean School'.[10] Martha Ornstein also asserted that early modern thinkers in Italy after Galileo gave prime importance to their experiments, and since universities refused to sponsor these experimentalist activities, they turned to scientific societies such as the Cimento: 'It is superfluous to say that they made every effort to foster the cause of experimental science. This was the keynote, the charter of their existence, the motive underlying their every activity'.[11] Finally, Rupert Hall, Roger Emerson, and Gaetano Pieraccini also described the use of Galilean or Baconian experimental method as the sole impetus behind the Cimento's work.[12]

These accounts, identifying the supposed birth of modern science within the practices of post-Galilean Tuscan natural philosophers, have remained virtually unchanged since the seventeenth century. For this reason we refer to them here as 'traditional' historiographical perspectives. Galileo is made out to be the first scientist to use experiments to counter Aristotelian physics, the first scientist whose work was essentially empirical—further, as the man who inspired and grounded experimental science, a practice that was perfected by his followers during the next fifty years and developed into a supposedly unbiased and objective research method.

Recently historians have largely broken away from this traditional positivist perspective. Writers have begun to suggest that the situation was more complex than the pure adaptation of an experimental method. More specifically, some historians have linked early modern Tuscan experimentalism to certain rules of behaviour and etiquette for investigating nature inside the princely court. In particular, for Jay Tribby, Paula Findlen and Mario Biagioli, the focus has shifted to the social and political aims and interests of the Medici Court, the patrons of the Accademia del Cimento.[13] Thanks largely to these authors, we can identify how Galileo and his followers, legitimated their work within the region's princely court. Inversely we can examine how the Court also used these new representatives of elite culture to

---

[8] Nelli placed particular emphasis on the role Evangelista Torricelli (1608–1647) supposedly played in advancing the cause of 'experimental science'. Torricelli was Galileo's successor to the position of Court Mathematician in Tuscany, and according to Nelli, practised experimental science when constructing what came to be known as the first barometer; Nelli 1759, p. 111.

[9] Fermi 1903, p. 87.

[10] Barbensi 1947, p. 19.

[11] Ornstein 1963, p. 259.

[12] Hall 1983, p. 38; Emerson 1990, p. 964; Pieraccini 1925, II, p. 603.

[13] Tribby 1995, p. 321; Biagioli 1992, pp. 11–54; Findlen 1993, pp. 39–40.

raise their own status as patrons of the latest intellectual activity.[14] According to Tribby, it was crucial that the experimental and natural philosophical work carried out by the Cimento project the political power and cultural identity of Florence and the Medici rulers.[15] However, it is easy to get carried away with such historiographies that are focused on courtly culture. They can be used effectively to examine the broad political context of early modern science, but, at times, these authors slip into the type of discussions typical of 'traditional' historiographies, that is, they wind up claiming that the birth of an experimental method occurred in institutions such as the Accademia del Cimento and the early Royal Society of London.

For example, in recent years, Steven Shapin and Simon Schaffer have produced extensive studies of the experimental life of the early London Royal Society.[16] Their works have particularly focused on the rise of 'the new empirical science of seventeenth century England' in place of traditional natural philosophical interests.[17] According to Shapin and Schaffer, Robert Boyle (1627–1691) made it clear to his colleagues in England that the only certain way of acquiring knowledge was through a 'programme' of experimental fact-making. Furthermore, Shapin and Schaffer claim that the success of this experimental science depended on the trustworthiness of the experimenters to produce matters of fact: that they reported their experimental findings to each other according to codes of civil and honest gentlemanly behaviour and discourse. All players in this gentlemanly, courtly game could trust and build on each other's reports. This reporting of matters of fact supposedly replaced natural philosophical concerns that were previously pursued in the early seventeenth century.

In all fairness to Tribby, Findlen and Biagioli, their intentions were not to present this type of origin story for the Accademia del Cimento. They claimed not to be interested in seeking the origins of experimental science as a product of courtly culture. In fact, they openly and strongly constructed their arguments on the basis of their interest in 'cultural' history, the political interests of early modern courts and their involvement in natural philosophy. Nevertheless, while we may be willing to praise these writings for their erudite work on 'cultural' history, Tribby and Findlen still make some allusions to the history of science and the supposed birth of atheoretical experimental knowledge inside the Cimento. For example, Findlen often describes Francesco Redi's career inside the Tuscan Court as the beginnings of an 'experimental method', or 'scientific method'.[18] Meanwhile, Tribby describes

> ... the emergence of a new vocational category within the court, that of the experimenting courtier who, in contrast to the philosophising courtier, relies on these new, narrowly conceived activities known as *esperienze* to keep his feet—and his

---

[14] Tribby 1995, p. 324.

[15] *Ibid.*, p. 321.

[16] The best known of these works include: Shapin and Schaffer 1985; Shapin 1988 and 1994.

[17] Shapin 1994, p. xxi.

[18] Findlen 1993, pp. 43–45.

thoughts— ... far away from the speculative work that had ruined the career of another
Medici courtier, Galileo, just a few decades earlier.[19]

Regardless of his insistence that he does not deal with science or philosophising in his analysis of the Tuscan Court, Tribby still makes the point in this passage that the 'experimental life' inside the Court in the mid to late seventeenth century, signalled a rupture from the theoretical and speculative work produced by natural philosophers such as Galileo. The new 'experimenting courtiers', as Tribby calls them, all of a sudden began producing atheoretical knowledge claims thanks to a perfected inductive method of research. This is where the works by Tribby and Findlen carry some serious implications. While they claim to be doing 'cultural' history, any spin-offs from their arguments into history of science could mean once again slipping into stories about the origins of modern experimental science of the type produced either by 'traditional' historians or by Shapin and Schaffer. An example is Marco Beretta's recent work on the Accademia del Cimento.[20]

Uninterested in the Accademia's links to Renaissance culture and politics, Beretta clings to the type of history of the birth of experimental science to which Findlen and Tribby have alluded and that has evidently survived since the eighteenth century. With the assistance of Shapin and Schaffer's analysis of early modern 'experimental life', Beretta arrived at the conclusion that the Cimento academicians were, like the members of the Royal Society, breaking away from natural philosophical theorising in order to produce factual experimental knowledge. Beretta claimed that the Cimento's emphasis on experimental science signalled the emergence of a society completely different to the Renaissance academies: 'As a matter of fact', states Beretta, 'the foundation of the Accademia del Cimento sanctions the birth of a new way of confronting science'.[21] Beretta does not go so far as to mention the role of the Tuscan Court's gentlemanly culture in maintaining a 'matter of fact' investigation of nature. Nevertheless, he clearly insists that the Accademia broke away from traditional natural philosophy, including Galileo's emphasis on mechanics and mathematics, to be the first institution to practise experimental science, providing 'the birth of a new form of scientific knowledge'.[22] By taking as literal reporting the uncontroversial experimentalist rhetoric of the *Saggi di naturali esperienze*, Beretta claims that the 'Accademia del Cimento remained neutral, adhering faithfully to the mere description of facts'.[23] Furthermore, Beretta makes a loose reference to how 'unpublished manuscripts and laboratory diaries also confirm the general tendency of the academicians to proceed on a purely experimental ground'.[24] Unfortunately, he does not tell us to which manuscripts he is referring. The problem here is that the academicians' surviving

---

[19] Tribby 1992, p. 386.
[20] Beretta 2000, pp. 131–151.
[21] *Ibid.,* p. 134.
[22] *Ibid.,* p. 148.
[23] *Ibid.,* p. 141.
[24] *Ibid.,* p. 137.

correspondence and manuscripts actually provide crucial evidence showing that they were concerned with much more than producing purely atheoretical matters of fact.[25]

Marco Beretta, like Tribby and Findlen, does not take at all seriously the academicians' natural philosophical concerns, but while the latter two at least locate the supposedly atheoretical pursuits of the Accademia in their presumed cultural context, Beretta does not provide much indication that he appreciates the social and political value of experiments to the Tuscan Court. Therefore, Beretta's work seems to endorse the 'traditional' stories about experimental method that we have seen in the early twentieth-century writings. By doing this, Beretta leaves his account hostage to the following point, to be established during the course of this paper, a point that also weighs heavily against the 'traditional' historiography. Despite the experimentalist rhetoric of the *Saggi*, the manuscript evidence provides valuable clues regarding how the Accademia del Cimento constructed experiments and knowledge claims according to their natural philosophical concerns. We shall see that rather than search for the origins of 'experimental science' in the courtly traditions of civility and gentlemanly behaviour when accumulating factual accounts of nature, we should understand that the use of an experimental programme, and the gathering of so-called 'matters of fact', were not the central concerns of these early modern thinkers. In other words, while the decision-making and action-taking processes of the Cimento were undoubtedly linked with, and partially shaped by, the cultural and political interests and traditions of their Medici patrons, this does not mean that the thinkers employed by Grand Duke Ferdinando II and Prince Leopoldo de'Medici were willing to abandon the natural philosophical concerns they had been pursuing throughout their entire careers. Instead, besides their allegiance to the Court's social and political traditions and ambitions, they were also concerned with much deeper and pertinent issues related to natural philosophical debates which spanned the entirety of western Europe.

My argument is that the Accademia del Cimento had two distinct objectives during its ten-year history. Firstly, its members wished to resolve natural philosophical controversies by performing experiments and appealing to their purported 'results', and secondly, they aimed to promote the legitimacy of their work to their European colleagues. Because of this second aim, they were forced to suppress the controversial theoretical framing of their experiments when presenting them. That is, if they wished to be applauded for their work, they dared not upset religious and political authorities, and preferred to present their claims in an authoritative and persuasive manner—by narrating experiments that purportedly accumulated factual knowledge and not controversial theorising. In other words, the actual natural philosophising that the academicians practised and recorded in their diaries and letters was quite different from what they revealed in their official reports and publication.

With this in mind, it is now time to turn our attention to a case study of the Cimento to explicate the claim that both the 'traditional' and the 'cultural' historiographies have been misled by the façade of the Cimento—the

---

[25] This argument is made in Boschiero 2003a.

experimentalist rhetoric that the Grand Duke was using to advance his own status and reputation, as well as the status and reputation of his courtiers, across Europe. The academicians' work regarding the appearance of Saturn will reveal that Leopoldo and his academicians faced some serious political and religious pressures when working in the field of astronomy. The Cimento concerned itself heavily in resolving issues that were laden with natural philosophical implications, but the Medici patrons were carefully trying to avoid any controversy with the Catholic Church by not allowing the natural philosophical opinions of the academicians to be published in their own text, the *Saggi di naturali esperienze*. This attempt to avoid controversy also assisted Leopoldo in his endeavour to create an image of the Cimento as an unbiased and uncontroversial institution.

This case study is based on Albert van Helden's narration of events surrounding observations of Saturn between 1610 and 1660, but goes beyond van Helden's work by examining the political and religious issues that helped to shape the experimental rhetoric in the *Saggi*, a rhetoric that veiled the natural philosophical concerns with which the academicians' experiments were laden. This is also the rhetoric that so many 'traditional' historians, and even some recent 'cultural' historians, have mistakenly credited as being the beginnings of an unbiased, inductive, and atheoretical experimental method for accumulating matters of fact.

3.THE SATURN PROBLEM

In July 1610, Galileo made an astounding observation with his telescope: that Saturn is not just another planet of ordinary appearance, but is in fact, the composite of three spherical bodies. Galileo believed that he had observed two small stars sitting very close to either side of the much larger central globe. He may have even thought that he was observing two satellites of Saturn, much like the Medicean stars moving around Jupiter.[26] However, after having observed no changes at all in the positions of the smaller bodies for the following two years, it seems highly unlikely that Galileo would have felt comfortable with such an hypothesis. In fact, as he expressed in his first letter on sunspots, dated 4 May 1612, Galileo was quite certain that these two bodies on either side of Saturn were not like any other stars.[27]

By the end of 1612, a sudden and dramatic change in Saturn's appearance justified Galileo's suspicion that these were not ordinary satellites. On 1 December of that year, when he wrote the third of his published letters on sunspots, he mentioned that Saturn was apparently no longer triple-bodied, but could instead be seen as a single perfectly spherical planet, 'without its customary supporting stars'.[28] Regardless of this sudden and strange change in the planet's appearance, Galileo predicted, with some degree of caution and reservation, that the two small bodies on either side of Saturn that he had earlier observed with his telescope, would return to sight by the northern-hemisphere summer of 1613 for only two months, before

---

[26] Van Helden 1974a, p. 106; Galileo 1957, p. 74.
[27] Favaro 1890, V, p. 110.
[28] 'senza l'assistenza delle consuete stelle'; *ibid.*, V, p. 237.

reappearing in the winter solstice of 1614 for another brief period, and finally again in the summer solstice of 1615. According to Galileo, they would then remain in view for many years.[29]

It is not clear what type of model Galileo was using as the basis of his predictions or how he believed the three apparent spheres were moving.[30] In any case, those predictions were fairly accurate. At the very least, the three-bodied appearance of Saturn was confirmed to have returned briefly in 1613, and again in 1615.[31] Despite these predictions, Galileo could not have possibly expected the observations he was to make of Saturn later, in 1616. Around August that year, in a letter addressed to Federico Cesi in Rome, Galileo stated that Saturn no longer appeared to consist so clearly of three separate bodies, or even one body on its own. Instead, now only one middle globe could be observed, with 'two half eclipses', or 'handles', on either side.[32] Any confidence that Galileo may have gained from accurately predicting some of Saturn's phases, could have possibly been destroyed by this latest observation.

Nevertheless, he continued to observe Saturn's movements until he lost his sight towards the end of the 1630s. During this time, as van Helden stated, Galileo's observations of Saturn's appearance, and his early predictions about its phases, became the most trusted available authority on the subject. Few other astronomers made any effort during this time to improve on Galileo's work. Only Pierre Gassendi (1592–1655) and Francesco Fontana (1585–1656), a Neopolitan instrument maker, recorded several more observations of Saturn during the 1630s. While Gassendi observed the 'handled' appearance in 1633 and described it in the posthumously published *Opera omnia* (1658), Fontana used his own telescopes, more powerful than Galileo's, to view the handles and to depict them in a manuscript in 1638.[33] But it wasn't until August 1642 that interest in the Saturn problem began to increase across Europe. It was at this point that Gassendi observed the planet without its 'handles' and discussed this apparent change in Saturn's appearance with colleagues. Following this, several other astronomers around Europe began to make further observations of the planet and to contribute to the resolution of the problem regarding its strange phases. So within the following two decades, several publications were released on the topic and various theories were

---

[29] *Ibid.*

[30] Van Helden 1974a, p. 107.

[31] A letter from a correspondent of Galileo in Rome, Giovanni Battista Agucchi, confirmed the return of Saturn's satellites in July 1613. Agucchi congratulated Galileo on the accuracy of his predictions thus far. Favaro 1890, XI, p. 532.

[32] 'Non voglio restare di significare a V.E. un nuovo et stravagante fenomeno osservato da me da alcuni giorni in qua nella stella di Saturno, li due compagni del quale non sono più due piccoli globi perfettamente rotondi, come erano già, ma sono di presente corpi molto maggiori, et di figura non più rotonda, ma come vede nella figura appresso, cioè due mezze ecclissi con due triangoletti oscurissimi nel mezzo di dette figure, et contigui al globo di mezzo di Saturno'. As cited by Giovanni Faber, an associate of the Accademia dei Lincei, in a letter to his friend Federigo Borromeo in Milan, dated 3 September 1616: Favaro 1890, XII, p. 276. Galileo also included a diagram of his observations of the 'handles' in *The Assayer* in 1623. See Favaro 1890, VI, p. 361.

[33] Van Helden 1974a, pp. 112–113.

proposed.[34] However, despite this sudden increase in interest in Saturn's strange appearance, nobody devised an hypothesis that could be agreed upon by most astronomers, to account plausibly for each of the planet's phases. Wren summarised the situation in his 1658 treatise about the planet:

> For Saturn alone stands apart from the pattern of the remaining celestial bodies, and shows so many discrepant phases, that hitherto it has been doubted whether it is a globe connected to two smaller globes or whether it is a spheroid provided with two conspicuous cavities or, if you like spots, or whether it represents a kind of vessel with handles on both sides, or finally, whether it is some other shape.[35]

It was at this point that Christiaan Huygens (1629–1695) put forward his hypothesis of a ring in his *Systema Saturnium* in 1659. In this text, dedicated to Prince Leopoldo de'Medici, Huygens confidently claimed that what he could see through his telescope was a solid, rigid and thick ring surrounding Saturn, and at the same time, completely detached from the planet. He believed that this ring created the illusion of handles or satellites that had been observed by other astronomers during the previous fifty years. Such illusions, claimed Huygens, were assisted by the use of telescopes inferior in quality to his own.[36] In fact, in what he believed to be a demonstration of the superiority of his telescope, which he had himself constructed, over those used by his colleagues in other parts of Europe, Huygens also claimed that in 1656 he had seen a small satellite of Saturn, never before spotted by anyone else.

So, from the combination of these two observations, the new satellite and the ring, Huygens was convinced that not only had he constructed the best telescope in Europe, but that he had also solved the puzzle regarding Saturn's various appearances. However, the situation was hardly that simple and the problem of Saturn was not so easily resolved. Huygens' work carried some serious natural philosophical, religious and political implications for the traditional Aristotelian astronomers still dominating the Jesuit schools, especially in Rome, and who were assessing the validity of Huygens' ring theory.

4.HUYGENS VERSUS FABRI AND DIVINI: RELIGION, REPUTATIONS AND NATURAL PHILOSOPHICAL COMMITMENTS ON THE LINE

The implications arising from Huygens' work were numerous. In the first place he was rejecting Galileo's claim that Saturn is the composite of three spherical bodies. Secondly, and more importantly, Huygens was providing a definitive claim against the Aristotelian notion that planets, since they belong to the celestial realm, are perfectly spherical and incorruptible. Furthermore, Huygens' confidence in the

---

[34] These publications include: Fontana 1646; Hevelius 1647; Gassendi 1649; Riccioli 1651; and Wren 1651. Most of these writings had important contributions to make to the discussions about Saturn, For more information on the opinions of Fontana, Hevelius, Gassendi, Riccioli and others regarding Saturn, see: van Helden 1974a, pp. 105–121.

[35] See van Helden 1968, p. 220.

[36] Van Helden 1970, p. 37.

validity of his hypothesis was based on his openly expressed heliocentric Corpernican commitments. In a clearly illustrated diagram used to explain Saturn's phases, Huygens was suggesting that the appearance of Saturn from Earth depended upon the illumination of the ring from the centrally located Sun [**Fig. 20**].

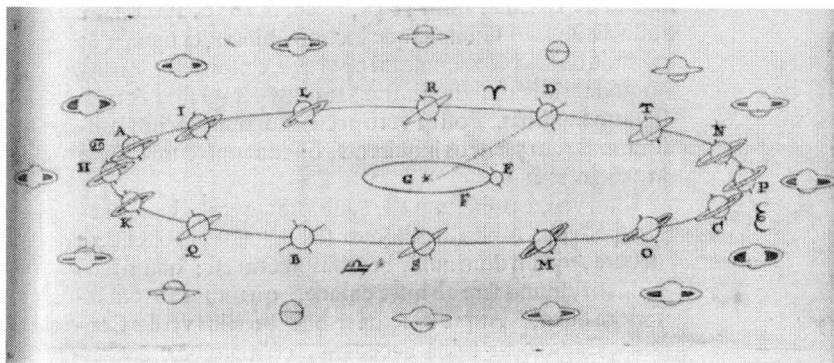

Figure 20. C. Huygens, Oeuvres completes de Christiaan Huygens, *1888–1950, 15, p. 309.*

In fact, Huygens openly expressed his support for Copernicanism by clearly stating in the dedicatory letter of *Systema Saturnium* that Saturn, like Earth, orbits the Sun.[37] He even suggested that Saturn and Earth were quite alike, contrary to scholastic belief in the uniqueness of the Earth: he claimed that both have only one satellite, and both have the same degree of inclination.[38]

This type of Copernican framing for an astronomical hypothesis was certainly not unusual for a Protestant astronomer in the 1650s, educated in early modern, rather than scholastic, astronomy; but, from the perspective of traditional scholastic astronomers, Huygens was re-igniting the same issues that had seen the condemnation of Galileo by the Catholic Church in 1633 regarding the teaching of Copernican and anti-Aristotelian cosmology as truth. Once Huygens put forward a claim that was strongly tied to Copernican and anti-Aristotelian beliefs and based on telescopic observations that he claimed were the most accurate so far, natural philosophical implications and conflict with rival instrument makers would be unavoidable.

[37] 'Unum hoc inanimadversum eos praeterire nolim; nempe quam non leve argumentum ad astruendum pulcherrimum illud mundi universi ordinem qui a Copernico nomen habet Saturnius hic mundus adferat'; Huygens 1888–1950, 3, p. 433, as cited by Galluzzi 1981, p. 826.
[38] Van Helden 1974b, p. 163.

So rather than resolve the problem with Saturn, Huygens' claims actually helped to increase the contention surrounding its phases and appearances. In particular, since Huygens dedicated his text to Prince Leopoldo, the issue was especially fraught with political, religious, and natural philosophical dangers for the Tuscan Court and its members, including of course, the Accademia del Cimento. This is, therefore, where we begin to examine the reasons why the academicians' firstly became involved in this issue, and why their astronomical work was omitted from the *Saggi*.

After Huygens published and distributed his text to his friends and colleagues across Europe, criticisms of the ring theory were immediately raised. Some astronomers were sceptical of the validity of Huygens' hypothesis, because Saturn sometimes appeared to be unaccompanied by any ring or the illusion of satellites. That is, if the ring were as thick as Huygens proposed, then it should be visible all the time.[39] These types of criticisms were additionally aimed against Huygens' claim that he was using a telescope superior in power and quality to anyone else's in Europe. This statement obviously did not sit well with other highly respected instrument makers. Johannes Hevelius (1611–1687), as an example, in a letter to Ismael Boulliau (1605–1694) in December 1659, insisted that his telescopes were not inferior to Huygens'. Hevelius stated that he had made the same observations as Huygens ten years earlier, only he was too 'careless', as he put it, to speculate upon whether the new satellite of Saturn that Huygens claimed to have discovered, could be anything other than a fixed star. As for Huygens' proposed ring, Hevelius was also sceptical of this notion on the basis of its supposed thickness. In any case, he claimed to have already carefully annotated each of Saturn's apparent phases before Huygens even suggested the existence of a ring. Hevelius concluded to Boulliau: 'Thus, on this point I don't concede anything to him'.[40]

Meanwhile, Eustachio Divini (1610–1685), a Roman manufacturer of telescopes was also quite annoyed by Huygens' claims. In a letter he wrote to Prince Leopoldo on 10 July 1660, in which he mentioned how he had only recently received and read the *Systema Saturnium*, Divini gave his estimation of Huygens' work: 'I found that he placed too much faith in some things, in himself, and in his lenses'.[41] As van Helden pointed out, it was to be expected that Divini, whose livelihood depended on his reputation as a worthy manufacturer of telescopes, would be critical of an opponent who claimed to be making the best telescopes in Europe, and believed that he was carrying out far more accurate observations of Saturn than anyone else.[42] Divini was therefore unimpressed by Huygens' claims and set about trying to discredit them.[43]

[39] Ismael Boulliau and Christopher Wren were particularly critical of Huygens' theory on these grounds; *Ibid.*, p. 163.

[40] Bibiotheque Nationale, MSS. Collection Boulliau', fols 89v–90r. As cited by van Helden 1970, p. 38.

[41] '... trovai ch'in qualche cosa troppo egli si sia fidato, e di sé, e delli suoi occhiali'. Biblioteca Nazionale Centrale di Firenze, MS. Gal. 276, fol. 33r.

[42] Van Helden 1970, p. 38.

[43] Both Divini's and Hevelius' objections to Huygens' observations and claims illustrate some of the sociological issues involved in the historiography of this case study. The significance of an observation was being challenged and negotiated, yet the task for historians is not to discuss who had

Divini's letter to Leopoldo accompanied a text, *Brevis annotatio in systema Saturnium*, which was also dedicated to the Tuscan Prince, and which was intended to be a public response to Huygens' claims. Not surprisingly, this writing also contained criticism of Huygens' overwhelming faith in his telescopes. But the discrediting of Huygens as an instrument maker was not the only aim of this text. The author of *Brevis annotatio* also suggested replacing Huygens' theory with an hypothesis that was more of a compromise between Galileo's earliest observations of a triple-bodied Saturn, and traditional Ptolemaic astronomy.

This theory suggested that there were four stars near Saturn, apart from the satellite discovered by Huygens. These stars did not orbit Saturn, but rather two points behind that planet. Since two of the stars reflected light and the other two did not, an illusion was supposedly created in which, when the light-reflecting stars were partially obscured by the dark stars, two half eclipses could be observed from Earth. In addition, at their greatest elongation, the light-reflecting satellites could be seen in their entirety sitting closely by the sides of Saturn, but when they were hidden from view behind the planet, obviously only the sphere of Saturn could be seen [**Fig. 21**].

---

the 'best' instrument and who was making the 'right' observations. Indeed, rather than make 'Whiggish' statements about this case, we should come to understand that knowledge claims were being debated by rival telescope makers with social and political concerns extending well beyond who had the best theory. The efficacy of one's instrument was grounds upon which to be critical of a rival theory, and to be supportive of one's own intellectual, political, and religious commitments. So these astronomers were challenging each other's observations on the basis of their own social and natural philosophical agendas. We will find that these were concerns that extended to the Cimento in their involvement in this topic in 1660, and their decisions regarding the presentation of their work. This discussion recalls the sociological analysis of scientific knowledge pioneered by Harry Collins and Trevor Pinch: Collins 1975, pp. 205–224; Pinch 1985, pp. 3–36.

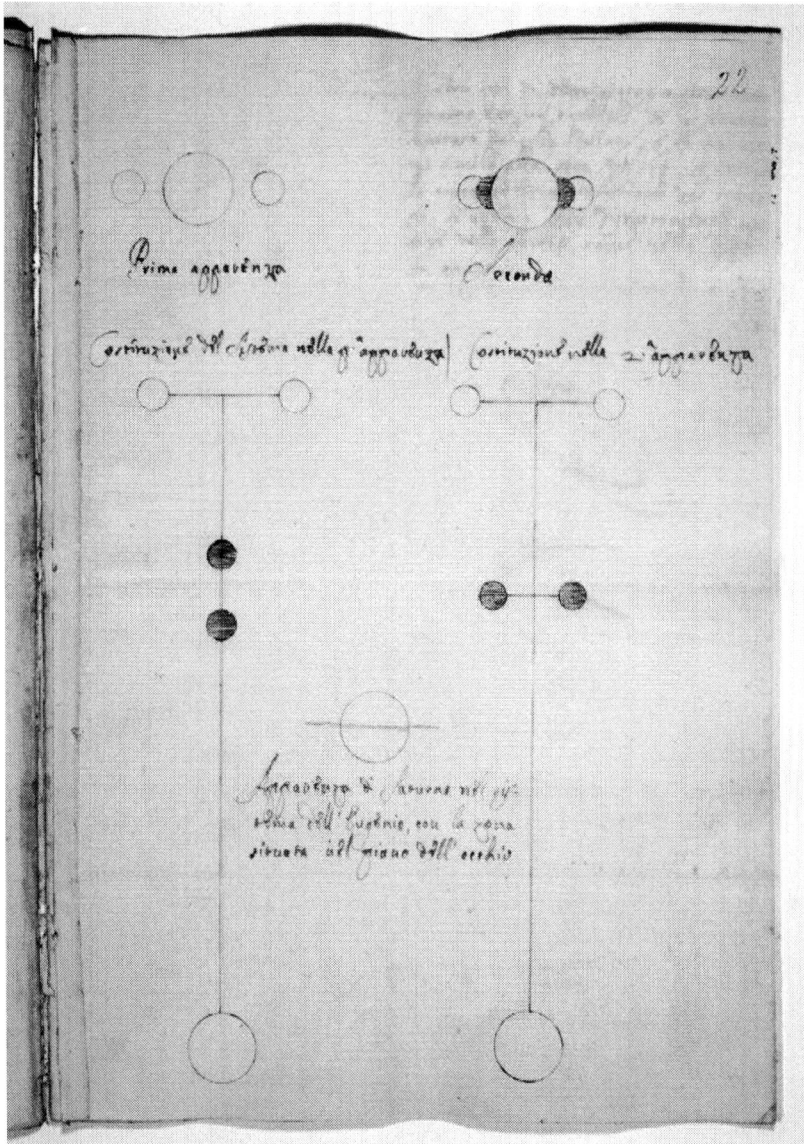

Figure 21. Saturn's appearance according to the hypothesis in Divini's Brevis annotatio. *This illustration is in a manuscript in Lorenzo Magalotti's handwriting, dated July 1660 and entitled* Osservazioni delle stele di Saturno. *Note that Saturn is also drawn here according to Huygens' system. MS. Gal. 271, fol. 22r. Courtesy of the Biblioteca Nazionale Centrale di Firenze.*

So despite making the same telescopic observations as Huygens, Divini's publication proposed a theory very distant from the suggestion of the existence of a ring around Saturn. Significantly, it did not include any references to Copernicanism. In fact, it insisted that the phases of Saturn were as viewed from the centrally located Earth, and it rejected the notion that any imperfect solid could be suspended around Saturn. This theory even validated Galileo's observations of separate spherical bodies surrounding Saturn, and framed those observations within an acceptable geocentric Tychonic model of the universe. As the first of the 23 propositions outlined in *Brevis annotatio* states, the heliocentric system could not be acceptably applied to the movements of any planet, since, according to Aristotle, 'the earth is immobile at the centre of the world and ... the celestial spheres turn around it'. This is an opinion, continues the text, that the author defended 'with tenacity, judging it to conform at the same time to the Catholic decrees, to the Sacred Scriptures, to the phenomena observed, and to sane reason'.[44]

Let us be clear, this was not a theory of Saturn that was strongly and logically tied to a geocentric universe or traditional natural philosophy, and it is even possible that any such ties were being overplayed by its author, but it was still designed to put forward an alternative hypothesis that rejected Huygens' Copernican-based model, and retained scholastic astronomy. Therefore, should this theory be sanctioned by a natural philosophical authority such as the patron of the Cimento, Divini's reputation would not only be restored, but traditional cosmological and astronomical observations and values would also be kept intact. Leopoldo and his academicians were being led into a minefield of political, religious, and natural philosophical issues.

One other point remains to be mentioned before we analyse exactly how the Cimento became involved in this debate and what decisions they made regarding the two competing theories. Although *Brevis annotatio* was published under Divini's name, in the above-mentioned letter to Leopoldo from 10 July 1660, Divini hinted at the possibility that the text was actually written by a correspondent of the Cimento, Father Honoré Fabri (1607–1688), a French Inquisitor with the Holy Office and Jesuit mathematician in the Roman College. Divini stated that he called upon Fabri to assist him in making this response to Huygens' work available in Latin, since Divini was himself only experienced in writing in Italian, and as he confessed to Leopoldo, a tract in Italian 'would only be of service to a few'.[45] This would seem to suggest that Fabri simply translated Divini's manuscript. But given that this public response to Huygens was littered with the type of religious and anti-Copernican rhetoric that we could expect from a Jesuit astronomer such as Fabri, it is likely that Fabri's contribution went much further than a mere translation. In fact, from the moment the tract arrived in Tuscany, it was considered to be Fabri's work.[46]

---

[44] Huygens 1888–1950, 15, p. 422, as cited by van Helden 1973, p. 242. See also Galluzzi 1981, pp. 826–827.

[45] 'ad alchuni pochi serviriano'; BNCF, MS. Gal. 276, fol. 33r.

[46] When summarising the accompanying letter from Divini to the Prince, Magalotti, the Cimento's secretary, wrote: 'Eustachio Divini manda a S.A. il suo libro contro l'Ugenio. Dice essere stato disteso dal Padre Fabri col fondamento d'alcune poche particolarità notate da lui nel libro dell'Ugenio'; BNCF, MS. Gal. 276, fol. 33v.

According to van Helden, Divini probably provided Fabri with some doubts about Huygens' telescopic observations and Fabri, with his own agenda to disprove the heretical arguments put forward by the Dutch Protestant, Huygens, compiled the *Brevis annotatio* himself.[47]

Further evidence suggesting that Fabri composed this text lies in the fact that all the subsequent references in letters and manuscripts to the theory opposing Huygens' ring hypothesis mentioned Fabri as the innovator. For example, Michelangelo Ricci, who was regularly dispatching news to the Tuscan Court from Rome, made mention on several occasions of his discussions with Fabri, 'talking to him about his system of Saturn'.[48] So it is likely that Fabri was the central figure behind the scholastic objections from Rome regarding Huygens' claims, and as we have already seen, his aim was to defend traditional Aristotelian cosmology and the Scriptures, 'with tenacity'.[49]

The involvement of Fabri and his use of such scholastic rhetoric against the Protestant astronomer, Huygens, certainly elevated the stakes in these seventeenth-century studies of Saturn. Supporting Copernicanism did not seem to be the central focus of Huygens' *Systema Saturnium*. Instead, as we have just seen, he was far more concerned with boasting about the superiority of his telescope over all others and advancing his ring theory. Copernican astronomy was, nevertheless, the basis of his description of Saturn's phases and this left him open to criticism from Catholic authorities and scholars still determined to have Copernicanism taught as nothing more than hypothetical. Therefore, Huygens could not avoid having to defend the anti-Aristotelian implications in his work. Once he received Fabri's and Divini's tract against the ring theory in August 1660, he immediately composed a reply, *Brevis assertio systematis Saturni*, once again dedicated to Prince Leopoldo.

This work was eagerly anticipated in Rome and probably also in Florence, where Leopoldo was continually receiving news from Michelangelo Ricci about his conversations with Fabri. But the anticipation surrounded not so much the technicalities of Huygens' argument, such as his beliefs regarding the inclination of Saturn or the thickness of the ring, as the cosmological framing of his work. In a letter from Rome dated 13 September 1660, Ricci mentioned to Leopoldo his expectations of Huygens' reply to the publication made in Divini's name and against the ring theory. Ricci revealed the dangers that he believed Huygens faced and the restraint and caution that Huygens should practise when compiling his response to the criticisms of the *Systema Saturnium*.

> A friend of mine sent Eustachio's book to Huygens. I said to him that Huygens should write carefully without insulting anyone, or touching on the motion of the Earth or anything else that could give the Congregation in Rome reason to prohibit him, impeding the book from being seen and also prejudicing the reputation of the cause.[50]

This was not a warning about Huygens' safety, but more so about the threat his theory was posing for scholastics in the Jesuit colleges, the Courts close to Rome,

---

[47] Van Helden 1970, p. 39.

[48] 'Parlandogli io di quel suo sistema di Saturno'. 26 July 1660; BNCF, MS. Gal. 276, fols 42r–v.

[49] See note 44, above.

[50] Fabroni 1775, 2, p. 97; see also Galluzzi 1981, p. 827.

and in the Catholic Church, those who remained determined to uphold traditional cosmology and Aristotelian natural philosophical beliefs. In any case, if Huygens actually received this warning from Ricci, he completely disregarded it. In his *Brevis assertio*, Huygens was critical of Fabri's use of Aristotelianism and defended the Copernican basis of his work by maintaining that Copernicus' model was closer to the truth than Ptolemy's, or even Tycho's. This, so Huygens believed, was even widely accepted by many Catholic astronomers.[51] These statements compounded the practical problems that Huygens adduced in Fabri's four-satellite theory. According to Huygens, the apparent 'handles' were not circular, as should be the case in the theory about the dark stars eclipsing the light-reflecting satellites. Instead, they were clearly elliptical. Furthermore, Huygens criticised Fabri and Divini for failing to provide a model for predicting the phases of these satellites moving behind Saturn.[52]

Clearly then, aside from Divini's own concern about his reputation, there was a strong natural philosophical and religious perspective at stake in these three treatises mentioned so far: Huygens' first work on Saturn, Fabri's and Divini's criticism, and Huygens' response to that criticism, all published in 1659–1660. As is revealed in these writings, as well as in the unpublished letters between the central figures in this debate, natural philosophical and religious beliefs were crucial to the acceptance or rejection of the opposing theories and the instrumentation used by the rival astronomers. So this was certainly not about who was using a correct method of observation, but rather how the rival instrument makers and astronomers protected their careers, and pursued their social and political concerns. In other words, what religious, political and natural philosophical aims they were each trying to achieve.

So how did the Cimento become involved in such a potentially volatile situation?[53] When a copy of the *Systema Saturnium* arrived in Tuscany with the dedication to the Medici Prince in July 1659, Leopoldo delayed his reply to Huygens for over one year. The reason for this, as van Helden suggested, may have been partly due to Huygens' failure to send Leopoldo an accompanying letter with the text.[54] But we may be willing to believe that Leopoldo was concerned with far more than a mere lack of communication from the Dutch astronomer. As Galluzzi claimed, such a long delay was quite out of the ordinary and reflected the Prince's extreme caution when facing the possibility of giving his approval to a Protestant astronomer with openly Copernican beliefs.[55] Furthermore, not only did Huygens' Copernicanism count against him, but Leopoldo may have also been aware of the

---

[51] Galluzzi 1981, p. 827.

[52] Van Helden 1974b, p. 165.

[53] The Prince and his academicians were acutely aware of potentially controversial topics. There sensitivity to such topics is discussed further in: Boschiero 2002, pp. 383–410; Galluzzi 1981, pp. 788–844.

[54] It is not even clear why Huygens decided to dedicate this work to the Tuscan Prince since he had never met or even written to Leopoldo prior to August 1660. It is a possibility that Hevelius or Boulliau, both mutual acquaintances of Huygens and the Prince may have given Huygens the idea to write the dedication to Leopoldo, given the Prince's interest in natural philosophy. In any case, as van Helden suggested, Huygens did not include an accompanying letter, perhaps for the reason that he did not wish to give Leopoldo the impression that he was seeking patronage from the Tuscan Court; van Helden 1973, p. 240.

[55] Galluzzi 1981, p. 826.

practical concerns with the ring theory that Boulliau and others had been expressing, especially Huygens' proposed thickness of the ring. So the Medici Prince, being so close to Rome and so eager to avoid any controversy, understandably hesitated in providing his approval of a theory that lacked credibility according to many astronomers, and more importantly, had the potential of being highly controversial because of its Copernican content. If this was indeed the reason for Leopoldo's seemingly cautious response to having *Systema Saturnium* dedicated to him, his judgement was not misplaced, considering Divini's and Fabri's objections to Huygens' claims.

It was in fact, Fabri's and Divini's public criticism of Huygens that forced Leopoldo to take some action regarding the Saturn problem. Forming an assessment of Huygens' work was unavoidable once Fabri and Divini also dedicated their publication to the Prince. This was not only because Leopoldo was the recipient of both dedications by the opposing astronomers, but also because Divini even specifically pleaded with the Prince to act as mediator in the debate between him and Huygens. Divini asked Leopoldo 'to inquire into which of us got it right and if the glasswork from Holland is more perfect than ours'.[56] Furthermore, in *Brevis annotatio* Fabri and Divini again appealed to Leopoldo's 'very good censure' and his 'enlightened judgement' to adjudicate between Huygens' theory and the hypothesis from Rome.[57]

5.LEOPOLDO TAKES CONTROL

In July and August 1660, Leopoldo called upon his Cimento academicians to assist him in the resolution of this controversy between Huygens and Fabri. The following actions taken by the Cimento were indicative of the natural philosophical concerns pursued by the academicians, led by Giovanni Borelli (1608–1679), and their social, political and religious concerns when presenting their work. Leopoldo was forced to choose between a Dutch Protestant who was not afraid to express the same views that had seen the condemnation of Galileo by the Catholic Church, and a Jesuit mathematician in Rome who was also an Inquisitor for the Holy Office, no less. As a result, the academicians were aware that their work on this topic was going to be anticipated in several parts of Europe, particularly Rome, where scholastics may have felt the most threatened by the Copernican content of Huygens' work. This was, therefore, a topic that demanded the attention of Europe's scholarly community, especially in Rome, since it was concerned with traditional conceptions about natural philosophical macro-structures, such as Aristotle's cosmology. This was to have an impact upon the way in which they would decide to carry out their observations, how they would choose to discuss their natural philosophical concerns, and how they would present their work to their colleagues. In other words, they

---

[56] This appeal to Leopoldo was made by Divini in his previously mentioned letter written on 10 July. '… esplorare chi di noi habbia accertato e se li vetri d'Ollanda siano più perfetti della nostra Italia'; BNCF, MS. Gal. 276, fol. 33r.

[57] Huygens 1888–1950, 15, p. 436. As cited by van Helden 1973, pp. 241–242.

knew about the controversial nature of this case when they approached it, and this influenced how they carried out their work. We can look forward, therefore, to the decisions and actions that Leopoldo and his academicians were to take, keeping in mind the deep concern the Prince had for the public image of the academy under his control and his relationship with ecclesiastical authorities.

As they began to investigate the contrasting theories regarding Saturn's appearance, the situation was not looking favourable for Fabri and Divini. On 17 July 1660, when the academicians had only just read through the work that had been published under Divini's name, that is, before they had even agreed on a course of action, Magalotti, the Cimento's secretary, recorded in the group's diary that Borelli was already making some remarks in Huygens' defence.[58] Furthermore, during late July, while the academicians were planning their approach to the problem, the correspondence to Leopoldo from Rome and France regarding the choice that Leopoldo had to make between the competing theories, was also quite favourable to Huygens.

Firstly, on 26 July 1660, Ricci wrote to Leopoldo giving his own judgement of Fabri's theory. According to Ricci, whose opinion was highly respected by Leopoldo, Fabri's work was certainly worthy of praise. However, upon closer investigation of his theory, Ricci claimed that Fabri did not provide a satisfactory account of the phases and movements of the planet and its apparent satellites. In fact, so sceptical was Ricci of the validity of Fabri's hypothesis, that he believed the Jesuit astronomer was only concerned with doing everything possible to defend traditional scholastic beliefs: 'It only seems so far that the Father introduces many changes in order to save only one ancient opinion of Saturn moving around the Earth'.[59] This reinforces the idea mentioned earlier that the tenuous ties between Fabri and Divini's hypothesis and their broader natural philosophical concerns about geocentricity, were being emphasised and even exaggerated in order to counter Huygens' Copernican-based theory.

Several days later, on 9 August 1660, as he was awaiting news from Florence about the Cimento's observations, Ricci again wrote to the Prince expressing more criticism of Fabri's hypothesis and expecting a similar sceptical report from the academicians. Ricci wrote that although Fabri had heard several criticisms of his theory, including Ricci's own thoughts, Fabri was convinced that he would be able to formulate a suitable response. Ricci concluded: 'I doubt that perhaps things will not come out for him as easily as he believes'.[60]

In the meantime, Huygens' first letters to the Prince were arriving from France. One of those, written on 16 August, stated Huygens' eager anticipation of the Prince's judgement. In particular, Huygens expressed his confidence that the academicians would decide in his favour, especially since observations from

---

[58] 'Si lesse tutto il libro del Divini scritto contro il sistema Saturnico di Christiano Eugenio, et in esso quello, che ha inventato il Padre Fabri Gesuita. Si sentirono alcune annotazioni fatte dal Sig. Borelli sopra a detto Libro in difesa dell'Eugenio, e si stabilirono alcune esperienze in questo istesso proposito'; BNCF, MS. Gal. 262, fols 93r–v.

[59] 'Quel che appare fin ora è che'l Padre introduce molte novità per salvare una sola antica opinione di Saturno mosso intorno la terra'; BNCF, MS. Gal. 276, fol. 42v.

[60] '... dubbito forse non sia per riuscirli così facilmente come si crede'; BNCF, MS. Gal. 276, fol. 49r.

England, such as those made by Christopher Wren, could be used to support the ring theory, or at least falsify Fabri's hypothesis, even though Wren himself argued that Huygens' theory was problematic.[61] By this time, the academicians had already performed an experiment to assist them in their assessments of the opposing hypotheses. But, we may see from the above-mentioned correspondence from Rome and Paris addressed to Leopoldo, that Fabri's and Divini's work was not gaining a great deal of favourable publicity in Florence, despite its conformity with scholastic beliefs, and despite also the criticisms that were aimed against the credibility of Huygens' theory. In addition to this, we must not forget that most of the academicians, including the Prince, were supportive of Galileo and his work on Copernican astronomy. All this reflected the unlikely advantage that Huygens held over his Roman colleagues. Indeed, as we shall now see, although they did not dare to express openly any anti-scholastic sentiments, especially in astronomy, the Cimento's work on Saturn was far from favourable for Aristotelian natural philosophers such as Fabri. We shall also see that the judgement reached by his academicians placed Leopoldo in a difficult position with regard to the Catholic Church. That is, he now had to try to negotiate the credibility of his judgement against Fabri in the face of traditional religious and natural philosophical pressures. This meant that Leopoldo was also fighting to preserve the credibility of the academy under his control, and its reputation as an institution producing reliable and uncontroversial knowledge.

6.EXPERIMENTING WITH MODELS

The Cimento diary entry for 20 July 1660, mentioned how the academicians discussed their options for investigating the appearance of Saturn. Since, they agreed, observing Saturn's phases would require years of telescopic observations, they would have to devise other ways of arriving at a quick resolution of this topic, including the construction of models.[62] So, during the weeks that followed this meeting, the academicians made two models of Saturn, one with Huygens' ring, and the other with Fabri's satellites [**Fig. 22**].[63] How the first of those models was set up,

---

[61] BNCF, MS. Gal. 276, fols 51r–v. Wren did not believe that what he saw around Saturn was a ring unattached from the planet, but rather an elliptical corona that touched Saturn at two opposite ends. Although not agreeing with Huygens' theory, Wren was certainly far from being supportive of Fabri's hypothesis about four satellites of different light-reflecting capabilities. For a summary of Wren's corona theory, see van Helden 1974b, p. 160.

[62] 'Si consultò il modo, e il tempo da farvi le osservazioni di Saturno con l'occhiale del Divini, perciò si discorsero diverse maniere di macchine per addopperare con facilità il Telescopio'; BNCF, MS. Gal. 262, fol. 93v. As Borelli stated in his letter to Leopoldo cited below, and published by Targioni Tozzetti, regular observations of Saturn's phases would have taken 'eight or nine years'. For this reason they decided to make the models and record their conclusions as soon as possible; BNCF, MS. Gal. 271, fol. 3r.; Tozzetti 1780, 2, p. 740.

[63] A sketch of the ring-theory model was sent to Huygens and was published in: Huygens 1888–1950, 3, pp. 154–155. Rough sketches of the model, as well as drawings of the ring around Saturn, and even of Fabri's proposed satellites can also be found in BNCF, MS. Gal. 271, fols 34r–47r.

and how the following observations were carried out, was described in a letter
Borelli wrote to Leopoldo in August 1660.

*Figure 22. Illustration of the model used by the Cimento to test Huygens' ring-
theory. MS. Gal. 289,81r. Courtesy of the Biblioteca Nazionale Centrale di
Firenze.*

Borelli described how the academicians set up their model of Huygens' system
of Saturn in a long gallery, probably in the Pitti Palace, at a distance of about 37
braccia (75 metres) away from the two telescopes, a powerful and large one, and
another one smaller and inferior to the first. Four torches were set up to illuminate
the model, but were hidden from the observer's field of vision. The initial
observations of these models through the telescopes were favourable for Huygens'
theory, since the powerful telescope observed the ring clearly, while the inferior
instrument created the illusion of two small satellites on either side of Saturn. This
strengthened the suggestion that the ring theory explained the strange appearances
and phases of Saturn better than any other hypothesis produced in the seventeenth
century. But, having constructed the model themselves, the academicians recognised
that they were of course already aware of its dimensions before observing it through

the telescopes, and were therefore not unbiased viewers. Not trusting their own senses, they called upon neutral observers, 'who had not seen the shape of this device from nearby', to look through the smaller telescope.[64] Besides some men who, for several possible reasons, recorded rather odd observations, 'it was obvious that the appearance that they almost all drew was the disc of Saturn in the middle of two little round balls and separated from it by a sensible distance'.[65] That is to say that Galileo's first observation of Saturn was recreated through the Cimento's experiment showing that the observations of the supposed satellites were simply illusions created by the ring and by the imperfections of those telescopes inferior in quality and power to Huygens' instruments. So the experiment was a resounding success from Huygens' point of view. Indeed, Borelli reported in his letter to the Prince that he had even observed what appeared to be Saturn's shadow being cast upon the ring. With this final observation, 'it would seem that a very efficacious argument', so Borelli claimed, 'could be deduced in Sig. Huygens' favour'.[66]

The only doubt that remained for the academicians was the same one for which Huygens was criticised by other French and English astronomers: in their observations of the model through the stronger of the two telescopes, it seemed that some trace of the ring was always apparent, meaning that Huygens' theory could not explain how in reality Saturn occasionally appeared only as a single sphere with no accompanying 'handles'. The only argument that Huygens made here in defence of his theory was that the edge of the ring was made out of a material that did not reflect light and was therefore sometimes invisible from Earth. This type of ad hoc claim was understandably not very well received, since it also meant that the academicians would have to accept Fabri's and Divini's suggestion that two of the hypothetical satellites behind Saturn could also be made out of this material.[67] In any case, the academicians' experiments proved to be far more successful for Huygens than for Fabri. A model of Fabri's theory with the satellites provided the three-bodied appearance of Saturn and the single sphere, but the 'handled' appearance was never achieved.[68] So while both Huygens and Fabri had practical problems with their hypotheses, according to the Cimento academicians, the observations carried out by

---

[64] Borelli did not actually state what telescopes, if any, these observers used. But judging from their observations and Borelli's comments about the results, cited below, it would seem that these independent participants could have been asked to observe the model only either with the inferior of the two instruments, or even with no telescope at all.

[65] BNCF, MS. Gal. 271, fol. 7v; Tozzetti 1780, 2, p. 742. 'Per chiarire adunque la verità di questa apparenza furono chiamati molti, fra quale anche delle persone idiote, e che non avessero veduta da presso la struttura di quella macchina, ad osservarla e fatta gliela vedere dalla detta distanza di 37 braccia, e disegnare ciascuno a parte ciò che se gli appresentasse, fu così patente l'apparenza che disegnarono quasi tutti il disco di Saturno in mezzo a due palline rotonde, e distaccate per sensibile spazio di essa', as translated by van Helden 1973, p. 245.

[66] '... pare che possa dedursene argumento molto efficace a favore del Sig. Ugenio'; BNCF, MS. Gal. 271, fol. 13r; Tozzetti 1780, 2, p. 745.

[67] See BNCF, MS. Gal. 271, fol. 10v.

[68] This observation was not mentioned in the above-cited letter from Borelli to Leopoldo. But it was described in the final report Borelli wrote in August, 1660, one of two reports written by the Cimento; BNCF, MS. Gal. 289, fols 15r–19v. See also van Helden 1973, p. 248.

Huygens were far more acceptable and less obviously flawed than those made by
Fabri.

This signalled the end of the academicians' observational work on the topic, but
it only marked the beginning of their religious and political concerns when
presenting their results—the type of concerns that guided the Cimento's public
rhetoric. The Tuscan Court faced a problem when Huygens' Copernican-based work
was dedicated to the Prince, but now that problem intensified as Leopoldo was
hearing recommendations from his courtiers in favour of Huygens' controversial
theory. Therefore, Leopoldo had to decide as to how the Cimento's work should be
presented to the public. Now that Huygens' work was not rejected, but in fact
supported by the academicians, there was still plenty of room for the type of
controversy with ecclesiastical authorities that Leopoldo was anxious to avoid for
the sake of preserving the uncontroversial status and reputation of his Court and his
academy.

On 17 August 1660, two reports on the experiment and its outcomes, one written
by Borelli, and the other by another academician, Carlo Dati (1619–1676), were sent
to Rome. They were addressed to Ricci, but the accompanying letters by Leopoldo
and Lorenzo Magalotti, the Cimento's secretary, were both intended for Fabri.[69] As
we saw from Borelli's earlier letter to Leopoldo describing the experiment, the
Cimento's leading contributor was quite supportive of the ring theory, despite
Huygens' doubtful claims about the ring's thickness. Borelli's official report
reflected similar sentiments against Fabri, and in favour of Huygens and his
Copernican-based theory of Saturn. Meanwhile, Dati was seemingly more impartial
in his assessment, suggesting that there could be grounds to dismiss either of the
competing theories. Nevertheless, he agreed that from the observations performed
by the Accademia, Fabri's hypothesis was the less likely to be true.

In the meantime, a different report, also written by Borelli, was sent to Huygens
via Heinsius. A letter from Dati to Heinsius also accompanied the report. After
presenting the same arguments supporting Huygens and based on the Cimento's
observations of the models, Borelli praised the Dutch astronomer for his
observations and interpretation of Saturn's phases. This was finally, as van Helden
pointed out, the approval of his *Systema Saturnium* that Huygens had been seeking
from the Tuscan Court when he dedicated his publication to the Prince in 1659.[70]

Fabri's response to the academicians' conclusions can be gauged firstly from
Ricci's 22 August letter to Leopoldo. Ricci himself was delighted with the
Cimento's 'ingenious' experiment and suggested that Huygens' theory was clearly
shown to be the more accurate of the two. However, Ricci warned that Fabri was far
from convinced and would seek Divini's collaboration to analyse the Cimento's
claims and to compile a defence of their theory.[71] Indeed, on 30 August, Ricci sent

---

[69] Copies of the letters and the reports are preserved in BNCF, MS. Gal. 289, fols 6r–9r.; 15r–21v.

[70] Van Helden 1973, pp. 250–251.

[71] 'Non vedo però che fin ora si possa dir altro se non che'l Sig. Ugenio non sia convinto dal P.ʳᵉ Fabbri di
falsità, ma che ne meno ci costi esser vero il di lui sistema, restandovi pur assai da smaltire. Gran
diletto ha poi recato all'animo mio l'esperienza che mostra la fascia intorno il globo formato a
simiglianza di Saturno, ora in forma di due globi separati, ora nella sua natural figura: pensiero de' più
ingegnosi e pellegrini ch'i'udissi mai. Lo dissi al P.ʳᵉ Fabbri prima di consegnargli il piego del Sig.ʳ

Leopoldo a manuscript, again written under Divini's name, which continued to defend the quality of Divini's telescopes.[72] In this apologia, entitled *Pro sua annotatione in Systema Saturnium*, Fabri also defended his theory by adding two more light-reflecting satellites to his system of Saturn that could create the elliptical shape of the 'handles'. Nevertheless, Fabri was still careful not to dismiss Huygens' claims completely. He conceded, for example, that while he disagreed with the ring theory, Saturn could appear to have a ring surrounding it. As van Helden suggested, Fabri was allowing himself the opportunity to retreat gracefully should any more criticisms be aimed against him.[73] Once again this demonstrates how observations of natural phenomena, including the instrumentation used, were laden with social, political, and natural philosophical commitments. Fabri and Divini were willing to go to any length to defend scholastic principles in which they had been trained and that formed a cornerstone of traditional intellectual endeavours, and, in the process, to show that Divini's telescopes were not inferior to those of Huygens. Yet, realising that their claims were attracting criticism from highly respected sources, such as the Cimento, they still manoeuvred to present their work in a way that could offer them an escape and save their own careers and reputations.

In the meantime, since the Cimento was now overwhelmingly in favour of the validity of Huygens' theory, and had even made their support clear in the reports they sent to the rival astronomers, Leopoldo had to ensure that no accusations of heresy could possibly be made against him or his academy. For this reason, firstly he made a request to Fabri and to Huygens that they not refer to the Cimento's work on this topic in their writings. In what is probably the clearest demonstration of the academicians' intention to keep well away from any type of conflict that could harm their reputation and relations with other courts, especially the Papal Court, Carlo Dati wrote the following message to Heinsius, intended for Huygens, in August 1660:

> For the moment it is desired that no public mention is made of it. For one thing this is because these men are very cautious in affirming anything, not wishing to commit themselves without much consideration and repeated trials …, and for another thing because, having written some rather severe censures against Father Fabri, they would not wish to commit themselves and to be held by the world to be impassioned and partial … In this I commit myself to your prudence.[74]

This appeal to Huygens' discretion would certainly save the academicians from being exposed by the Dutchman, but they still needed to convince Fabri that they did not intend to promote Copernicanism as the truth. This would require some subtle diplomatic manoeuvring. Leopoldo decided to publish, in Florence, Huygens' *Brevis Assertio*, the Dutch astronomer's reply to Fabri's and Divini's critical analysis of the *Systema Saturnium*. This text, as was mentioned, defended the ring theory largely on

---

Lorenzo Magalotti, e mi rispose che 'l Divini avrebbe voluto provar tutto questo e per quel che m'imagino ambidue s'armano alla difesa'; BNCF, MS. Gal. 267, fols 55r–v.

[72] Fabroni 1775, 2, pp. 94–95.

[73] Van Helden 1973, pp. 256–258.

[74] Huygens 1888–1950, 3, pp. 149–150, as cited by van Helden 1973, p. 250. This message was in the letter Dati sent to Heinsius accompanying Borelli's report to Huygens of the academicians' Saturn experiment.

the basis that Copernicus' system was true, and criticised its Roman detractors for their scholastic beliefs. Obviously the complete publication of this tract, including Huygens' anti-Aristotelian rhetoric, would hardly have aided the Prince's reputation in Rome, so in order to avoid any controversy and appease Fabri's natural philosophical and religious concerns, Leopoldo simply omitted Huygens' references to Copernicanism. This piece of editing did not detract greatly from Huygens' ring-hypothesis, nor did it deny the Copernican basis of his work, which most astronomers outside of Rome probably still would have inferred from having read, or heard about, Huygens' previous publications and utterances on the topic. But it did maintain the academicians' impartial reputation by showing that they were willing to listen to the varying claims and make an assessment without becoming participants in controversial speculations and debates about theory.

Furthermore, the experiment with a model, although an unusual approach to an astronomical question, provided Leopoldo and his courtiers with an opportunity to maintain a purely experimental, and therefore seemingly factual and atheoretical, approach. This awareness of the value of the academicians' experimental work was reflected in Borelli's report of the experiment to Leopoldo:

> In this matter too we have inviolably observed the custom of the Academy of Your Highness, which is to search out the truth through many experimental proofs, to a degree, however, in which it can be adapted to things so far removed from our senses, and we have fully and dispassionately examined the opinions of Mr. Huygens and those of the adversaries who oppose him, in the meeting before Your Serene Highness.[75]

This is the type of rhetoric that appealed to the Prince's political aims and interests. Through the construction of a model, the academicians managed to give the impression that they were avoiding the natural philosophical controversy surrounding the issue. This was an experimental approach and rhetoric that was intended to build their reputations as reliable producers of natural knowledge. In fact, this was the reputation that Leopoldo advertised to Huygens on 14 September 1660. The Prince wrote to Huygens about the unbiased quality of the Cimento's work and their ability to carry out an impartial judgement of the competing theories of Saturn. In this letter, Leopoldo firstly praised Huygens for his 'great desire to recognise the truth in everything' and then claimed that this same search for truth was also 'the most important maxim of an academy of many virtuous men, who gather together before me almost every day without impassioning themselves to the opinions of others, or even to their own'.[76]

By the time Leopoldo had written this letter, the academicians' reports about their observations of the models had already been sent to Huygens. Dati had also

---

[75] BNCF, MS. Gal. 271, fols 3v–4r.; Tozzetti 1780, 2, p. 740. 'Noi però altrimenti, secondo il costume dell'Accademia di Vostra A.S., che è d'investigare il vero per via di riprove sperimentali, l'abbiamo inviolabilmente osservato anche in questo affare, per quella parte però che può ridursi ad Esperienza di cose tanto remote da' nostri sensi, et esaminando per ultimo nei Congressi tenuti d'avanti all'A.V, disappassionatamente i Concetti dell'Ugenio, e quei degl'Avversari che gli oppongono, vi sono cadute alcune Riflessioni', as translated by van Helden 1973, p. 244.

[76] 'un desiderio grande di riconoscere la verità in ciascheduna cosa, come ho determinato che sia la principal massima di un'Accademia di molti Virtuosi, che quasi ogno giorno si radunano avanti di me, senza appassionarsi non solo alle opinioni altrui, ma nemmeno alle proprie'; Tozzetti 1780, 1, p. 382.

already appealed to the Dutch astronomer's discretion to preserve the Cimento's uncontroversial image. Now Leopoldo was personally advertising this image to Huygens by insisting that his courtiers were only searching for the truth and did not purposefully set out to support or reject any opinions or speculations. The academicians' reputation as unbiased knowledge makers was, therefore, undoubtedly quite important to Leopoldo and reflected the Cimento's censorship policy when it came to the presentation of their work.

Soon afterwards, Lorenzo Magalotti suggested to Leopoldo that since a censored copy of Huygens' work was to be published in Florence, the academicians could also publish censored versions, without Copernican theorising, of their own reports that they had sent to Huygens and Fabri. In a single page memorandum to the Prince, Magalotti proposed that the academicians' observations of Saturn should be recorded, but that any statements by Borelli made in support of Copernicanism should also be omitted, in order 'to avoid difficulties'.[77]

Magalotti's proposal was never accepted, but through the Prince's other efforts to avoid controversy, by the end of this debate the Cimento came out with its reputation high amongst European astronomers and intact amongst ecclesiastical authorities and Jesuit thinkers such as Fabri.[78] Although the work carried out by his academicians was in favour of the Dutch Protestant instead of the Jesuit Inquisitor, remarkably Leopoldo had managed to keep himself and his academy away from any controversy, and even reinforced their image as unbiased experimentalists. In other words, by refusing to publish or make public the natural philosophical skills, commitments, and agendas of the academicians, as they were expressed in the reports and in letters, they could not possibly have been condemned by the Catholic Church. They could support Huygens without publicly acknowledging his belief in Copernicanism. This then relieved them of being threatened by any accusations from Fabri that the ring theory was heretical. In the meantime, the Cimento was to boost its reputation across Europe for producing of reliable experimental knowledge claims, creating the status which the Medici Prince longed for as a protector of truth and knowledge in the Tuscan Court, and strengthening the reputations and careers of the academicians.

7.CONCLUSION

The Cimento's rhetoric in this case study was crucial to the political and religious concerns of its patron and members. As Galluzzi has pointed out, there is little evidence that the Prince was implementing a formal policy of censorship of natural philosophical expression. Nevertheless, Leopoldo seems to have adopted 'self-censorship' for the sake of attaining a respectable status and reputation for his

---

[77] 'Si è pensato di metter in sicuro tutto quello che l'anno 1660 si speculò, e si operò nell'Accademia di V.A. intorno a Saturno ... Bisognerà però che il Sig. Borelli, si contenti di ridurre fuori del sistema Copernicano quelle sue dimostrazioni per sfuggir difficultà': BNCF, MS. Gal. 271, fol. 16r; Tozzetti 1780, 1, p. 385.

[78] Van Helden 1973, p. 254.

academy.[79] While they could get away with publishing their work on the vacuum, air-pressure, and the effects of heat and cold with clear natural philosophical underpinnings, astronomy was a different story. Leopoldo could dare to publish experiments that hinted at a belief in micro-structures such as atomism and the existence of the vacuum, as long as such beliefs were never openly supported.[80] But astronomy was still based on the analysis of the macro-structures that not only formed the basis of Aristotelian cosmology, but were also used to support Catholic biblical teachings. To make such immensely controversial anti-Aristotelian statements in seventeenth-century Italy, the academicians would have been exposing themselves to the same type of religious scrutiny that resulted in the condemnation of their hero, Galileo.

The academicians' reputation for impartiality could not have been achieved if they had made public their work on Saturn and risked exposing the controversial Copernican interests of some of their members. This is why the Prince did not allow that he or his Court members should be publicly seen as participants in this debate. It is also the reason why Magalotti's proposal to publish the academicians' work was not acceptable, and why not a single word of the Cimento's role in this controversy was mentioned in the *Saggi*. In fact, so effective was Leopoldo's strategy to maintain this reputation in this case study, that even twentieth-century historians have found themselves marvelling at the academicians' approach to the Saturn problem, including the construction of models for experimenting, and extolling the virtues of the so-called 'Experimental Method'.

The Saturn episode is probably the most impressive of the Cimento's history because of the academicians' rather innovative experimental approach. They were intent on building models of the rival theories of Saturn's appearance and movements and seemingly relied purely on this astronomical experiment in order to resolve the controversy. Indeed, not trusting their own bias, they called upon naïve passers by to describe what they saw through the telescope to ensure that the results were achieved objectively. This is a perfect example of how they were supposedly avoiding controversial theorising by basing their work simply on their experiments and observations. Indeed, even Albert van Helden could not resist the temptation to provide a sweeping generalisation in conclusion to this case study. Van Helden provided an excellent account of the theoretical issues that tormented the academicians as they prepared to solve the dispute between Fabri and Huygens. But such was van Helden's admiration for the academicians' skills in observation and experimenting that he still believed that the most important part of this case study was the Cimento's demonstration of their 'mastery of the experimental method'.[81] So, despite Leopoldo's political maneuvering to suppress the publication of the natural philosophical opinions of his courtiers regarding this issue, van Helden could not resist concluding that the academicians' experiments were simply an 'illustration

[79] Galluzzi 1981, pp. 823–832.
[80] See: Boschiero 2002, pp. 383–410; Boschiero 2003b, pp. 329–349.
[81] Van Helden 1973, p. 247.

of the height of sophistication to which the experimental method had risen in Florence by 1660'.[82]

It is not difficult to understand how van Helden arrived at such a conclusion when considering the academicians' efforts to use so-called objective and independent observers, and their intention to examine the merits of both competing theories about Saturn's movements. In his report to Leopoldo about the Saturn observations carried out by the Cimento, Borelli even stated that they simply searched for the truth 'through many experimental proofs', and that they had 'fully and dispassionately examined the opinions of Mr. Huygens and those of the adversaries who oppose him'.[83] This also led Middleton to state, in his brief analysis of the Saturn problem, that the academicians' observations were an example of the Cimento's 'experimental psychology'.[84]

There is no doubt that the academicians were ardent and at times talented experimentalists, but despite the rhetoric in the *Saggi* and Borelli's statement above, is it fair to describe the mere use of experiments as a belief in, and mastery of, a putative modern experimental method? If by 'experimental method' we are to understand the type of fact-gathering and inductive reasoning that so many other traditional historians imply existed in the seventeenth century, then Middleton's and van Helden's statement could not be further from the truth. The observations of Saturn during the late 1650s and in 1660, especially the Cimento's work on the topic, were laden with natural philosophical concern and contention. Borelli in particular took a special interest in supporting the Copernican view represented by Huygens against the scholastic beliefs defended by Fabri. In fact, these observations were so heavily laden with natural philosophical contention that Leopoldo refused to have them published in the *Saggi*, or anywhere else for that matter, in order to maintain the Cimento's uncontroversial and unbiased image. So, while some historians evidently take this episode to be an example of some type of atheoretical experimental programme, equating with some purported modern scientific method, upon closer examination we find that the academicians never relied on such a practice when actually carrying out and interpreting their observations. Experiments played a crucial part of the culture of natural philosophising in mid to late seventeenth-century Tuscany by adding authority and persuasiveness to the academicians' work, but there is no indication that the Cimento was following an atheoretical experimental method that is tantamount to so-called modern experimental science.

In order to avoid this type of mistaken historiographical perspective, this case study has shown the need to return to the contextual approaches looking at theory-laden experimentation, that were first put forward by the proponents of the Sociology of Scientific Knowledge. In this case study, the significance of an observation was being challenged and negotiated, not according to gentlemanly standards of atheoretical experimental fact-making. But rather, the theories and

---

[82] *Ibid.*, p. 259.

[83] '... per via di riprovi esperimentali'; '... esaminando per ultimo ... dispassionatamente i concetti del Sig. Ugenio e quei degli avversari che se gl'oppongono'; BNCF, MS. Gal. 271, fols 3v–4r.

[84] Middleton 1971, p. 262.

instruments in this case were being questioned by rival telescope makers and astronomers with natural philosophical, religious, and political skills, commitments, and agendas extending well beyond questions about courtly experimental etiquette or accumulation of so-called 'matters of fact'. This should serve as a basis upon which to recognise the culture of natural philosophising that existed throughout mid to late seventeenth-century Europe.

REFERENCES

Accademia del Cimento (1667) *Saggi di naturali esperienze*, Firenze.
Barbensi, G. (1947) *Borelli*, Trieste: Zigiotti Editore.
Beretta, M. (2000) 'At the source of western science: the organisation of experimentalism at the Accademia del Cimento (1657–1667)', *Notes and Records of the Royal Society of London*, 54 (2), pp. 131–151.
Bermingham, A. and Brewer, J. eds (1995) *The Consumption of Culture, 1600–1800*, London: Routledge.
Biagioli, M. (1992) 'Scientific revolution, social bricolage, and etiquette' in *The Scientific Revolution in National Context*, eds R. Porter and M. Teich, Cambridge: Cambridge University Press, pp. 11–54.
Boschiero, L. (2002) 'Natural philosophising inside the late seventeenth-century Tuscan court', *British Journal for the History of Science*, 35 (4), pp. 383–410.
— (2003a) *Natural Philosophy Inside the Mid to Late Seventeenth-Century Tuscan Court: the History of the Accademia del Cimento*, Doctoral dissertation University of New South Wales.
— (2003b) 'Natural philosophical contention inside the Accademia del Cimento: the properties and effects of heat and cold', *Annals of Science*, 60, pp. 329–349.
Collins, H. M. (1975) 'The Seven Sexes: a study in the sociology of a phenomenon or the replication of experiments in physics', *Sociology*, 9, pp. 205–224.
Cropper, E., Perini, G. and Solinas, F. eds (1992) *Documentary Culture Florence and Rome From Grand Duke Ferdinand I to Pope Alexander VII: Papers from a Colloquium held at the Villa Spelman, Florence, 1990*, Bologna: Nuova Alfa Editoriale.
Dear, P. (1995) *Discipline and Experience: The Mathematical Way in the Scientific Revolution*, Chicago: University of Chicago Press.
Emerson, R. (1990) 'The organisation of science and its pursuit in early modern Europe' in *Companion to the History of Modern Science*, eds R. C. Olby, G. N. Cantor, J. R. R. Christie and M. J. S. Hodge, London: Routledge, pp. 960–979.
Fabroni, A. (1775) *Lettere inedite d'uomini illustri*, 2 vols, Florence: G. Barbèra Editore.
Favaro, A. ed. (1890), *Le Opere di Galileo Galilei, Edizione Nazionale*, 20 vols, Florence: Barbèra.
Fermi, S. (1903) *Lorenzo Magalotti. Scienziato e Letterato (1637–1712)*, Piacenza: Stab. Tip.-Lit. Bertola & C.
Findlen, P. (1993) 'Controlling the experiment: rhetoric, court patronage and the experimental method of Francesco Redi', *History of Science*, 31, pp. 35–64.
Fontana, F. (1646) *Novae coelestium terrestriumque rerum observations*, Naples.
Galileo, G. (1957) *Discoveries and Opinions of Galileo*, ed. and trans. S. Drake, New York: Doubleday.
Galluzzi, P. (1981) 'L'Accademia del Cimento: 'gusti' del principe, filosofia e ideologia dell'esperimento', *Quaderni Storici*, 48, pp. 788–844.
Gassendi, P. (1649) *Animadversiones in decimum librum Diogenis Laertii*, Lyons.
— (1658) *Opera omnia*, Lyon.
Hall, A. R. (1983) *The Revolution in Science 1500-1700*, 3rd edn, London, New York: Longman.
Henry, J. (2002) *The Scientific Revolution and the Origins of Modern Science*, 2nd edn, New York: Palgrave.
Hevelius, J. (1647) *Selenographia*, Gandsk.
Huygens, C. (1888–1950) *Oeuvres complètes de Christiaan Huygens*, 22 vols, The Hague: Nijhoff.
Middleton, W. E. K. (1971) *The Experimenters: a study of the Accademia del Cimento*, Baltimore: Johns Hopkins Press.
Nelli, G. B. C. (1759) *Saggio di storia letteraria fiorentina del secolo XVII scritta in varie lettere*. Lucca: Patrizio Fiorentino.

Olby, R. C., Cantor, G. N., Christie, J. R. R. and Hodge, M. J. S. eds (1990) *Companion to the History of Modern Science*, London: Routledge.

Ornstein, M. (1963) *The Role of Scientific Societies in the Seventeenth Century*, 3rd edn, Hamden: Archon Books.

Pieraccini, G. (1925) *La stirpe de' Medici di Cafaggiolo*, 3 vols, Florence.

Pinch, T. (1985) 'Towards an analysis of scientific observation: the externality and evidential significance of observational reports in physics', *Social Studies of Science*, 15, pp. 3–36.

Porter, R. and M. Teich, M. eds (1992) *The Scientific Revolution in National Context*, Cambridge: Cambridge University Press.

Riccioli, G. (1651) *Amalgestum novum*, Bologna.

Schuster, J. A. and Taylor, A. B. H. (1997) 'Blind trust: the gentlemanly origins of experimental science', *Social Studies of Science*, 27, pp. 503–536.

Segre, M. (1989) 'Viviani's Life of Galileo', *Isis*, 80, pp. 207–231.

Shapin, S. (1988) 'The house of experiment in seventeenth-century England', *Isis*, 79, pp. 373–404.

— (1994) *A Social History of Truth: Civility and Science in Seventeenth-Century England*, Chicago: University of Chicago Press.

Shapin, S. and Schaffer, S. (1985) *Leviathan and the Air-Pump*, Princeton: Princeton University Press.

Tozzetti, G. Targioni (1780) *Notizie degli aggrandamenti delle scienze fisiche accaduti in Toscana nel corso di anni LX del secolo XVII*, 3 vols, Florence.

Tribby, J. (1992) 'Of conversational dispositions and the *Saggi*'s Proem' in *Documentary Culture Florence and Rome From Grand Duke Ferdinand I to Pope Alexander VII: Papers from a Colloquium held at the Villa Spelman, Florence, 1990*, eds E. Cropper, G. Perini, and F. Solinas, Bologna: Nuova Alfa Editoriale, pp. 379–390.

— (1995) 'Dante's Restaurant: the cultural work of experiment in early modern Tuscany' in *The Consumption of Culture. 1600–1800*, eds A. Bermingham and J. Brewer, London: Routledge, pp. 319–337.

Van Helden, A. (1968) 'Christopher Wren's *De Corpore Saturni*', *Notes and Records of the Royal Society of London*, 23, pp. 213–229.

— (1970) 'Eustachio Divini versus Christiaan Huygens: a reappraisal', *Physis*, 12, pp. 36–50.

— (1973) 'The Accademia del Cimento and Saturn's ring', *Physis*, 15, pp. 237–259.

— (1974a) 'Saturn and his Anses', *Journal for the History of Astronomy*, 5, pp. 105–121.

— (1974b) '"Annulo Cingitur": the solution of the problem of Saturn', *Journal for the History of Astronomy*, 5, pp. 155–174.

Wren, C. (1651) *De Corpore Saturni*, London.

PETER R. ANSTEY

EXPERIMENTAL VERSUS SPECULATIVE NATURAL PHILOSOPHY

1.INTRODUCTION

This chapter discusses an undeservedly neglected distinction in the discussions of method in natural philosophy in early modern England. It is the distinction between experimental and speculative natural philosophy. The chapter makes no attempt to analyse the modes of deployment of this distinction within the method discourse and practice of early modern natural philosophy. Rather it merely seeks to establish its presence, importance and historical development within this discourse. It is evident that the distinction between experimental and speculative natural philosophy was deployed within rhetorical, heuristic and philosophical contexts during the period, but I do not discuss these uses here.[1] It will be enough in this paper to establish its widespread incidence and to trace its development.

A distinction between speculative and experimental natural philosophy is found in many different English writers in the latter half of the seventeenth century. For example, John Dunton's *The Young-Students-Library* (1692), which, as its title suggests, is addressed to a student audience, divides natural philosophy as follows:

> Philosophy may be consider'd under these two Heads, Natural and Moral: The first of which, by Reason of the strange Alterations that have been made in it, may be again Subdivided into *Speculative* and *Experimental*.[2]

Very roughly, and a degree of imprecision is important here, speculative natural philosophy is the development of explanations of natural phenomena without prior recourse to systematic observation and experiment. By contrast, experimental natural philosophy involves the collection and ordering of observations and experimental reports with a view to the development of explanations of natural phenomena based on these observations and experiments. Needless to say, it was experimental natural philosophy that was favoured by almost all natural philosophers in early modern England.[3] Indeed, the distinction is normally invoked

---

[1] For general discussions of method discourse in natural philosophy in the early modern period see: the Introduction to Schuster and Yeo 1986, pp. ix–xxxvii; Schuster 1986; and Dear 1998.

[2] Dunton 1692, p. vi.

[3] Hobbes was a notable exception. See Shapin and Schaffer 1985, chap. 4. For an early critique of this preference for experimental natural philosophy see Margaret Cavendish's *Observations upon*

P. R. Anstey and J. A. Schuster (eds.), The Science of Nature in the Seventeenth Century, 215-242.
© 2005 *Springer. Printed in the Netherlands.*

to affirm, to vindicate and to recommend the experimental methodology or to criticise those who indulged in speculative natural philosophy. Experimental philosophy was definitely 'in' and speculative philosophy was 'out'. Quoting Dunton again,

> We must consider, the distinction we have made of *Speculative* and *Experimental*, and, as much as possible, Exclude the first, for an indefatigable and laborious Search into Natural Experiments, they being only the Certain, Sure Method to gather a true Body of Philosophy, for the Antient Way of clapping up an entire building of Sciences, upon pure Contemplation, may make indeed an *Admirable Fabrick*, but the Materials are such as can promise no lasting one.[4]

Now this distinction is easily passed over or dismissed as one peruses early modern writings about natural philosophy. It could simply be just part of the rhetoric of those promoters of Baconian natural philosophy of the kind practised by the early Royal Society. However, I want to argue that not only is it a fundamental distinction in the characterising of natural philosophical method in the latter half of the seventeenth century, but that it is *the* fundamental dichotomy in discussions of natural philosophical methodology during the period. I will claim that these were the defining terms of reference for any practitioner of natural philosophy. In fact, I will argue for five strong claims regarding this distinction:

1. this distinction is in evidence, in some form or other, from the late 1650s until the early decades of the eighteenth century
2. this distinction provides the primary methodological framework within which natural philosophy was interpreted and practised in the late seventeenth century
3. this distinction is independent of disciplinary boundaries within and closely allied to natural philosophy
4. this distinction crystallised in the 1690s when opposition to hypotheses in natural philosophical methodology intensified
5. this distinction provides the terms of reference by which we should interpret Newton's strictures on the use of hypotheses in natural philosophy.

2. THE EXPERIMENTAL/SPECULATIVE DISTINCTION IS IN EVIDENCE, IN SOME FORM OR OTHER, FROM THE LATE 1650S UNTIL THE EARLY DECADES OF THE EIGHTEENTH CENTURY.

The origins of the distinction are not entirely clear. Certainly it is adumbrated in some form by Francis Bacon in *De dignitata et augmentis scientiarum* where he distinguishes between speculative (*speculativa*) and operative (*operativa*) natural

---

Experimental Philosophy (1666) and in particular the chapter entitled 'Ancient Learning Ought Not to be Exploded, nor the Experimental Part of Philosophy Preferred Before the Speculative' (2001, pp. 195–197).
[4] Dunton 1692, pp. vi–vii.

philosophy.[5] The former was further divided into physic (or physics) and metaphysics; physic being founded upon natural history and metaphysics being founded upon both natural history and physic.[6] Furthermore, the third of Bacon's tripartite division of the subjects of natural history, namely Arts, which is the history of artefacts and the manipulation of nature, is called '*Mechanical* and *Experimental*'. This is regarded by Bacon as the most important and yet the most neglected form of natural history and is naturally seen as an antecedent to the experimental philosophy.[7] But none of Bacon's distinctions is co-extensive with the experimental/speculative distinction as found later in the century, for, at least from the 1650s, it was 'physic' (a synonym for 'natural philosophy') which was divided into the speculative or the experimental. Bacon famously opposed idle speculation and promoted the derivation of natural knowledge from experiment. The idols of the theatre and the critique of the dogmatists who spin webs out of themselves are familiar themes in the *Novum Organum* and are picked up by the promoters of experimental philosophy.[8] (A discussion of Bacon's distinction between speculative and productive philosophers is found in Stephen Gaukroger's chapter.) There was also a distinction between speculative philosophy and other types of philosophy in some scholastic divisions of the sciences in the late Renaissance. Thus, Toletus claims that philosophy has three principal parts: speculative, practical and factive. The speculative part was further to be divided into physics, metaphysics and mathematics.[9]

So there was a precedent for the experimental/speculative terminology and for the making of divisions within natural philosophy. And we find the term 'Experimental Philosophy' used as early as 1635 in Samuel Hartlib's *Ephemerides*.[10] Yet one cannot claim that a definitive division of the science of natural philosophy was bequeathed to the first generation of serious English experimental philosophers in the mid-seventeenth century. It does appear to be adumbrated in the early writings and opinions of William Petty, who seeks to 'explode' the 'meerly phantasticall' and promote experimental learning.[11] But it is clearly not present in the eclectic methodological views of Hartlib. However, whatever its origins, once the distinction between experimental and speculative natural philosophy reached the form in which it is found in the sixth decade of the century, it is not hard to find.

---

[5] Bacon 1859, IV, p. 343 = *ibid.*, I, p. 547.

[6] 'And so of Natural Philosophy the basis is Natural History; the stage next the basis is Physic; the stage next the vertical point is Metaphysic', Bacon 1859, IV, p. 362.

[7] 'Description of the Intellectual Globe', Bacon 1996, p. 102/3.

[8] See *New Organon*, I, 61–65 and 95, Bacon 1859, IV, pp. 62–66 and 92–93. See for example Glanvill 1668, p. 5 where he denies that the aims of the Royal Society are not 'little Projects of *serving* a *Sect*, or *propagating* an *Opinion*; of *spinning* out a *subtile Notion* into a *fine thread*, or forming a *plausible System* of *new Speculations*'. See also Nedham 1665, pp. 234–235.

[9] See Wallace 1988, p. 210.

[10] Hartlib Papers 29/3/37B.

[11] Petty 1647, p. 2.

It is certainly evident in Boyle's methodological writings of the late 1650s and early 1660s.[12] Indeed some of Boyle's recently published manuscript notes for his *Usefulness of Natural Philosophy* furnish us with an explicit link to Bacon. Boyle tells us

> I shall ... do what is requisite to commend Experimental Learning to you, if I be so happy as to make it out, that Experiments considered in the Lump, or one with another, may very much assist the speculative Phylosopher, that is sollicitous about the causes and reasons of Naturall things; and that the speculative Phylosopher so assisted, may (on the other side) very much improve the Practical part of Physick. And consequently, that both of them may very happily conspire to the Establishing & Advancement of a Solid usefull Naturall Philosophy.[13]

He then goes on, alluding to *De augmentis* and referring to *Novum Organum*,

> before I proceed to handle these things distinctly, I must advertise you, that I forget not that our excellent Verulam has mentioned a *Scala ascensoria & descensoria*; the former from Experiments to Axioms, the latter from Axioms to Experiments, as designed parts of his *Novum Organum* ...[14]

Boyle even composed a work entitled 'Of Usefulnes of Speculative & Experimental Philosophy to one another', though this is no longer extant.[15] The distinction is also evident in Boyle's statement of the rationale of his *Spring of the Air* (1660). He tells us of this work that

> It was not my chief Design to establish Theories and Principles, but to devise Experiments, and to enrich the History of Nature with Observations faithfully made and deliver'd; that by these, and the like Contributions made by others, men may in time be furnish'd with a sufficient stock of Experiments to ground *Hypotheses* and *Theorys* on. ... I propos'd my Thoughts but as Conjectures design'd ... to excite the Curiosity of the Ingenious, and afford some hints and assistance to the Disquisitions of the Speculative.[16]

Likewise it is implicit in the preface to Henry Power's *Experimental Philosophy* when he tells us that 'this I am sure of, That without some such Mechanical assistance, our best Philosophers will but prove empty Conjecturalists, and their profoundest Speculations herein, but gloss'd outside Fallacies ...'.[17] And in the preface to Hooke's *Micrographia* (1665) we find

> The *real*, the *mechanical*, the *experimental* Philosophy, which has this advantage over the Philosophy of *discourse* and *disputation*, that whereas that chiefly aims at the subtilty of its Deductions and Conclusions, without much regard to the first ground-

---

[12] See for example Boyle 'Proemial Essay' in *Certain Physiological Essays*, Boyle 1999–2000, 2, pp. 23–25, which was written in 1657.

[13] Boyle 1999–2000, 13, p. 351, underlining added.

[14] *Ibid.*

[15] See Boyle 1999–2000, 14, p. 342.

[16] *Defence against Linus* (1662), Boyle 1999–2000, 3, p. 12, underlining added. See also *Experiments and Considerations touching Colours* (1663), *ibid.*, 4, p. 5.

[17] Power 1664, Preface [c3v]; Power calls Bacon 'that Patriark of Experimental Philosophy', *ibid.*, p. 82.

EXPERIMENTAL VERSUS SPECULATIVE NATURAL PHILOSOPHY

work, which ought to be well laid on the Sense and Memory; so this intends the right ordering of them all, and the making them serviceable to each other.[18]

We also find it in Sprat's *History of the Royal Society* of 1667.

> *Experimental Philosophy* will prevent mens spending the strength of their thoughts about *Disputes*, by turning them to *Works* … And indeed of the usual titles by which men of business are wont to be distinguish'd, the *Crafty*, the *Formal*, and the *Prudent*; … The *Formal* man may be compar'd to the meer *Speculative Philosopher*: For he vainly reduces every thing to grave and solemn general *Rules* … the *Prudent* man is like him who proceeds on a constant and solid cours of *Experiments*.[19]

Indicative of just how widespread was the appreciation of this distinction is the fact that it is found in the literary responses to the new science. In Shadwell's *The Virtuoso*, Sir Formal (the name, of course, alludes to Sprat's 'formal man') sings the praises of Sir Nicholas Gimcrack, who embodies both the speculative and experimental,

> Trust me, he is the finest speculative Gentleman in the whole World, and in his Cogitations the most serene Animal alive: Not a Creature so little, but affords him great Curiosities.

Sir Nicholas, when later found emulating a swimming frog, tells us 'I content my self with the Speculative part of Swiming, I care not for the Practick'.[20]

It is also worth citing some critics and opponents of the new natural philosophy. Margaret Cavendish is clearly working with the distinction in her *Observations upon Experimental Philosophy* (1666). In her preface she claims that 'as I have had the courage to argue heretofore with some famous and eminent writers in speculative philosophy; so have I taken upon me in this present work, to make some reflexions also upon some of our modern experimental and dioptrical writers'.[21] Later in the century, John Sergeant, in his *The Method to Science* (1696), sets the problem of the method of science up as follows

> The *METHODS* which I pitch upon to examine, shall be of two sorts, viz. that of *Speculative*, and that of *Experimental* Philosophers; The Former of which pretend to proceed by *Reason* and *Principles*; the Later by *Induction*; and both of them aim at advancing *Science*.[22]

Finally, the Scottish virtuoso George Sinclair compares his method in hydrostatics with that of Archimedes using what would have been a familiar trope, 'His way is more *Speculative*: this is more *Practical*'.[23]

Now, it would be going too far to say that this distinction is completely ubiquitous, but it is extremely common. The experimental philosophy quickly

---

[18] Hooke 1665, Preface [a3].

[19] Sprat 1667, p. 341, underlining added. See also p. 257 where Sprat claims the method of the members of the Royal Society 'to be chiefly bent upon the *Operative*, rather than the *Theoretical Philosophy*'. For Sprat on method see Wood 1980.

[20] Shadwell 1997, pp. 9 and 30. Gimcrack is really a composite character displaying all of the features of the virtuosi that Shadwell seeks to ridicule, including a keen interest in experiment.

[21] Cavendish 1666 (2001, p. 10). See also note 3 above.

[22] Sergeant 1696, Preface [b6r–v], underlining added.

[23] Sinclair 1683, Epistle to the Reader.

emerged as, far and away, the dominant form of natural philosophy. Not surprisingly then, the term 'experimental philosophy' became the key descriptor for the kind of natural philosophy that members of the Royal Society practised and promoted. It is important to note however, that the term 'experimental philosophy' was not co-extensive with 'natural philosophy' because natural philosophy could be practised in a speculative way. In fact, one of the consequences of the impact of the distinction was that the term 'speculative' in philosophical and natural philosophical contexts often had a pejorative connotation. Thus Henry Oldenburg could tell a correspondent that the Royal Society 'aimes at the improvement of all usefull Sciences and Arts, not by meer speculations, but by exact and faithfull Observations and Experiments'.[24]

3. THE EXPERIMENTAL/SPECULATIVE DISTINCTION PROVIDES THE PRIMARY METHODOLOGICAL FRAMEWORK WITHIN WHICH NATURAL PHILOSOPHY WAS INTERPRETED AND PRACTISED IN THE LATE SEVENTEENTH CENTURY.

The distinction between experimental and speculative natural philosophy provided the terms of reference for virtually all methodological reflection and practice of natural philosophy in England from the late 1650s to the end of the early modern period. Yet we should not simply use the distinction to classify natural philosophers as belonging to one camp or the other without further analysis. For, there is a cluster of epistemological issues that underlie the terms 'experimental' and 'speculative' and as these are unpacked it becomes clear that there was actually a range of natural philosophical methodologies within experimental natural philosophy, some of which incorporated elements normally attributed to speculative philosophy. It is therefore imperative that we examine that cluster of epistemological issues which were associated with the distinction and use them to shed light on the distinction's utility and the spectrum of methodologies which fell under this rubric.

The first point to stress as we unpack the distinction is that the epistemology of this period was in a state of flux.[25] Many notions that we now take for granted in the philosophy of science were emerging and some were receiving serious philosophical reflection for the first time. The notions of hypothesis, probability, induction, laws of nature, testimony, experimental replication and so on, were all being discussed and incorporated into accounts of natural philosophical method. Now what is important to stress here, is that this flux of ideas and notions is reflected in a certain vagueness or indeterminacy in the distinction between speculative and experimental natural philosophy. The distinction became a kind of demarcation criterion whose terms were never fully spelt out or clearly defined. Indicative of this is the fact that many natural philosophers—and those who reflected upon natural philosophical

[24] Oldenburg to Norwood, 10 February 1667/8, Oldenburg 1965–1986, IV, p. 168, quoted in Hunter 1989, p. 47.
[25] See B. Shapiro 1983, chap. 2.

methodology—in the latter half of the seventeenth century, upheld methodological precepts which were not always consistent or easily reconcilable. Thus we find the juxtaposition of a stress on Baconian natural histories and an ideal of a demonstrative science of nature with talk of probability, moral certainty. We even find criticism of hypotheses and the deployment of hypotheses in the same writer and sometimes in the very same work.

Indeed, there has been a tendency in some quarters to speak of natural philosophical methodology in this period as if it was constituted by a relatively coherent form of 'probabilism' and 'empiricism' and as if some practitioners were consciously employing a form of hypothetico-deductive method.[26] However, while the claim cannot be fully substantiated here, it seems rather that the natural philosophical methodologies of this period are better characterised as underdeveloped, tentative and sometimes internally inconsistent. To be sure, there are partial adumbrations of what was to come, but a careful perusal of, say, the methodological writings of Boyle, Hooke or Locke do not yield anything closely resembling modern scientific methodology.[27] This should not surprise us because some of the issues were relatively new (such as laws of nature) and those issues that had a lengthy genealogy in philosophical reflection were often in the process of reinterpretation in the light of the strongly polemical context in which the new philosophy was being forged.[28]

This brings us to a second point, namely that a strong polemical agenda underlies the origins and use of the distinction between speculative and experimental natural philosophy. It was the new natural philosophers, and in particular those aligned with the newly formed Royal Society and its precursor groups, who first used the distinction. They did this not simply to emphasise the fact that they were experimentalists or saw an indispensable need for experimentation, but also to distance themselves from the old speculative way of proceeding in physics or physiology (as natural philosophy was often called). And, of course, the old speculative way was that of the schools; that of the Aristotelians, who indulged in hypothetical and metaphysical speculations which were often untestable or which cluttered the ontological furniture of the world. Such entities as inexplicable occult qualities, substantial forms, virtual extension, sympathies and antipathies were paradigm cases of speculative indulgence in natural philosophy.[29] Thus we find

[26] See for example Laudan 1981, chaps 4 and 5.

[27] For example, Boyle's manuscript notes on the evaluation of hypotheses (Boyle 1999–2000, 13, pp. 270–272, Boyle 1991, p. 119) need to be assessed in conjunction with the very important exposition of the method of natural history in his letter to Oldenburg of 13 June 1666, Boyle 2001, 3, pp. 170–175. For recent assessments of Hooke's methodological views see: Lynch 2001, chap. 3; Hunter 2003; and for Locke see Anstey 2002 and 2003b.

[28] There is still no consensus as to the emergence and development of some of these notions in early modern natural philosophy, and, to my knowledge, they have never been interpreted in the light of the experimental/speculative distinction. For discussions of early modern notions of probability and certainty see: B. Shapiro 1983; Patey 1984; Daston 1988 and 1998; and Franklin 2001.

[29] There has been extensive work done on the critique of early modern Aristotelianism(s) and such notions as occult qualities and the theory of forms. See for example Grant 1987; Mercer 1993; Hutchison 1982 and 1991.

Joseph Glanvill's *Plus Ultra* (1668) is written as a defence of the experimental philosophy against the criticisms of an Aristotelian. He tells us that,

> the *Modern Experimenters* think, That the *Philosophers* of elder Times, though their *Wits* were excellent, yet the way they took was not like to bring much *advantage* to *Knowledge* or any of the *Uses* of *humane Life*; being for the most part *that* of *Notion* and *Dispute*, which still runs round in a *Labyrinth* of *Talk*, but *advanceth nothing*. And the *unfruitfulness* of those *Methods* of *Science*, which in so many *Centuries* never brought the World so much *practical, beneficial Knowledge* as would help towards the *Cure* of a *Cut finger*, is a palpable Argument, That they were *fundamental Mistakes*, and that the *Way* was not *right*.[30]

Nor did this polemical tone diminish as the century wore on and as the experimental philosophy became more entrenched. William Molyneux's dedicatory letter to 'the Illustrious The Royal Society' in his *Dioptrica nova* of 1692 exemplifies this,

> I cannot omit expressing my Sence of that excellent Method of <u>Experimental Philosophy</u>, which now, by your Example and Incouragement, does so universally prevail ... 'Tis wonderful to consider, how the Schools were formerly overrun with a senseless kind of Jargon, which they call'd *Philosophy*; ... The Commentators on *Aristotle*, ... have rendred *Physicks* an heap of froathy Disputes, managing the whole Knowledge of Body and Motion ... by <u>Hypothetical Conjectures</u>, confirm'd by plausible Arguments of Wit and Rhetorick, ordered in a Syllogistical form; and answering Objections in like manner: But never studied to prove their Opinions by Experiments.[31]

But the so-called speculative philosophers were not confined to the Aristotelians. Boyle lists Leucippus, Epicurus, Aristotle, Telesius and Campanella as 'speculative Devisers of new Hypotheses'.[32] Furthermore, there was a general sense that the experimental natural philosophy was, in contrast to the philosophy of the ancients, something novel or new. In particular, it was new in the way it emphasised the role of the senses in acquiring knowledge. Naturally this was often emphasised in the context of the deployment of the new instruments such as the telescope and microscope which had opened up new vistas of knowledge by extending human senses. Thus we find Hooke in the preface to *Micrographia* stressing how the members of the Royal Society 'have begun anew to correct all Hypotheses by sense',

> And I beg my Reader, to let me take the boldness to assure him, that in this present condition of knowledge, a man so qualified, as I have indeavoured to be, ... may venture to compare the reality and the usefulness of his services, towards the true

---

[30] Glanvill 1668, pp. 7–8. Glanvill's earlier defence of the Royal Society, *Scepsis scientifica* (1665), was also motivated by an anti-Aristotelian polemic. See also Boyle's *Excellency of Theology*, Boyle 1999–2000, 8, pp. 87–88 for the claim that, unlike the schools, some of the ancients did practice natural philosophy by experiment and observation. See also Boyle's *The Christian Virtuoso, I*, Boyle 1999–2000, 11, pp. 292 and 296.

[31] Molyneux 1692, dedicatory epistle [A1–3], underlining added.

[32] *Certain Physiological Essays*, Boyle 1999–2000, 2, p. 24.

Philosophy, with those of other men, that are of much stronger, and more acute *speculations*, that shall not make use of the same method by the Senses.[33]

Third, it is important to note the relation between the experimental/speculative distinction and the widespread emphasis on the construction of Baconian natural histories amongst English natural philosophers in the early decades of the Royal Society. Much has now been written on the importance, indeed the centrality, of natural history to the natural philosophical method of the early Royal Society.[34] Baconian natural histories were vast collections of facts pertaining to particular objects or qualities which were to be gathered by observation, experiment, travelers' reports and any other means. These were to be assembled and arranged by natural philosophers who would then use this data in order to develop explanations of natural phenomena. Much of the work of the Society's experimenters was conceived in terms of the development of natural histories and Henry Oldenburg, the Society's indefatigable first secretary, conceived of his role as intelligencer for the Society in terms of the promotion and realisation of this method.[35] The erection of these histories is quite naturally seen as the organising principle of the task of experiment and observation which constituted the method of the experimental natural philosopher. This is not to say that all English natural philosophers from the period saw the practice of experiment and observation exclusively in these terms. For there was much debate over the relation between the construction of natural histories on the one hand, and hypotheses, principles, causes and induction on the other. But there were very few active natural philosophers in England who did not conceive of their task in terms of the construction of natural histories or whose method does not reflect the influence of natural histories.[36] Nor was this methodological emphasis restricted to England. Oldenburg's championing of Baconianism had an immediate impact on many of his Continental correspondents. And some early members of the Académie Royale conceived of their natural philosophical method in similar terms. Thus the young Christiaan Huygens could say, 'The principal work and most useful occupation of this Assembly should be, in my opinion, to work on the Natural History, somewhat according to the plan of Verulam'.[37]

If natural history was a central component of experimental natural philosophy until the end of the century, all were agreed that hypotheses were the province of the speculative philosopher. And this brings us to the fourth and most important issue associated with the experimental/speculative distinction, namely the epistemic status of hypotheses. 'Hypothesis' in early modern natural philosophy could refer to a causal explanation, a metaphysical principle or maxim, what we would call an inductive generalisation, or even a theory or system of doctrines such as the

---

[33] Hooke 1665, Preface [b1]. See also Hooke 1661, pp. 41–43 and *idem.* 1705, pp. 4–18; Glanvill 1668, pp. 52–53; and Cavendish 1666, (2001, p. 196).

[34] Hunter 1989 and 2003; Anstey 2002; Findlen 1997; Levine 1983; Cook 1993.

[35] For a sampling see Oldenburg 1964–1986, II, pp. 143–144, 146; III, p. 537; IV, pp. 315, 422 and 451.

[36] While the point cannot be established here, I have argued elsewhere (Anstey 2004) that Newton's use of queries has its roots in the lists of queries and heads for the writing of natural histories.

[37] Quoted from Sabra 1967, p. 171. See *ibid.*, pp. 170–174 for a discussion of Huygens on Bacon's method of natural history.

corpuscular hypothesis or the Copernican hypothesis. The word was also used as a synonym for conjecture, speculation and so on.

The core epistemological issue relating to hypotheses was the extent and manner in which they were related to observation and experiment. It is important to note that observation and experiment were not normally conceived as standing in an evidential relation to hypotheses. Rather, hypotheses were subservient to experience; hypotheses illustrated, explained, were deduced from or shed light on experiments and observations. As Boyle says in his dialogue on 'The Requisites of a Good Hypothesis', 'an hypothesis is a *supposition* ... that men have pitchd upon, or devis'd, as a Principle, by whose help the Phænomeno[n] wherto it is to be applyd may be *explicated'.*[38] It is this inversion of the salient epistemic relation between hypotheses and experience which can seem so foreign to a twenty-first-century reader. A speculative philosopher then, was one who indulged in hypotheses without recourse to observation and experiment at all, or only as an afterthought in order to save the phenomena or in order 'to adapt them to their Hypothesis'.[39] Speculative philosophers either failed to admit any relation between hypotheses and experience or subordinated experience to the hypothesis at hand.

Now such was the disdain amongst some English natural philosophers for hypotheses of any sort that some practitioners who were honest enough to realise that they actually employed them, admitted this very self-consciously as a kind of confession. Hooke for example, after affirming in his *Micrographia* that the experimental philosophy aimed at 'avoiding *Dogmatizing*, and the *espousal* of any *Hypothesis* not sufficiently grounded and confirm'd by Experiments', confesses to the Royal Society that 'I may seem to condemn my own Course in this Treatise; in which there may perhaps be some *Expressions*, which may seem more *positive* then YOUR Prescriptions will permit' and that he desires 'to have them understood only as *Conjectures* and *Quæries* (which YOUR Method does not altogether disallow)'.[40] Others, such as Glanvill and Sprat, who saw a role for hypotheses in experimental philosophy, were acutely aware that this case needed to be argued for and could not be taken as a given.[41] If the core issue was the relation between hypotheses and experience, there was a degree of systematic confusion as to the relation between hypotheses and causal explanations, hypotheses and inductive generalisations and hypotheses and metaphysical first principles. Indeed, one needs to take early modern English discussions of hypotheses, and they are legion, on a case by case basis because the term is not always used consistently, not even by the same author.

[38] Boyle 1999–2000, 13, p. 271. See also Newton to Oldenburg for Pardies, 10 June 1672, Newton 1959–1977, 1, p. 164.
[39] Quoting Hooke from 'A General Scheme, or Idea of the Present State of Natural Philosophy', Hooke 1705, p. 4.
[40] Hooke 1665, To the Royal Society [A2v]. Auzout soon took Hooke to task on an example of this and Hooke's reply is telling, 'I could wish that this worthy Person had rectified my mistakes, not by speculation, but by experiments', Oldenburg 1965–1986, II, p. 383. See also Brouncker's cautionary comment made upon licensing the work, 'though they [the Royal Society] have licensed it, yet they own no theory, nor will be thought to do so: and that the several hypotheses and theories laid down by him therein, are not delivered as certainties, but as conjectures', quoted in Hunter 2003 from Birch 1756–1757, I, p. 491.
[41] See for example Sprat 1667, pp. 107 and 257.

Part of the problem with hypotheses was the fact that some natural philosophers maintained the ideal of a demonstrative natural philosophy. Inevitably, as we can now see with hindsight, any form of proto-hypotheticalism in early modern natural philosophy was tied both to the newly emerging probabilism and an acceptance of a central role for inductive reasoning (in the modern sense). This ran counter to the goal of a demonstrative science of nature and, for those committed to such a goal, undermined the epistemological status of hypotheses. For example, this is clearly a problem for Locke in his critical discussions of hypotheses in the *Essay* and elsewhere.[42] Another issue lay in the manner in which many speculative hypotheses were contrived merely to save the phenomena. The underdetermination of hypotheses by observational data came to be regarded by many experimental natural philosophers as a fundamental flaw in the method of hypothesis. Thus we find Newton defending his view of colours in the following terms

> For what I shall tell concerning them is not an Hypothesis but most rigid consequence, not conjectured by barely inferring 'tis this because not otherwise or because it satisfies all phænomena (the Philosophers universall Topick,) but evinced by ye mediation of experiments concluding directly & wthout any suspicion of doubt.[43]

Now the disdain that some natural philosophers felt for hypotheses led not only to a kind of justificatory pose for those experimentalists who advocated their use in natural philosophy, but also to a whole vocabulary of dismissive and pejorative terms. Hypotheses were castles in the air, mere speculations, fancies, phantasms, chimeras, and so on.[44] Indeed, this kind of invective is even found amongst Continental writers. Thus Huygens, who was a mechanist and upholder of the Cartesian vortex theory, could still claim that 'Descartes has only spread idle fancies' and given out 'conjectures in the guise of truths'.[45]

It is also important to appreciate that while there were experimental natural philosophers who saw the utility of hypothetical reasoning for natural philosophical methodology, there was always a vocal group who opposed their use except under the most stringent of conditions. Sir Robert Moray expressed this outlook when he wrote of the Royal Society that

> This Society will not own any Hypothesis, systeme, or doctrine of the principles of Naturall philosophy, proposed, or maintained by any Philosopher Auncient or Moderne, nor the explication of any phaenomenon, where recourse must be had to Originall causes, … Nor dogmatically define, nor fixe Axiomes of Scientificall things, but will question and canvas all opinions[,] adopting nor adhering to none, till by mature debate & clear arguments, chiefly such as are deduced from legittimate experiments, the trueth of such positions be demonstrated invincibly.[46]

---

[42] See *An Essay concerning Human Understanding*, IV. xii. 13, Locke 1975, p. 648. For further references see below and for a recent analysis of Locke on hypotheses see Anstey 2003b.

[43] Newton 1959–1977, 1, pp. 96–97. See also Cotes' Preface to the 2nd edition of Newton's *Principia*, Newton 1999, p. 393. For an early expression of this complaint about underdetermination see Childrey 1661, Preface to the Reader [b2v–b3r].

[44] See for example Sydenham 1848, II, p. 173.

[45] Quoting from Westman 1980, p. 98. See Huygens 1888–1950, 10, p. 405.

[46] Transcribed in 'The early Royal Society and the shape of knowledge' in Hunter 1995, p. 173.

It should now be clear why we cannot simply take the dichotomy between speculative and experimental natural philosophy and use it to divide up early modern natural philosophers. If we take the dichotomy and group those natural philosophers, commentators and propagandists for the new science, we find a diversity of positions. There were those like Boyle and Sprat who were of a 'reconciling disposition' who conceived of their work within the dichotomy, but who sought to find a role for hypotheses in experimental natural philosophy. Then there were those who like Oldenburg repeatedly emphasised the centrality of natural history to the enterprise and whose stress on history as a foundation for natural philosophy effectively precluded any serious consideration of the epistemological problems associated with incorporating the methodological emphases of 'the other side'. There were those like Hooke who strongly identified with experimental philosophy, but who were conscious that they used speculations and that their causal explanations of phenomena could be branded as speculative. They were normally quite self-conscious in their hypothesising and at times even apologetic. Finally, there were those who regarded the experimental approach as the only legitimate way forward in natural philosophy and who were strongly opposed to speculation and the method of hypothesis. This diversity of 'Baconianisms' is not only evident amongst the members of the early Royal Society,[47] where it was sometimes expressed in a rather schematic and embryonic way, but also amongst its later established practitioners and theorisers. What is striking however, is that this range of views is always expressed in terms that are consistent with the fundamental dichotomy of speculative and experimental methods.

4. THE EXPERIMENTAL/SPECULATIVE DISTINCTION IS INDEPENDENT OF DISCIPLINARY BOUNDARIES WITHIN AND CLOSELY ALLIED TO NATURAL PHILOSOPHY.

It is important to emphasise that the distinction between speculative and experimental natural philosophy was not restricted to practitioners of the experimental philosophy or to those who reflected on natural philosophical methodology. It was not the special province of the members of the Royal Society and those within its ambit, nor indeed of the English. But neither was it more commonly found amongst those working in a particular branch of natural philosophy, say, mechanics or astronomy, chymistry or pneumatics, physiology or hydrostatics. It is a distinction which transcended disciplinary boundaries and which was applicable to any form of natural philosophy whatsoever. Thus we find it appears in apologists for the Royal Society such as Samuel Parker. Parker tells us,

> The chief reason therefore, why I prefer the Mechanical and Experimental Philosophie before the *Aristotelean*, is not so much because of its so much greater certainty, but because it puts inquisitive men into a method to attain it, whereas the other serves only to obstruct their industry by amusing them with empty and insignificant Notions. And therefore we may rationally expect a greater Improvement of Natural Philosophie from

[47] See Hunter and Wood 1989.

the *Royal Society*, (if they pursue their design) then it has had in all former ages; for they having discarded all particular *Hypotheses*, and wholly addicted themselves to exact Experiments and Observations, they may not only furnish the World with a compleat *History of Nature*, (which is the most useful part of *Physiologie*) but also laye firm and solid foundations to erect *Hypotheses* upon, (though perhaps that must be the work of future Ages:) at least we shall see whether it be possible to frame any certain *Hypotheses* or no, which is the thing I most doubt of, because, though the *Experiments* be exact and certain, yet their Appliction to any *Hypotheses* is doubtful and uncertain; so that though the Hypothesis may have a firm *Basis* to bottome upon, yet it can be fastned and cemented to it no other way, but by conjecture and uncertaine (though probable) applications, and therefore I doubt not but we must at last rest satisfied with true and exact Histories of Nature for use and practice; and with the handsomest and most probable *Hypotheses* for delight and Ornament.[48]

One cannot help but note the anti-Aristotelian polemic, the reference to natural history, the talk of discarding all 'particular hypotheses' and being addicted to observation and experiment. Glanvill's 'An Adress to the Royal Society' in his *Scepsis scientifica* repeats the same themes,

Nor are these all the *advantages* upon the Account of which we owe *acknowledgments* to *Providence* for your *erection*; since from your *promising* and *generous endeavours*, we may hopefully expect a considerable inlargement of the *History* of *Nature*, without which our *Hypotheseis* are but *Dreams* and *Romances*, and our *Science* meer *conjecture* and *opinion*. For while we frame *Scheames* of things without consulting the *Phænomena*, we do but *build* in the *Air*, and *describe* an *Imaginary World* of our *own making*; that is but little a kin to the *real one* that *God made*. And tis possible that all the *Hypotheseis* that yet have been contrived, were built upon too narrow an *inspection* of *things*, and the phasies of the *Universe*. For the advancing *day* of *experimental knowledge* discloseth such *appearances*, as will not lye *even*, in any *model* extant.

Yet Glanvill and Parker are the natural places to look for this kind of experimentalist rhetoric. It is important to note therefore, that from the late 1650s a number of physicians were calling for reforms to medicine that would parallel the changes taking place in natural philosophy. Thomas Willis expressed the wish that the successors of Hippocrates 'had betaken themselves to Observations only, and Experiments' for

without doubt the Art of Physick had been advanced to a greater perfection and fineness, and with much more advantage to the sick. But that which presently shut out the light which had been at first set up, and dimmed the eyes of posterity, was the preposterous endeavour of those men, who hastily, and in a manner after their own Phantasie, framed the Art of Physick into a general Method, after the fashion of some Speculative Science.[49]

But what of progressive medical practitioners who had no direct association with the Royal Society and who were not natural philosophers as such? Marchamont Nedham, in his *Medela medicinæ* (1665), explicitly links the need for reform of

[48] Parker 1666, pp. 45–46.
[49] Quoting Nedham's translation (1665, p. 238) of Willis 1659, *De febribus*, Preface [H3v–H4]. See also Starkey's *Pyrotechny* (1658) which opens with the question 'What profit is there of curious speculations, which doe not lead to real experiments? To what end serves Theorie, if not appplicable unto practice', p. 1. See also p. 3.

medicine with reform in natural philosophy, appealing to the authority of Bacon.[50] If we turn to the writings of, say, Thomas Sydenham our familiar themes are also in evidence. For example, in *On Dropsy* Sydenham asserts,

> however much hypotheses based upon the speculations of philosophy may be wholly futile — and futile they will be until men become endued with such intuitive knowledge as shall enable them to find foundations for these superstructures — hypotheses directly derived from the facts themselves, and arising from those observations only which are suggested by practical and natural phenomena, are stable and permanent; so much so that, although the practice of medicine to one who looks at the arrangement of writers only, appears as if it arose out of hypotheses, the truer view is that the hypotheses themselves, so far as they are true and genuine, themselves originated in practice ... Had I begun with my hypotheses, I should have shown the same want of wisdom that a builder would show who began with the roof and tiles, and ended with the basement and foundation. But it is only those who build castles in the air [Aere *Castella*] that may begin at either end indifferently.[51]

The familiar characterisation of unfounded hypotheses as 'castles in the air', the building metaphor, the reference to the 'speculations of philosophy' are all indicative of the experimental/speculative dichotomy. As for natural histories, the construction of histories of diseases was one of Sydenham's key *desiderata* for medicine and one for which he appealed to the authority of Francis Bacon.[52]

5. THE EXPERIMENTAL/SPECULATIVE DISTINCTION CRYSTALLISED IN THE 1690S WHEN OPPOSITION TO HYPOTHESES IN NATURAL PHILOSOPHICAL METHODOLOGY INTENSIFIED.

Of course Sydenham's influence on the Molyneux brothers and Locke in these matters is well documented: not least in Locke's correspondence where we find Thomas Molyneux agreeing with Locke on speculative theories in medicine,

> I perfectly agree with you concerning general theories, that they are for the most part but a sort of waking dreams, with which men have warm'd their own heads ... beginning at the wrong end, when men lay the foundation in their own phansies, and then endeavour to sute the Phænomena of diseases, and the cure of them, to those phansies. I wonder that, after the pattern of Dr. Sydenham has set them of a better way, men should return again to that romance way of physick. But I see it is easier and more natural for men to build castles in the air of their own, than to survey well those that are to be found standing [i.e. natural histories of diseases]. ... Upon such grounds as are the establish'd history of diseases hypotheses might with less danger be erected.[53]

In fact, it appears from a close reading of the sources during the 1690s that there was something of a 'ratcheting up' of the opposition to hypotheses and speculative methodology. Locke's correspondence is revealing here as well, particularly with

---

[50] See Nedham 1665, pp. 234–235. See also Simpson 1669, Preface.

[51] Sydenham 1848, II, p. 173 = Sydenham 1683, pp. 165–166. See also *idem*. 1848, I, p. 14 and 'De Arte Medica' (possibly by Locke) in Dewhurst 1966, p. 81 and Royal College of Physicians of London MS 572, fol. 9b, Meynell 1991, p. 17.

[52] Sydenham 1848, I, pp. 12 and 21.

[53] 20 January 1693, Locke 1976–, IV, pp. 628–629. See also William Cole to Locke, *ibid.*, p. 91.

reference to his exchange with William Molyneux over Richard Blackmore's *King Arthur*. Locke writes to William on 15 June 1697,

> I have always thought, that laying down, and building upon hypotheses, has been one of the great hindrances of natural knowledge; and I see your notions agree with mine in it. And though I have a great value for Sir R. Blackmore, on several accounts, yet there is nothing has given me a greater esteem of him, than what he says about hypotheses in medicine, in his preface to K. Arthur.[54]

Turning to Blackmore's preface we find the following,

> the raising of an Hypotheses in Philosophy obtains little more Credit with me, than the erecting a Scheme in Astrology; and the Judgments and Decisions that are given upon them seem to me alike Precarious and uncertain. I was once enamour'd with the *Cartesian System*, but the warmth of my Passion is quite extinguish'd. It may indeed make a Man capable of entertaining and amusing others, but not of quieting and satisfying himself. All Knowledge is valuable according to it's degree of Usefulness, as it do's more or less promote the benefit of Mankind, and for this Reason 'tis a great mortification to consider how little the Pains and Time I have bestow'd in Philosophical Enquirys, have contributed to my knowledge in Curing Diseases. I am now inclin'd to think, that 'tis an Injury to a Man of good sense and natural Sagacity, to be hamper'd with any Hypothesis before he comes to the Practice of Physic. For this prepossession obstructs the Freedom of his Judgement, puts a strong Byass on his Thoughts, and obliges him to make all the Observations that occur to him in his Practise, to comply with, and humour his pre-conceived Opinions; whereas in Reason, his Observations on Nature should be first made, before any Hypotheses should be establish'd. A clear and penetrating Understanding, Cultivated and Matur'd by repeated, Diligent Observation, will in my Opinion, make a more able and accomplish'd Physitian, than any *Philosophical Scheme* that has yet obtain'd in the World.[55]

Note that the Cartesian system is used as an example of a hypothetical system. This is a point to which we will return below. The pertinent issue here, however, is that the occurrence of this opposition to hypotheses in a literary work, albeit in a preface and by a poet and a physician who had come under the influence of Thomas Sydenham,[56] is indicative of just how widespread this phenomenon had become in the 1690s. Not surprisingly therefore, we also find this anti-hypotheticalism in the writings of the theologian and polemicist John Toland who in a very poignant remark claimed, 'since PROBABILITY is not KNOWELG, I banish all HYPOTHESES from my PHILOSOPHY'.[57]

Now the two really explicit statements of the division in natural philosophy between the speculative and the experimental quoted above are also from the 1690s. Dunton's student manual and Sergeant's *Method to Science* may well reflect a

---

[54] Locke 1976–, VI, p. 144. See also Locke to Thomas Molyneux, 1 November 1692, 'I hope the age has many who will follow his [Sydenham's] example, and by the way of accurate practical observation, as he has so happily begun, enlarge the history of diseases, and improve the art of physick, and not by speculative hypotheses fill the world with useless, tho' pleasing visions', Locke 1976–, IV, p. 563 and John Baron to Locke, *ibid.*, VI, p. 471.

[55] Blackmore 1697, Preface, pp. ix-x.

[56] Dewhurst 1966, p. 49.

[57] Toland 1696, p. 15, quoted in Feingold 1988, p. 305 who wrongly claims that it alludes to Newton's *Principia*. The 'hypotheses non fingo' appeared in the 2nd edition of 1713. See also Toland 1696, The Preface, p. viii.

'suring-up' or crystallising of the methodological precepts of the new science, such that natural philosophy can now definitively be divided into two quite distinct types. But what is certain is that the discussion of hypotheses in this decade is filled with more invective and ridicule and is far less concessive than in previous decades.

It is difficult to explain the causes of this phenomenon. I have one suggestion as to the underlying process by which the distinction between speculative and experimental natural philosophy became so widely accepted and deployed and by which the anti-hypotheticalism in English natural philosophy strengthened. It is a striking change in the polemical context. As the century progressed Descartes' natural philosophy, and in particular his vortex theory, (which is discussed in the chapters by Schuster, Dear and Hattab) came to be regarded as the archetypal form of speculative natural philosophy. It was not simply the substantial forms of the scholastics, but Descartes' threaded screws and whirlpools which became objects of ridicule. And his identification of matter and extension and consequent denial of the possibility of a vacuum had come increasingly to be regarded as metaphysical speculations. One can plot an increasing discomfort with Descartes' natural philosophy amongst natural philosophers in England from the mid 1660s.[58] Power and Glanvill both seem quite sanguine about Cartesian natural philosophy and methodology in the 1660s. But both Boyle and Hooke, while being influenced by the general cast of Descartes' mechanism, were critical of particular Cartesian doctrines, either because they were untestable or because they were not founded upon observation and experiment.[59] To be sure, many English natural philosophers continued to speak of the solar system as 'our vortex',[60] however, increasingly Descartes' views were used as examples of speculative natural philosophy. And by the early 1680s the tide had turned against Descartes and the Cartesians.

By 1680 Locke was sceptical of the vortex theory. He comments rather sarcastically in his correspondence on the size of giant hailstones, 'I doubt whether the Cartesians can have any contrivances to help in this matter, and whether the occult qualities of the Peripatetics may not break under such a load'.[61] More significant though is the shift in Newton's attitudes. Newton appears to have accepted Cartesian vortices until the early 1680s.[62] However, his 'De gravitatione et aequipondio fluidorum', which B. J. T. Dobbs has recently argued was composed in

---

[58] It should be noted that in the late 1640s William Petty attacked Cartesian natural philosophy on the grounds that it was too speculative and not founded on enough experiments. See his exchange with Henry More, who was quite sanguine about the experimental support that Descartes had rendered for his system. It is interesting to note, however, that Petty's attack seems not to have had any significant repercussions for the early acceptance of Cartesianism in England and that More's later rejection of Cartesian mechanism was made on independent grounds. See More to Hartlib 11 Dec 1648, Hartlib Papers, 18/1/38B and William Petty to Hartlib?, *ibid.*, 7/123/1A–2A. For further discussion see Webster 1969 and Gabbey 1982. For Isaac Barrow's opposition to Descartes and its possible influence on Newton see Gascoigne 1985, p. 409.

[59] See for example Boyle's *Origin of Forms and Qualities* (1666), Boyle 1999–2000, 5, p. 353 and Hooke 1665, pp. 46 and 54–61.

[60] See Boyle's *Notion of Nature* 1686, Boyle 1999–2000, 10, p. 508, for a reference to 'our *Vortex*'.

[61] Locke 1976–, II, p. 176. See also Keill 1698, pp. 11–18.

[62] See Wilson 2002, p. 206 and especially Dobbs 1991, pp. 122–129.

the mid 1680s,[63] is a strongly anti-Cartesian essay in metaphysics in which Newton attempts to do away with Descartes' fictions (*figmenta*).[64] Newton's attack on Descartes was to culminate in his demolition of the vortex theory in the *Principia*. And by the 1690s the casting of Descartes as the archetypal speculative philosopher was complete. Thus, in Locke's *Second reply to Stillingfleet* (1699) we find a typical example of this criticism of Descartes,

> 'That Des Cartes, a mathematical man, has been guilty of mistakes in his system.'
> Answ. When mathematical men will build systems upon fancy, and not upon
> demonstration, they are as liable to mistakes as others. And that Des Cartes was not led
> into his mistakes by mathematical demonstrations, but for want of them, I think has
> been demonstrated by some of those mathematicians who seem to be meant here.[65]

Similar criticisms are even found in Huygens who remained a mechanist and advocate of the vortex theory after reading the *Principia*. His notes on Baillet's *Life of Descartes* (1691) describe Cartesian hypotheses as 'conjectures and fictions'. And his reaction on reading the *Principia* was to claim 'vortices destroyed by Newton'.[66]

The point here is that Descartes was a modern. He was a contributor to the new science and a reflector on natural philosophical method, but increasingly he, like the scholastics, came to represent the wrong way to do natural philosophy. Perhaps a compounding factor here was the battle of books which raged in the early 1690s. For, one line of argument in favour of modern learning was that the moderns applied a new experimental method, unlike the ancients who indulged in gross speculation. It is the application of the new method which accounts for the recent gains in natural knowledge. Descartes however was in some respects an awkward exception, for he was in many ways just as guilty of speculative natural philosophy as the ancients. Compounding this was the fact that a steady stream of truly speculative Cartesian cosmological systems was flowing from the presses on the Continent and in England.[67] It was as if Descartes' principles had gone to seed. Very few of these works were related in any way to observation or experiment and many of them were speculative in the extreme. Indeed before long we find other offenders being singled out. Thus William Wotton claimed in his *Reflections upon Ancient and Modern Learning* (1694),

> I do not here reckon the several *Hypotheses* of *Des Cartes*, *Gassendi*, or *Hobbes*, as
> Acquisitions to real Knowledge, since they may only be Chimæra's and amusing
> Notions, fit to entertain working Heads. I only alledge such Doctrines as are raised upon

---

[63] Dobbs 1991, pp. 143–146. McGuire (2000, p. 271) finds this redating of 'De gravitatione et aequipondio fluidorum' 'persuasive if not decisive', whereas Stein (2002, p. 303 n. 39) claims that 'uncertainty remains' and Hall rejects it (2002, pp. 412–421)

[64] Published with an English translation in Hall and Hall 1962, pp. 89–156.

[65] Locke 1823, IV, p. 427.

[66] Quoted from Koyré 1965, p. 117. For discussion of the status of hypotheses in French Cartesian natural philosophy from the 1660s see Clarke 1989, especially chap. 5.

[67] See for example Mallement de Messange 1679, Barin 1686, Fontenelle 1686 and Burnet 1681. For attacks on the English cosmogonies see for example Keill 1698 who dismisses Burnet's theory as 'a philosophical romance', p. 26. Of Burnet's *Telluris theoria sacra* (1681) Locke claimed 'I imagine, if I should trouble you with my fancies, I could give you an hypothesis would explain the deluge without half the difficulties, which seem to me to cumber this', Locke to James Tyrrell, 14/24 February 1687?, Locke 1976–, III, p. 140.

faithful Experiments, and nice Observations; and such Consequences as are the
immediate Results of, and manifest Corollaries drawn from, these Experiments and
Observations.[68]

I present this anti-Cartesianism as only one instance of the kind of forces which
led to a heightened anti-hypotheticalism in the late seventeenth century and what
was arguably a more rigid characterising of natural philosophical method as either
speculative or experimental. It is in this context that we need to evaluate the most
notorious of all comments on hypotheses in the early modern period, the 'hypotheses
non fingo' of Isaac Newton.

6. THE EXPERIMENTAL/SPECULATIVE DISTINCTION PROVIDES THE TERMS OF REFERENCE BY WHICH WE SHOULD INTERPRET NEWTON'S STRICTURES ON THE USE OF HYPOTHESES IN NATURAL PHILOSOPHY.

There are many studies of Newton on hypotheses, but few which deal with the
broader context of the acceptance of hypotheses in early modern natural philosophy.
For instance, Ernan McMullin asks of Newton's letter to Oldenburg of 8 July 1672
'Why this opposition to hypothesis, which had, by the 1670s, become common coin
in natural philosophy?'.[69] But apart from noting Newton's aversion to Cartesian-
style speculative hypotheses, he ignores the 'common coinage' entirely. Yet a
careful perusal of Newton's many discussions of the place of hypotheses in natural
philosophical method clearly reveals that his terms of reference are the
experimental/speculative distinction. Perhaps the most explicit statement of this is in
a draft of a letter to Roger Cotes from March 1713.

> Experimental Philosophy reduces Phænomena to general Rules & looks upon the Rules
> to be general when they hold generally in Phænomena. It is not enough to object that a
> contrary phænomenon may happen but to make a legitimate objection, a contrary
> phenomenon must be actually produced. Hypothetical Philosophy consists in imaginary
> explications of things & imaginary arguments for or against such explications, or
> against the arguments of Experimentall Philosophers founded upon Induction. The first
> sort of Philosophy is followed by me, the latter too much by Cartes, Leibniz & some
> others.[70]

Note the contrast between the experimental and hypothetical (speculative)
philosophy and the claim that the latter is practised by Descartes. Note too the claim
that the hypothetical philosophy consists of 'imaginary explications' and 'imaginary
arguments'. I. B. Cohen has characterised this kind of comment in Newton as his
'insistence on maintaining a sharp distinction between empirical science and

---

[68] Wotton 1694, p. 244.

[69] McMullin 1990, p. 69.

[70] Newton 1959–1977, 5, pp. 398–390, underlining added. See also Newton 1715, p. 224 where Newton
contrasts his experimental natural philosophy with Leibniz' in the following terms, 'The one
[Newton] proceeds upon the Evidence arising from Experiments and Phænomena, and stops where
such Evidence is wanting; the other [Leibniz] is taken up with Hypotheses, and propounds them, not
to be examined by Experiments, but to be believed without Examination'.

philosophy'.[71] But a more nuanced interpretation arises from our foregoing discussion such that we can characterise it as an insistence on the distinction between experimental and speculative natural philosophy and a clear acceptance of the former.

Now, a comprehensive analysis of Newton's discussions of hypotheses in the light of the experimental/speculative distinction is beyond the scope of this chapter, however, we can get a taste for just how illuminating such a study would be by examining some of the claims made about Newton and hypotheses in the secondary literature. We will address claims made by Barbara Shapiro, Larry Laudan, Zev Bechler, George Smith and Mordechai Feingold.

Barbara Shapiro claims '[a] modest yet positive view of hypothesis thus was characteristic of the English scientific community at least until Newton's more critical views were proclaimed'.[72] Larry Laudan makes a similar claim to the effect that in the mid-seventeenth century there was a well-developed 'method of hypothesis' which emerged in the writings of Boyle and Glanvill under the influence of Descartes. However, Laudan regards Descartes' mechanism, as encapsulated in the clock metaphor, as sowing the seeds of the downfall of the 'method of hypothesis'.

The demand for a hypothesis-free science, which was widely circulated after Newton, could never have gathered such enthusiastic adherents if the probabilism of Descartes, Boyle, and Glanvill had not died such a quick and needless death at the hands of those who thought nature's clock had no secrets which man's instruments could not seek out and know with certainty. As it happened the method of hypothesis went into virtual eclipse after 1700 ... [73]

Both Shapiro and Laudan see Newton as derailing the advances made in the method of hypothesis earlier in the century. However, if we take account of the ever-present critique of hypotheses from the 1660s on and the ideal for a demonstrative natural philosophy in writers like Hobbes and Locke, and if we interpret these phenomena in the light of the experimental/speculative dichotomy, a different picture emerges. It becomes clear that rather than Newton causing the derailment of the 'method of hypothesis', his comments on the role of hypotheses in natural philosophy are entirely consistent with one committed to the experimental philosophy and opposed to the speculative philosophy with its conjectures and fancies. Thus, rather than instigating the decline of the 'method of hypothesis', Newton's comments are indicative of the pre-existing terms of reference by which the discipline of natural philosophy was understood. And if Cohen is correct in claiming that Newton's attitude hardened against hypotheses in the 1690s,[74] we should see this as typifying the trend in natural philosophical method of the last decade of the seventeenth century.

Newton made significant changes to the hypotheses in the first edition of the *Principia* in the 1690s when he was in relatively close relations with Locke and

---

[71] 'A Guide to Newton's *Principia*', Newton 1999, p. 62.

[72] Shapiro 1983, p. 54.

[73] Laudan 1981, p. 48.

[74] Cohen 1966, p. 179. See also McGuire 1970, pp. 28–29.

when, as we have seen, the fortunes of hypotheses in natural philosophy were taking a turn for the worse.[75] His changing of the famous hypotheses in the first edition to the rules of reasoning in the second edition is best interpreted as indicative of the hardening against speculative natural philosophy in this decade and not as a development initiated solely by his own internal ruminations on the nature of natural philosophical method.[76] See, for example, his elaborative comments to Rule III where he says, 'Certainly idle fancies ought not to be fabricated recklessly against the evidence of experiments'.[77]

Furthermore, if my thesis about the consolidation of the experimental/speculative dichotomy in the 1690s is correct, Newton's 'hypotheses non fingo'[78] of 1713 is also better regarded as a response to a generalised and widespread denigrating of speculative natural philosophy, rather than as the cause of it. Thus, when Newton says in the General Scholium that 'hypotheses, whether metaphysical or physical, or based on occult qualities, or mechanical, have no place in experimental philosophy' we should be aware of the clear connotation of Newton's first published use of the term 'experimental philosophy'.[79] Newton here is identifying himself with the experimental philosophy in opposition to the speculative. His terms of reference are identical to those of other natural philosophers we have discussed above. His comments are entirely consistent with anyone who favours the experimental side of the experimental/speculative dichotomy and are consistent with other earlier comments by Newton himself. This is reinforced when we consider that in the third edition he changed the word 'experimental' to 'natural' in the General Scholium in his comment that 'to treat of God from phenomena is certainly a part of experimental philosophy'. For Newton there was experimental philosophy and there was natural philosophy and the two were not co-extensive. For Newton, as for Locke,[80] the study of God pertained to natural philosophy, but not to experimental philosophy.

Indeed the whole General Scholium is something of a declaration of Newton's commitment to the experimental philosophy and opposition to speculative philosophy. As such he begins by recounting his rejection of the archetypal

---

[75] See Smith 2002, p. 161.

[76] For the details of Newton's changes see Cohen 1966.

[77] Newton 1999, p. 795.

[78] It is perhaps worth pointing out the Baconian origin of the notion of 'feigning from nature'. Bacon uses the expression *neque enim fingendum, aut excogitandum, sed inveniendum, quid Natura faciat aut ferat* in the *Novum Organum*, Bk II, Aphorism X (Bacon 1859, I, p. 236) which appears on the title pages of Boyle's *Colours* and *Cold*, works with which Newton was familiar from the mid-1660s.

[79] This has recently been stressed by Alan Shapiro who has found no use of the term by Newton before a draft of Query 23 for the Latin translation of the *Opticks* in 1706: Shapiro 2004, pp. 186–189. We should also be aware that Newton is using the term 'hypothesis' in a very specific sense to mean a proposition that is independent of observation or experiment. See for example Newton to Cotes, 28 March 1713, 'the word Hypothesis is here used by me to signify only such a Proposition as is not a Phænomenon nor deduced from any Phænomena but assumed or supposed without any experimental proof', Newton 1959–1977, 5, p. 397.

[80] 'The end of this [natural philosophy], is bare speculative Truth, and whatsoever can afford the Mind of Man any such, falls under this branch, whether it be God himself, Angels, Spirits, Bodies ...', *An Essay concerning Human Understanding*, IV. xxi. 2, Locke 1975, p. 720.

speculative system, Descartes' 'hypothesis of vortices' and ends in the penultimate paragraph with the famous claim that 'I do not feign hypotheses'.[81] This scholium is best interpreted as the culmination of a long struggle that Newton engaged in as he attempted to situate and understand his own natural philosophical methodology in relation to the distinction between experimental and speculative methods, a struggle in evidence as early as his first reports on the properties of light in 1672.

An awareness of the distinction between speculative and experimental natural philosophy also enables us slightly to recast Zev Bechler's interpretation of the controversy in the early 1670s between Newton, Hooke and Huygens over Newton's optical experiments.[82] Bechler finds the tension between Newton and the others as resulting from his inability to appreciate the fallibilism of Hooke and Huygens and his uncompromising commitment to a deductive, even dogmatic, approach to his experiments and their analysis. Bechler sees this as the first such incident which marked the ushering in of a 'period of the blind spot'; a period in which natural philosophers were unable to see the efficacy of hypotheses or the way of hypothesis. However, if we interpret the exchanges between Newton and Hooke and Newton and Huygens in the light of the distinction between speculative and experimental philosophy, it is clear that Newton saw himself very much as an experimental philosopher and that dogmatism (of which he is accused by Bechler) was the province of the speculative natural philosopher, as were hypotheses. Little wonder that he took umbrage at Pardies' description of his view of colour as an hypothesis.[83] We need also to fine-tune George Smith's recent claim that 'From the beginning of his work in optics in the 1660s, Newton had always distrusted the hypothetico-deductive approach, arguing that too many disparate hypotheses can be compatible with the same observations'.[84] Rather, it is better to say that Newton distrusted speculative natural philosophy, for it is clear that what this amounted to is not what we call the hypothetico-deductive method. That cluster of epistemological issues in natural philosophical methodology surrounding probability, induction and hypotheses had not yet congealed into what we might, somewhat anachronistically, call the hypothetico-deductive method.

What Newton was unequivocally committed to was the experimental philosophy, with its rejection of dogmatism and speculation.[85] Clearly he was not partial to the sort of ill-formed probabilism which was to be found amongst a number of the members of the Royal Society and which was to find its clearest articulation in Huygens' preface to the *Treatise on Light* and *Discourse on the Cause of Gravity*.

---

[81] The anti-speculative/Cartesian polemic is also present in Cotes' Preface to the 2nd edition of the *Principia*. See Newton 1999, p. 393 and Newton 1959–1977, 5, p. 391.

[82] Bechler 1974.

[83] See the exchange between Newton and Pardies, mediated by Oldenburg, that was initiated by Pardies' letter to Oldenburg of 30 March 1672, Newton 1959–1977, 1, pp. 130ff.

[84] Smith 2002, p. 154.

[85] Newton wrote anonymously in 1715 'The Philosophy which Mr. *Newton* ... has pursued is Experimental; and it is not the Business of Experimental Philosophy to teach the Causes of things any further than they can be proved by Experiments. We are not to fill this Philosophy with Opinions which cannot be proved by Phænomena. In this Philosophy Hypotheses have no place, unless as Conjectures or Questions proposed to be examined by Experiments', Newton 1715, p. 222.

Yet, as we can see with hindsight, a coherent scientific methodology would have to embrace some form of hypothetico-deductive or retroductive inference. The problem was that Newton's strengths and successes lay in the realm of mathematical natural philosophy that most closely approximated the demonstrative ideal of Bacon, and that others committed to the experimental natural philosophy, such as Locke, believed to be at least a possibility in the light of Newton's achievements.[86]

This brings us to the recent claims of Mordechai Feingold in his discussion of some of the internal tensions between members of the Royal Society over the question of correct method in natural philosophy. Feingold argues that Newton's assertion about the certainty of his theory of colour in his 6 February 1671/2 letter to Oldenburg was a 'bombshell' and that Hooke took the comments as 'disparaging of the naturalist and experimental tradition of the Society'.[87] Furthermore, Feingold claims that Newton withdrew from engagement with the Society after his early optical controversies because 'he refused to abide by its moratorium on theoretical pronouncements'.[88] He finds an 'incessant rebuke of theory' amongst the Baconians of the early Royal Society and he claims that the emphasis on the primacy of mathematics in natural philosophy as found in the Newtonian party in the Royal Society from the 1680s on led not only to the deprecation of natural history but of the experimental philosophy itself.

However, if we regard the distinction between speculative and experimental natural philosophy as the backdrop to the controversies which Feingold discusses, I suggest that we can arrive at a more nuanced interpretation of them. For this backdrop enables us to see that the opposition between promoters of natural history and mathematical natural philosophers is a conflict internal to the experimental philosophy itself. To be sure, Oldenburg may have found Newton's assertion of certainty too dogmatic or presumptuous for publication in the *Philosophical Transactions*, but part of Newton's motive in being so forthright was to reason from his optical experiments without reverting to hypotheses, a motive with which Oldenburg could concur. If we take the railing against hypotheses, so prevalent from the 1660s, as a central facet of the stance against speculative natural philosophy (and remember Newton himself was one of the most severe critics of speculative hypotheses), we can see that there was no moratorium on theorising *per se*, but on the empty conjectures of the speculative natural philosophers. Nor was there any deprecation of the experimental philosophy.[89]

---

[86] See for example, Locke 1975, IV. ii. 9–13, pp. 534–536; IV. vii. 11, pp. 598–603; Locke 1989, p. 248; and Locke 1823, IV, p. 427.

[87] Feingold 2001, p. 83.

[88] *Ibid.*, p. 84.

[89] William Wotton draws a distinction between hypotheses and theories in his *Reflections*, claiming that theories are raised upon experiment and observation whereas hypotheses are not. Now, this distinction in Wotton was his own and is not a widely held view, moreover, it does emerge some three decades after the founding of the Royal Society. However, it is illustrative of the point that there was certainly no moratorium on theorising but rather on speculative hypothesising. See Wotton 1694, pp. 235 and 244. Hooke had claimed in his *Micrographia* that axioms and theories are to be raised upon natural histories, Preface [b2]. While Glanvill claimed that the members of the Royal Society 'continually declare against the *establishment* of *theories*, and *Speculative Doctrines*' (1668, p. 89) it is clear from

Furthermore, the denigration of the method of natural history, which gathered force in the 1680s, should not be identified with, nor should it be thought to entail, a rejection of the experimental philosophy. For the method of natural history was not constitutive of experimental philosophy, but rather it was the most prominent early manifestation of it.[90] The fact that the method of natural history, in a sense, lost its way at the same time as an appreciation of Newton's achievement in mathematical natural philosophy was dawning on the natural philosophical community, may well have precipitated its decline. The sentiment against natural history is summed up in the comments of Steele in *The Spectator* of 24 March 1711, when he speaks of physicians who, for want of better things to do, only

> amuse themselves with the stifling of Cats in an Air Pump, cutting up Dogs alive, or impaling of Insects upon the point of a Needle for Microscopical Observations; besides those that are gathering weeds, and the Chase of Butterflies: Not to mention the Cockle-shell-Merchants and Spider-Catchers.[91]

Interestingly too, this rivalry issued in a rather ironic twist in the deployment of the experimental/speculative distinction in the early eighteenth century. For, some of the defenders of the method of natural history began to accuse the emulators of Newton's methodology of indulging in speculation themselves. In particular this accusation was levelled against the mathematisers of medicine such as Archibald Pitcairne.[92]

7.CONCLUSION

It is clear then that the experimental/speculative distinction is an important way of demarcating different approaches to method in English natural philosophy in the latter half of the seventeenth century. It functioned as a kind of general methodological rubric from the late 1650s until the early decades of the following century and was deeply ingrained in the methodological discourse of many practitioners, promoters and even critics of the new science. In fact, it even transcended disciplinary boundaries in so far as it also impacted upon the medical methodology of Sydenham who was quickly to become an exemplar of a more Hippocratic approach to medical practice. As natural philosophers became disillusioned with speculative systems such as the Cartesian vortex theory in the final decades of the century, the critical attitude towards hypotheses hardened and in the 1690s the experimental/speculative distinction appears to have become more firmly entrenched. This is reflected in the writings of Newton, whose changes to the

---

the context that he means speculative theories, for immediately preceding this comment he speaks of the need to raise axioms from experiments, *ibid.*, p. 87.

[90] Thus there is no need to follow Feingold (2001, p. 85 and p. 100 n. 18) who implies that William Molyneux privately criticised the experimental philosophy when he criticised the method of natural history but in public 'was a bit more circumspect'. Molyneux's endorsement of the experimental philosophy, including natural history, as we have seen, was unequivocal. See also W. Molyneux to Locke, 27 May 1697, Locke 1976–, VI, p. 134; and Molyneux 1686, Epistle Dedicatory.

[91] Steele and Addison 1888, p. 37.

[92] See Feingold 2001, pp. 88–90.

hypotheses of the *Principia* in this decade are indicative of the broader intellectual climate as reflected in the writings of theologians, poets and philosophers alike.

Now it may be tempting to view this experimental/speculative distinction as equivalent to the modern distinction between rationalism and empiricism. After all, the experimental philosophy emphasised the importance of the senses, constantly appealing to observation and experiment, and it decried the use of mere reason in generating hypotheses. However, while these are certainly tenets of empiricism, there are also marked discontinuities between the two. The wariness of, and at times outright opposition to, hypotheses as well as the preference of some for a demonstrative science of natural philosophy are features of early modern methodologies which are foreign to modern empiricist theories of knowledge. Some might still desire to foist the nineteenth century *historiographical* categories of rationalism and empiricism on the broad spectrum of approaches to natural philosophical knowledge in this period, but to my mind the early modern *historical* categories of speculative and experimental philosophy are more effective terms of reference for interpreting the diverse range of discussions of method in the period. These terms 'save the phenomena' of our historical data in a manner that is far more satisfactory than the 'fancies' of nineteenth and twentieth century historiographers. Indeed it may be that the very origins of the categories rationalism and empiricism are to be found in the philosophical deployment of this unduly neglected distinction.[93]

REFERENCES

Anstey, P. R. (2002) 'Locke, Bacon and natural history', *Early Science and Medicine*, 7(1), pp. 65–92.
— ed. (2003a) *The Philosophy of John Locke: New Perspectives*, London: Routledge.
— (2003b) 'Locke on method in natural philosophy' in *The Philosophy of John Locke: New Perspectives*, ed. P. R. Anstey, London: Routledge, pp. 26–42.
— (2004) 'The methodological origins of Newton's queries', *Studies in History and Philosophy of Science*, 35, pp. 247–269.
Bacon, F. (1859) *The Works of Francis Bacon*, 7 vols, eds J. Spedding, R. L. Ellis and D. D. Heath, London.
— (1996) *The Oxford Francis Bacon: Philosophical Studies c. 1611–c.1619*, ed. G. Rees, Oxford: The Clarendon Press.
Baillet, A. (1691) *De la vie de Monsieur Descartes*, Paris.
Barin, T. (1686) *Le Monde Naissant*, Utrecht.
Bechler, Z. (1974) 'Newton's 1672 optical controversies: a study in the grammar of scientific dissent' in *The Interaction between Science and Philosophy*, ed. Y. Elkana, New Jersey: Humanities Press, pp. 115–140.

---

[93] I would like to thank Mordechai Feingold, Peter Harrison, Michael Hunter, David Miller, John Schuster, Richard Serjeantson and Alan Shapiro for their comments on this paper. It was first read at the Australasian Association for the History, Philosophy and Social Studies of Science conference at the University of Melbourne in July 2003.

Bennett, J. Cooper, M., Hunter, M. and Jardine, L. eds (2003) *London's Leonardo: The Life and Work of Robert Hooke*, Oxford: Oxford University Press.

Birch, T. (1756–1757) *The History of the Royal Society of London*, 4 vols, London.

Blackmore, R. (1697) *King Arthur*, 2nd edn, London.

Bos, H. J. M., Rudwick, M. J. S., Snelders, H. A. M. and Visser, R. P. W., eds (1980) *Studies on Christiaan Huygens*, Lisse: Swets & Zeitlinger B. V.

Boyle, R. (1991) *Selected Philosophical Papers of Robert Boyle*, ed. M. A. Stewart, Indianapolis: Hackett.

— (1999–2000) *The Works of Robert Boyle*, 14 vols, eds M. Hunter and E. B. Davis, London: Pickering and Chatto.

— (2001) *The Correspondence of Robert Boyle*, 6 vols, eds M. Hunter, A. Clericuzio and L. M. Principe, London: Pickering and Chatto.

Buchwald, J. Z. and Cohen, I. B., eds (2001) *Isaac Newton's Natural Philosophy*, Cambridge MA: The MIT Press.

Burnet, T. (1681) *Telluris theoria sacra*, London.

Cavendish, M. (1666) *Observations upon experimental philosophy*, London. Reprinted in 2001, ed. E. O'Neill, Cambridge: Cambridge University Press.

Childrey, J. (1661) *Britannia Baconia*, London.

Clarke, D. M. (1989) *Occult Powers and Hypotheses*, Oxford: The Clarendon Press.

Cohen, I. B. (1966) 'Hypotheses in Newton's Philosophy', *Physis*, 8, pp. 163–184.

Cohen, I. B. and Smith G. E., eds (2002) *The Cambridge Companion to Newton*, Cambridge: Cambridge University Press.

Cook, H. (1993) 'The cutting edge of a revolution? Medicine and natural history near the shores of the North Sea' in *Renaissance and Revolution: Humanists, Scholars, Craftsmen and Natural Philosophers in Early Modern Europe*, eds J. V. Field and F. A. J. L. James, Cambridge: Cambridge University Press, pp. 45–61.

Daston, L. (1988) *Classical Probability in the Enlightenment*, Princeton: Princeton University Press.

— (1998) 'Probability and evidence' in *The Cambridge History of Seventeenth-Century Philosophy*, eds D. Garber and M. Ayers, Cambridge: Cambridge University Press, pp. 1108–1144.

Dear, P. (1998) 'Method and the study of nature' in *The Cambridge History of Seventeenth-Century Philosophy*, 2 vols, eds D. Garber and M. Ayers, Cambridge: Cambridge University Press, 1, pp. 147–177.

Dewhurst, K. (1966) *Dr. Thomas Sydenham (1624–1689): his Life and Original Writings*, London: Wellcome.

Dobbs, B. J. T. (1991) *The Janus Faces of Genius: The Role of Alchemy in Newton's Thought*, Cambridge: Cambridge University Press.

Dunton, J. (1692) *The Young-Students-Library*, London.

Elkana, Y. ed. (1974) *The Interaction between Science and Philosophy*, New Jersey: Humanities Press.

Feingold, M. (1988) 'Partnership in glory: Newton and Locke through the Enlightenment and beyond' in *Newton's Scientific and Philosophical Legacy*, eds P. B. Scheurer and G. Debrock, Dordrecht, Kluwer, pp. 291–308.

— (2001) 'Mathematicians and naturalists: Sir Isaac Newton and the Royal Society' in *Isaac Newton's Natural Philosophy*, eds J. Z. Buchwald and I. B. Cohen, Cambridge MA: The MIT Press, pp. 77–102.

Findlen, P. (1997) 'Francis Bacon and the reform of natural history in the seventeenth century' in *History and the Disciplines: The Reclassification of Knowledge in Early Modern Europe*, ed. D. R. Kelley, New York, University of Rochester Press, pp. 239–260.

Fontenelle, B. Le B. (1686) *Entretiens sur la pluralité de mondes*, Paris and Lyon.

Franklin, J. (2001) *The Science of Conjecture*, Baltimore: Johns Hopkins University Press.

Gabbey, A. (1982) 'Philosophia Cartesiana Triumphata: Henry More (1646–1671)' in *Problems of Cartesianism*, eds T. M. Lennon, J. M. Nicholas and J. W. Davis, Kingston and Montreal: McGill-Queens University Press, pp. 171–250.

Garber, D. and Ayers, M. (1998) *The Cambridge History of Seventeenth-Century Philosophy*, 2 vols, Cambridge: Cambridge University Press.

Gascoigne, J. (1985) 'The universities and the Scientific Revolution: the case of Newton and Restoration Cambridge', *History of Science*, 23, pp. 391–434.

Glanvill, J. (1665) *Scepsis scientifica*, London.

— (1668) *Plus Ultra*, London.
Grant, E. (1987) 'Ways to interpret the terms "Aristotelian" and "Aristotelianism" in Medieval and Renaissance natural philosophy', *History of Science*, 25, pp. 335–358.
Hall, A. R. (2002) 'Pitfalls in the editing of Newton's papers', *History of Science*, 40, pp. 407–424.
Hall, A. R. and Hall, M. B. eds (1962) *Unpublished Scientific Papers of Isaac Newton*, Cambridge: Cambridge University Press.
Hooke, R. (1661) *An attempt for the Explication of the Phænomena*, London.
— (1665) *Micrographia*, London.
— (1705) *The Posthumous works of Robert Hooke*, London.
Hunter, M. (1989) *Establishing the New Science: The Experience of the Early Royal Society*, Woodbridge: Boydell.
— (1995) *Science and the Shape of Orthodoxy: Intellectual Change in Late Seventeenth-Century Britain*, Woodbridge: Boydell.
— (2003) 'Robert Hooke: the natural philosopher' in *London's Leonardo: The Life and Work of Robert Hooke*, eds J. Bennett, M. Cooper, M. Hunter and L. Jardine, Oxford: Oxford University Press, pp. 105–162.
Hunter, M. and Wood, P. (1989) 'Towards Solomon's house: rival strategies for reforming the early Royal Society' in *Establishing the New Science: The Experience of the Early Royal Society* by M. Hunter, Woodbridge: Boydell, pp. 185–244.
Hutchison, K. (1982) 'What happened to occult qualities in the Scientific Revolution?', *Isis*, 73, pp. 233–253.
— (1991) 'Dormative virtues, scholastic qualities, and the new philosophies', *History of Science*, 21, pp. 245–278.
Huygens, C. (1690) *Traité de la lumière*, Leiden.
— (1888–1950) *Oeuvres complètes de Christiaan Huygens*, 22 vols, The Hague: Nijhoff..
Keill, J. (1698) *An Examination of Dr Burnet's Theory of the Earth*, Oxford.
Kelley, D. R. ed. (1997) *History and the Disciplines: The Reclassification of Knowledge in Early Modern Europe*, New York: University of Rochester Press.
Koyré, A. (1965) *Newtonian Studies*, Chicago: University of Chicago Press.
Laudan, L. (1981) *Science and Hypothesis*, Dordrecht: Kluwer.
Lennon, T. M., Nicholas, J. M. and Davis, J. W., eds (1982) *Problems of Cartesianism*, Kingston and Montreal: McGill-Queens University Press.
Levine, J. M. (1983) 'Natural history and the scientific revolution', *Clio*, 13, pp. 519–542.
Lindberg, D. C. and Westman, R. S., eds (1990) *Reappraisals of the Scientific Revolution*, Cambridge: Cambridge University Press.
Locke, J. (1823) *The Works of John Locke*, London.
— (1975) *An Essay concerning Human Understanding*, ed. P. H. Nidditch, Oxford: Clarendon Press.
— (1976–) *The Correspondence of John Locke*, 9 vols, ed. E. S. de Beer, Oxford: The Clarendon Press.
— (1989) *Some Thoughts concerning Education*, eds J. W. Yolton and J. S. Yolton, Oxford: Clarendon Press.
Lynch, W. T. (2001) *Solomon's Child: Method in the Early Royal Society of London*, Stanford: Stanford University Press.
McGuire, J. E. (1970) 'Atoms and the "analogy of nature": Newton's third rule of philosophizing', *Studies in History and Philosophy of Science*, 1, pp. 3–58.
— (2000) 'The fate of the date: the theology of Newton's Principia revisited' in *Rethinking the Scientific Revolution*, ed. M. J. Osler Cambridge: Cambridge University Press, pp. 271–295.
McMullin, E. (1990) 'Conceptions of science in the Scientific Revolution' in *Reappraisals of the Scientific Revolution*, eds D. C. Lindberg and R. S. Westman Cambridge: Cambridge University Press, pp. 27–92.
Mercer, C. (1993) 'The vitality and importance of early modern Aristotelianism' in *The Rise of Modern Philosophy*, ed. T. Sorell, Oxford: Oxford University Press, pp. 33–67.
Messange M. de (1679) *L'Ouvrage de La Creation. Traité Physique du Monde*, Paris.
Meynell, G. G. ed. (1991) *Thomas Sydenham's Observationes Medicae and his Medical Observations*, Folkestone: Winterdown Books.
Molyneux, W. (1686) *Sciothericum telescopicum; or A New Contrivance of Adapting a Telescope to an Horizontal Dial*, Dublin.
— (1692) *Dioptrica nova*, London.

Nedham, M. (1665) *Medela medicinæ*, London.

Newton, I. (1715) 'An account of the book entituled Commercium Epistolicum Collinii & aliorum, De Analysi Promota', *Philosophical Transactions*, 342, pp. 173–224.

— (1959–1977) *The Correspondence of Isaac Newton*, 7 vols, eds H. W. Turnbull, J. F. Scott, A. R. Hall and L. Tilling, Cambridge: Cambridge University Press.

— (1999) *The Principia*, eds I. B. Cohen and A. Whitman, Berkeley: University of California Press.

Oldenburg, H. (1965–1986) *The Correspondence of Henry Oldenburg*, 13 vols, eds A. R. Hall and M. B. Hall, Madison, Milwaukee and London.

Osler, M. J. ed. (2000) *Rethinking the Scientific Revolution*, Cambridge: Cambridge University Press.

Parker, S. (1666) *A free and impartial censure of the Platonick philosophie*, London.

Patey, D. L. (1984) *Probability and Literary Form: Philosophic Theory and Literary Practice in the Augustan Age*, Cambridge: Cambridge University Press.

Petty, W. (1647) *The Advice of W. P. to Mr. Samuel Hartlib for The Advancement of some particular Parts of Learning*, London.

Power, H. (1664) *Experimental Philosophy*, London.

Sabra, A. I. (1967) *Theories of Light from Descartes to Newton*, London: Oldbourne.

Scheurer, P. B. and Debrock, G., eds (1988) *Newton's Scientific and Philosophical Legacy*, Dordrecht: Kluwer.

Schmitt, C. B., Skinner, Q. and Kessler, E. (1988) *The Cambridge History of Renaissance Philosophy*, Cambridge: Cambridge University Press.

Schuster, J. A. (1986) 'Cartesian method as mythic speech: a diachronic and structural analysis' in *The Politics and Rhetoric of Scientific Method: Historical Studies*, eds J. A. Schuster and R. R. Yeo, Dordrecht: Reidel, pp. 33–95.

Schuster, J. A. and Yeo, R. R. eds (1986) *The Politics and Rhetoric of Scientific Method: Historical Studies*, Dordrecht: Reidel.

Sergeant, J. (1696) *The Method to Science*, London.

Shadwell, T. (1997) *The Virtuoso*, Cambridge: Chadwyck-Healey: 1st edn London, 1676.

Shapin, S. and Schaffer, S. (1985) *Leviathan and the Air-Pump: Hobbes, Boyle, and the Experimental Life*, Princeton: Princeton University Press.

Shapiro, A. E. (2004) 'Newton's "experimental philosophy"', *Early Science and Medicine*, 9 (3), pp. 185–217.

Shapiro, B. J. (1983) *Probability and Certainty in Seventeenth-Century England*, Princeton: Princeton University Press.

Simpson, W. (1669) *Hydrologia chymica*, London.

Sinclair, G. (1683) *Natural philosophy improven by new experiments*, Edinburgh.

Smith, G. E. (2002) 'The methodology of the *Principia*' in *The Cambridge Companion to Newton*, eds I. B. Cohen. and G. E. Smith, Cambridge: Cambridge University Press, pp. 138–173.

Sorell, T., ed. (1993) *The Rise of Modern Philosophy*, Oxford: Oxford University Press.

Sprat, T. (1667) *History of the Royal Society*, London.

Starkey, G. (1658) *Pyrotechny Asserted and Illustrated*, London.

Steele, R. and Addison, J., eds (1888) *The Spectator*, London: Routledge and Sons.

Stein, H. (2002) 'Newton's metaphysics' in *The Cambridge Companion to Newton*, eds I. B. Cohen and G. E. Smith, Cambridge: Cambridge University Press, pp. 256–307.

Sydenham, T. (1683) *Tractatus de podagra et hydrope*, London.

— (1848) *The Works of Thomas Sydenham M. D*, trans. R. G. Latham, London.

Toland, J. (1696) *Christianity not mysterious*, London.

Wallace, W. A. (1988) 'Traditional Natural Philosophy' in *The Cambridge History of Renaissance Philosophy*, eds C. B. Schmitt, Q. Skinner and E. Kessler, Cambridge: Cambridge University Press, pp. 201–235.

Webster, C. (1969) 'Henry More and Descartes: some new sources', *British Journal for the History of Science*, 4, pp. 359–377.

Westman, R. S. (1980) 'Huygens and the problem of Cartesianism' in *Studies on Christiaan Huygens*, eds H. J. M. Bos, Rudwick, M. J. S., Snelders, H. A. M and R. P. W. Visser, Lisse: Swets & Zeitlinger B. V., pp. 83–103.

Willis, T. (1659) *Diatribae duae*, London.

Wilson, C. (2002) 'Newton and celestial mechanics' in *The Cambridge Companion to Newton*, eds I. B. Cohen and G. E. Smith, Cambridge: Cambridge University Press, pp. 202–226.

Wood, P. (1980) 'Methodology and apologetics: Thomas Sprat's History of the Royal Society', *British Journal for the History of Science*, 13, pp. 1–26.
Wotton, W. (1694) *Reflections Upon Ancient and Modern Learning*, London.

Index

STUDIES IN HISTORY AND PHILOSOPHY OF SCIENCE

1. R. McLaughlin (ed.): *What? Where? When? Why?* Essays on Induction, Space and Time, Explanation. Inspired by the Work of Wesley C. Salmon. 1982
ISBN 90-277-1337-5
2. D. Oldroyd and I. Langham (eds.): *The Wider Domain of Evolutionary Thought*. 1983
ISBN 90-277-1477-0
3. R.W. Home (ed.): *Science under Scrutiny*. The Place of History and Philosophy of Science. 1983
ISBN 90-277-1602-1
4. J.A. Schuster and R.R. Yeo (eds.): *The Politics and Rhetoric of Scientific Method*. Historical Studies. 1986
ISBN 90-277-2152-1
5. J. Forge (ed.): *Measurement, Realism and Objectivity*. Essays on Measurement in the Social and Physical Science. 1987
ISBN 90-277-2542-X
6. R. Nola (ed.): *Relativism and Realism in Science*. 1988
ISBN 90-277-2647-7
7. P. Slezak and W.R. Albury (eds.): *Computers, Brains and Minds*. Essays in Cognitive Science. 1989
ISBN 90-277-2759-7
8. H.E. Le Grand (ed.): *Experimental Inquiries*. Historical, Philosophical and Social Studies of Experimentation in Science. 1990
ISBN 0-7923-0790-9
9. R.W. Home and S.G. Kohlstedt (eds.): *International Science and National Scientific Identity*. Australia between Britain and America. 1991
ISBN 0-7923-0938-3
10. S. Gaukroger (ed.): *The Uses of Antiquity*. The Scientific Revolution and the Classical Tradition. 1991
ISBN 0-7923-1130-2
11. P. Griffiths (ed.): *Trees of Life*. Essays in Philosophy of Biology. 1992
ISBN 0-7923-1709-2
12. P.J. Riggs (ed.): *Natural Kinds, Laws of Nature and Scientific Methodology*. 1996
ISBN 0-7923-4225-9
13. G. Freeland and A. Corones (eds.): *1543 and All That*. Image and Word, Change and Continuity in the Proto-Scientific Revolution. 1999
ISBN 0-7923-5913-5
14. H. Sankey (ed.): *Causation and Laws of Nature*. 1999
ISBN 0-7923-5914-3
15. R. Nola and H. Sankey (eds.): *After Popper, Kuhn and Feyerabend*. Recent Issues in Theories of Scientific Method. 2000
ISBN 0-7923-6032-X
16. K. Neal: *From Discrete to Continuous*. The Broadening of Number Concepts in Early Modern England. 2002
ISBN 1-4020-0565-2
17. S. Clarke and T.D. Lyons (eds.): *Recent Themes in the Philosophy of Science*. Scientific Realism and Commonsense. 2002
ISBN 1-4020-0831-7
18. E. Booth: *"A Subtle and Mysterious Machine"*. 2005
ISBN 1-4020-3377-X
19. P.R. Anstey and J.A. Schuster (eds.): *The Science of Nature in the Seventeenth Century*. Patterns of Change in Early Modern Natural Philosophy. 2005
ISBN 1-4020-3603-5

springeronline.com